The Theory of Thin Antennas and Its Use in Antenna Engineering

Authored By

Boris Levin

Holon Institute of Technology
Israel

CONTENTS

FOREWORD

B. Levin's eBook "The Theory of Thin antennas and Its Use in Antenna Engineering" is dedicated to the theory of thin perfectly conducting and impedance radiators. For the most part, it bases on original results of the author who has successfully been tackling the problems of thin antennas for about fifty years and has contributed hundred published papers and fifty inventions in this area of great interest. Particularly impressive is that the author succeeds in an eBook so small in volume (~ 200 pages), in treating at a high scientific level the most important aspects of the analysis and synthesis of thin radiators and probable fields of their application. In the manner of rendering the material and in the number of thin antenna variants considered by the author, the eBook comes close to a handbook, which makes the results set forth in it much easier to use. Undoubtedly B. Levin's eBook will be greeted with interest by the community of antenna professionals and students of higher radio engineering schools.

Prof. Yakov S. Shifrin
Honored Worker of Science and Technology of Ukraine
President of Ukrainian National Antenna Association
Life Fellow IEEE
Ukraine

PREFACE

The eBook deals with the methods of theoretical and experimental research of antennas, which are based on the electromagnetic (EM) theory. The theory of thin antennas underlies the antenna analysis, as they represent one of the main types of radiators and are extensively put into practice both as independent antennas and as elements of more complicated antennas.

Here, techniques for calculations of the electrical characteristics of thin linear antennas are described and analyzed consistently, in particular, the methods of Poynting vector and the induced electromotive force (emf) (first and second formulations) as well as the integral equation method for the antenna conductor current. The theorem of the oscillating power is shown to allow bringing the difficult answers of the antenna theory to light.

The eBook presents several new methods of antenna analysis and design, including the theory of electrically coupled lines, and the methods of complex potential, of the compensation, of the impedance line, and of the mathematical programming among others. New results are obtained. The complex potential method is generalized to inhomogeneous media and is employed in conical and parabolic problems. The theory of electrically coupled lines allows considering theoretical problems of multi-conductor cables and multi-radiator antennas. The compensation method permits creating a weak field area in the near region of a transmitting antenna. The method of impedance line is applied to antennas with loads, and the method of mathematical programming offers selecting loads to develop antennas with required characteristics.

It is shown how the methods of dipoles and monopoles calculation are generalized to the more complicated structures: an antenna with displaced feed point, with losses in the wire, with the lumped and distributed loads, multi-wires, multi-radiators, folded and multi-folded antennas, and antennas of parallel wires with different length. The duality principle provides the slot antenna analysis.

The methods described enable calculating designs of new types of antennas, particularly, the self-complementary ones. A self-complementary structure exhibits constant and purely resistive input impedance in a wide frequency range. In the general case, similar properties pertain to the three-dimensional self-complementary antennas, their variants being considered in the eBook.

Different types of arrays including reflect and adaptive arrays are treated in the eBook in detail. The reciprocity theorem for a reflect array permits to determine a field phase step in a signal reradiated by a reflect array element and to build a reflect array with any given radiation direction.

Close attention is paid to the field compensation in the near region of a transmitting antenna and to creation of the dark spot. The shape and dimensions of dark spots and the irradiation reduction factor are found. Also, a compensation method seeking to form a weak field area over a broad frequency band is considered. Broadband field compensation is shown to be achievable in various structures where the required anti-phase second field is established either by an auxiliary antenna of identical type or by radiators located at equal distances from the compensation point, or by flat reflectors.

The characteristics of metal-shielded two-wire lines (twisted pairs) are calculated using a rigorous method based on the theory of electrically coupled lines. The cross-coupling between lines in multi-conductor cables is shown to result in a kind of electromagnetic interference (crosstalk) in communication channels, while the asymmetry of excitation and loads is shown to result in appearance of common mode currents in the cable. Voltage values (interference) for loads placed at the beginning and the end of the adjacent line are found at a given power in the main line. The effect of loads connected between wires and shield is examined.

A modified method for solving integral equations, which permits to obtain an integral expression for the current at each point of antenna, is proposed. The distribution law for the current along a radiator with distributed and concentrated (lumped) loads is analyzed. The distribution is shown to be similar to that along an equivalent long line (an impedance line). Linear and V-antennas with lumped capacitive loads are described. The problem of load selection to develop antennas with optimal characteristics is considered. Special attention is devoted to the synthesis of wide-band loaded antennas as well as to the issues of producing a given current distribution along a dipole and to those of reducing the effects of re-radiators.

In addition, some other matters are discussed in the eBook: transparent antennas, the rectangular loop field, structural features of ship borne (onboard) antennas. For example, creating a uniform distribution of currents along the transparent radiator cross-section by means of a metal triangle is shown to allow obtaining a sufficiently efficient antenna, its undeniable advantage being wide frequency band. The azimuth and radial field components of a rectangular loop in far and near field are found. Loop dimensions and shape are demonstrated to have a substantial effect on the field magnitude, especially in the near region.

The proposed eBook is a natural addition to known monographs. It is intended for professionals, who are engaged in engineering electrodynamics and in development, deployment and operation of antennas. To benefit from it will also be lecturers (university-level professors), teachers, students, advisors, *etc.* in their profound study of fields radiated by antenna equipment. The contents of the eBook can be drawn on to a short university course.

ACKNOWLEDGEMENT

I want to express my deep gratitude my teachers - for the transferred to me knowledge of electromagnetic theory and principles of antenna engineering, employees at work - for his help in the calculations and experiments, my wife – for understanding.

CONFLICT OF INTEREST

The author(s) confirm that this eBook content has no conflict of interest.

Boris Levin

Holon Institute of Technology
Israel
E-mail: levinpaker@gmail.com

List of Abbreviations

ACW   = Adaptive Control of Weighting Coefficients

AGC   = Automatic Gain Control

AP   = Adaptive Processor

BAA   = Broadband Antenna Absorber

CST   = Computer Simulation Technology

EM   = Electromagnetic

emf   = Electromotive Force

FTO   = Fluorine-Tin-Oxygen

ITO   = Indium-Tin-Oxygen

LPDA   = Log-Periodic Dipole Antenna

PF   = Pattern Factor

PLL   = Phase Locked Loop

R   = Receiver

RG   = Reference Generator

SAR   = Specific Absorption Rate

SWR   = Standing Wave Ratio

TEM   = Transverse Electromagnetic

TWR   = Travelling Wave Ratio

UWB   = Ultra-Wideband

2

Send Orders for Reprints to reprints@benthamscience.net

CHAPTER 1

Theory of Thin Antennas

Abstract: The great value of the theory of thin antennas is substantiated. Models of a linear radiator shaped as a straight perfectly conducting filament with zero and finite radii and as a straight circular thin-wall cylinder are described. Methods of calculation, which were applied before resorting to integral equations, are presented, in particular the induced emf method, its first and second formulations. Results of its application to symmetrical dipoles, to radiators with displaced feed point, to radiators with constant and piecewise constant surface impedance and lumped loads, to folded and multi-radiators antenna are given.

Keywords: Antenna theory, Conducting filament, Constant impedance, Current derivative jump, Displaced feed point, First formulation, Folded antenna, Induced emf method, Lumped loads, Modified solution method, Multi-radiators antenna, Oscillating power, Piecewise constant impedance, Poynting's vector, Reactive power, Second formulation, Sinusoidal distribution, Stepped impedance long line, Surface impedance, Symmetrical dipole, Thin antennas, Thin-wall cylinder.

1.1. FIRST STEPS

Ronold King gives an account of first antennas and first steps of antenna engineering [1]. In twenty years after Maxwell formulated his famous equations [2] to become the foundation of the classical electromagnetic (EM) theory, Hertz proved experimentally the existence of wave effects predicted by the equations. He used a spark gap to excite attenuated oscillations in a wire of length 60 cm with metal plates at the ends [3, 4]. Hertz' experiment started the future stormy development of radio engineering.

The first two Maxwell's equations in differential form are written as

$$curl\vec{H} = \vec{j} + \frac{\partial \vec{D}}{\partial t}, \ curl\vec{E} = -\frac{\partial \vec{B}}{\partial t}, \tag{1.1}$$

where $\vec{H}$ is the vector of magnetic field strength, $\vec{j}$ is the vector of volume density of conduction current, $\vec{D}$ is the electric displacement vector, t is time, $\vec{E}$ is the vector of electric field strength, $\vec{B}$ is the vector of magnetic induction. Hereinafter the International System of Units is used.

The equations (1.1) are to be complemented with the equation of continuity

$$div\vec{j} = -\frac{\partial \rho}{\partial t},$$ (1.2)

where ρ is the volume density of the electrical charge.

Typically, two more equations are included into a system of Maxwell's equations:

$$div\vec{D} = \rho, \ div\vec{B} = 0,$$ (1.3)

but they follow from the equations (1.1) and (1.2) [5].

The equations (1.1) interconnect the electromagnetic fields and currents in free space. Here it would be wrong to consider the left- or the right-hand side of an equation as the cause and, accordingly, the other side as the consequence. Currents, the electric and magnetic components of fields exist jointly only. And none of the quantities are the prime cause of appearance of others.

Radiation of an antenna is result of the action of another medium. In order to take it into account, the set of equations should in accordance with the equivalence theorem have extraneous (impressed) currents and fields as the original sources of excitation. They are introduced as summands in quantities $\vec{j}$, $\vec{E}$ and $\vec{H}$. Their nature and location area depend on the model of the section close to a generator, which is commonly known as the excitation zone. The total electromagnetic field of an antenna adds up from the field produced by the excitation zone and the field produced by the wires' currents, which arise at switch-on of initial sources. As a rule, the first summand is substantially less than the second one far from the antenna and can be neglected.

The Maxwell's equations for electromagnetic field, which are complemented with boundary conditions on the surface of some or other antenna, allow writing the equation for the current in the antenna wire. Solving it and finding the current distribution along the wire, one can determine electrical characteristics of a radiator. But in the first decades after Hertz' works researchers were interested in other matters. Among engineers trying to solve the problem of signal reception, names of Marconi and Popov are best known.

In 1894, 23-year-old Rutherford made a device for receiving radio signals, which was based on demagnetizing of a bunch of needles, and even demonstrated it to Marconi, and the latter undertook to improve it. And to solve the transmission problem the invention of radio tubes was essential. The power of radio tubes began to grow from year to year.

Published in 1884, paper [6] was devoted to calculation of the radiated signal power. The paper introduced the Poynting's vector as

$$\vec{S} = [\vec{E}, \vec{H}].$$ (1.4)

Quantity $\vec{S}$ is the power flux density. Its projection onto the normal to the corresponding part of a closed surface is equal to the power flux density, outgoing from the volume, bounded by the surface.

Using the Poynting's vector, one can find the active component of antenna radiation impedance. The power flux passing through an antenna surface does not change in free space and is equal to the power flux in the far region. The vectors of electric and magnetic field strengths are mutually perpendicular. Here,

$$|H| = |E|/Z_0 \,,$$

where $Z_0 = 120\pi$ is the wave impedance of free space. Since the field strength of a vertical linear antenna in the spherical coordinate system is

$$E = E_m F(\theta, \phi),$$

where E_m is the field in the direction of maximal radiation, and $F(\theta, \phi)$ is the pattern, the radiation power of such antenna is equal to an integral of Poynting's vector

$$P_\Sigma = (1/Z_0) \int_{(S)} E_m^2 F^2(\theta, \phi) dS \,.$$ (1.5)

Integration is performed over the surface of sphere S of a great radius. The area of a sphere element is $dS = R_0^2 \sin\theta d\theta d\phi$. Taking the ratio of the radiation power to the square of the generator current, we obtain the antenna radiation impedance

$$R_\Sigma = \frac{1}{J^2(0)Z_0} \int_0^{2\pi} d\phi \int_0^{\pi} E_m^2 F^2(\theta,\phi) R_0^2 \sin\theta d\theta .$$

(1.6)

Let us calculate the electric field in the far region and the pattern of a radiator, considering the latter as a sum of simple electrical dipoles (Hertz' dipoles). The field of such dipole of length b with current I, located along z-axis, is

$$E_{\theta 0} = j(30kIb/R)\exp(-jkR)\sin\theta .$$

(1.7)

Here $k = \omega\sqrt{\mu\varepsilon}$ is the wave propagation constant in the surrounding medium, ω is the circular frequency of the signal, μ is the permeability, $\varepsilon = \varepsilon_r \varepsilon_0$ is absolute permittivity (ε_r is relative permittivity, ε_0 is the absolute permittivity of the air). The current distribution along a symmetrical dipole with arm length L is determined by the expression

$$J(z) = J(0)\frac{\sin k(L-|z|)}{\sin kL}.$$

(1.8)

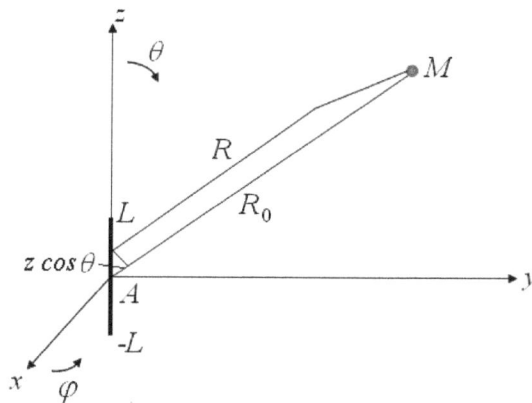

Figure 1: Field in the far region.

Setting $R = R_0 - z\cos\theta$ (see Fig. **1**), $I = J(z)$, $b = dz$ and neglecting small quantities, we obtain for the field of the symmetrical dipole

$$E_\theta = j\frac{30k\exp(-jkR_0)\sin\theta}{\varepsilon_r R_0} \int_{-L}^{L} J(z)\exp(jkz\cos\theta)dz ,$$

(1.9)

where R_0 is the distance from the dipole center to the observation point. Substituting (1.8) into (1.9), we find

$$E_\theta = j\frac{60J(0)}{\varepsilon_r \sin kL} \cdot \frac{\exp(-jkR_0)}{R_0} \cdot \frac{\cos(kL\cos\theta) - \cos kL}{\sin\theta}. \tag{1.10}$$

The last factor of the expression defines the dipole pattern. Substitution of (1.10) into (1.6) gives,

$$R_\Sigma = \frac{60}{\varepsilon_r^2 \sin^2 kL} \int_0^\pi \frac{\left[\cos(kL\cos\theta) - \cos kL\right]^2}{\sin\theta} d\theta. \tag{1.11}$$

Such method of calculating radiation resistance is commonly known as the Poynting's vector method. If the dipole length is small ($kL \langle\langle 1$), then limiting ourselves to the first terms of the function expansion in a series (at small x, $\cos x = 1 - x^2/2$), we find:

$$R_\Sigma = 20k^2 L^2 / \varepsilon_r^2. \tag{1.12}$$

For comparatively short antennas ($L < 0.3\lambda$, here λ is the wavelength), one can proceed from the expression

$$R_\Sigma = 20k^2 h_e^2 / \varepsilon_r^2, \tag{1.13}$$

where $h_e = \dfrac{2}{k}\tan\dfrac{kL}{2}$ is the effective length of the symmetrical dipole.

The next step in the theory of linear radiators was taken in the twentieth century. It is known as the induced emf method. But before proceeding to it we shall consider the field of the conduction current along a filament and a cylinder.

1.2. FIELD OF A FILAMENT AND A CYLINDER

With the Maxwell's equations written, we face their solution. This simplifies essentially, if auxiliary functions are introduced, called potentials. A vector potential (an auxiliary vector field) is introduced by comparing the second equation from (1.3) with the mathematical identity

$$divcurl\vec{A} = 0 \,,$$

where $\vec{A}$ is an arbitrary vector. This comparison shows that vector $\vec{B}$ can be represented as a curl of some vector $\vec{A}$:

$$\vec{B} = curl\vec{A} \,, \tag{1.14}$$

Yet, equation (1.14) is an ambiguous definition of vector $\vec{A}$. To define it unambiguously, one should also specify quantity $div\vec{A}$.

Substituting (1.14) into the second equation of the set (1.1) and using the mathematical identity

$$curl \; grad \; U=0,$$

where U is an arbitrary scalar function (scalar potential of field), we obtain

$$\vec{E} = -\frac{d\vec{A}}{dt} - gradU \,. \tag{1.15}$$

Substituting (1.14) and (1.15) into the first equation of the set (1.1) and taking account of the mathematical identity

$$curlcurl\vec{A} = graddiv\vec{A} - \Delta\vec{A} \,,$$

we find

$$\Delta\vec{A} - \mu\varepsilon\frac{d^2\vec{A}}{dt^2} - grad(div\vec{A} + \mu\varepsilon\frac{dU}{dt}) = -\mu\vec{j} \,. \tag{1.16}$$

Let us define now $div\vec{A}$ to simplify the last expression as far as possible. For this purpose, let

$$div\vec{A} = -\mu\varepsilon\frac{dU}{dt} \,. \tag{1.17}$$

This equality is known as the calibration condition, or as the Lorentz condition. In accordance with (1.16) and (1.17)

$$\Delta \vec{A} - \mu \varepsilon \frac{d^2 \vec{A}}{dt^2} = -\mu \vec{j} . \tag{1.18}$$

For harmonic field, varying with time as exponential function $\exp(j\omega t)$, the equation (1.18) takes the form

$$\Delta \vec{A} + k^2 \vec{A} = -\mu \vec{j} . \tag{1.19}$$

The equation (1.19) is called the vector wave equation. Its solution permits to find the vector potential $\vec{A}$, and then the electric and magnetic fields of antenna. Actually, accordingly to (1.14), (1.15) and (1.19)

$$\vec{E} = -\frac{\omega}{k^2}\left(grad\, div\vec{A} + k^2\vec{A}\right), \ \vec{H} = \frac{1}{\mu}curl\vec{A} . \tag{1.20}$$

If the electromagnetic field sources are distributed continuously in some region V, and the medium surrounding the region V is a homogeneous isotropic dielectric, the solution of the equation (1.19) for harmonic field appears as

$$\vec{A} = \mu \int_{(V)} \vec{j}G dV , \tag{1.21}$$

where $G = \exp(-jkR)/(4\pi R)$ is Green's function.

A similar expression for the scalar potential follows from (1.17), (1.21) and (1.2):

$$U = j\frac{1}{\omega \varepsilon} \int_{(V)} G div\vec{j} dV = \frac{1}{\varepsilon} \int_{(V)} \rho G dV . \tag{1.22}$$

Note that region V, where the electromagnetic field sources are located, may be multiply connected (if, *e.g.*, radiation of several antennas is considered, or metal bodies are located close to the antenna).

Further, consider the special case when the field source is the electrical currents that are in parallel to the z-axis in some region V and have the axial symmetry:

$$\vec{j} = j_z \vec{e}_z, \ j_z = j_z(z) = const(\phi) . \tag{1.23}$$

Here, the cylindrical system of coordinates (ρ, ϕ, z) with unit vectors $\vec{e}_\rho, \vec{e}_\phi, \vec{e}_z$ along theirs axes is used. As seen from (1.21), the vector potential in this case has component A_z only:

$$\vec{A} = A_z(\rho, z)\vec{e}_z,$$ (1.24)

i.e.,

$$div\vec{A} = \frac{\partial A_z}{\partial z}, \ graddiv\vec{A} = \frac{\partial^2 A_z}{\partial \rho \partial z}\vec{e}_\rho + \frac{\partial^2 A_z}{\partial z^2}\vec{e}_z, \ curl\vec{A} = -\frac{\partial A_z}{\partial \rho}\vec{e}_\phi,$$

and in accordance with (1.20)

$$E_z(\rho, z) = -\frac{j\omega}{k^2}\left(k^2 A_z + \frac{\partial^2 A_z}{\partial z^2}\right), E_\rho(\rho, z) = -\frac{j\omega}{k^2}\frac{\partial^2 A_z}{\partial \rho \partial z}, H_\phi(\rho, z) = -\frac{1}{\mu}\frac{\partial A_z}{\partial \rho},$$

$$E_\phi = H_z = H_\rho = 0.$$ (1.25)

Obviously, given the distribution of current $J(z)$ along the radiator, one can calculate the electromagnetic field of the current with the help of the presented formulas. If the radiator is excited by a generator with concentrated emf e at some point (*e.g.*, $z = 0$), the antenna impedance at the driving point is

$$Z_A = e/J(0),$$ (1.26)

and knowing the current magnitude at the corresponding point is enough to define it. When calculating power absorbed in the load of a receiving antenna, the current magnitude is needed also. Like that the current distribution along the antenna is its important characteristic.

As a model of a vertical linear radiator, one can use a straight, perfectly conducting filament, coinciding with *z*-axis (Fig. **2a**), with conductance current *J(z)* running in it. Current density $\vec{j}$ is related to this current by

$$\vec{J}(z) = \int_{(S)} \vec{j}dS,$$

where *S* is the filament cross-section. With the help of the relationship we obtain from (1.21) and (1.25):

$$A_z(\rho,z) = \mu \int_{-L}^{L} J(\varsigma) G_1 d\varsigma, \quad E_z(\rho,z) = \frac{1}{j\omega\varepsilon} \int_{-L}^{L} J(\varsigma) \left(k^2 G_1 + \frac{\partial^2 G_1}{\partial z^2} \right) d\varsigma. \qquad \textbf{(1.27)}$$

Here $G_1 = \exp(-jkR_1)/(4\pi R_1)$, distance R_1 from observation point M to integration point P being $\sqrt{(z-\varsigma)^2 + \rho^2}$.

In the considered model, the radiator radius is zero. The model of a radiator shaped as a straight circular thin-wall cylinder with radius a (Fig. **2b**) has finite dimensions. For the current to have longitudinal components only as before, both ends of the cylinders are left open, without covers. The surface current density along the cylinder is $J_S(z) = J(z)/(2\pi a)$. Since a volume element in the cylindrical coordinates system is equal to $dV = \rho d\rho d\phi dz$, and $\rho = a$ on the cylinder surface, so, in accordance with (1.21) and (1.25),

$$A_z(\rho,z) = \frac{\mu}{2\pi} \int_{-L}^{L} J(\varsigma) \int_{0}^{2\pi} G_2 d\phi d\varsigma, \quad E_z(\rho,z) = \frac{1}{j\omega\varepsilon} \int_{-L}^{L} J(\varsigma) \left(k^2 G_2 + \frac{\partial^2 G_2}{\partial z^2} \right) d\varsigma, \qquad \textbf{(1.28)}$$

where $G_2 = \exp(-jkR_2)/(4\pi R_2)$, with distance R_2 from observation point M to integration point P being $\sqrt{(z-\varsigma)^2 + \rho^2 + a^2 - 2a\rho\cos\phi}$. In particular, if the observation point is located on the radiator surface, $R_2 = \sqrt{(z-\varsigma)^2 + 4a^2\sin^2(\phi/2)}$.

Sometimes the dipole model shaped as a filament with finite radius a, *i.e.*, expression (1.27) is used for A_z and E_z, but distance R_2 from the observation point to the integration point is believed equal to $R_3 = \sqrt{(z-\varsigma)^2 + \rho^2 + a^2}$.

Obtained expressions for the vector potential and the vertical component of the electrical field strength produced by different models of a radiator confirm the stated above assertion on the significance of the current distribution along the radiator. An assumption that this distribution has sinusoidal form played a great role in the antenna theory. This was based partly on measurements results, but mainly on a simple thought, that the current distribution along conductors of two-wire open-end line does not change when the conductors are separated from each other by moving them apart. Later on, upon derivation and solution of integral equations for currents in radiators, the sinusoidal distribution was shown rigorously to be the first approximation to the true current distribution. Thus, its use is quite founded.

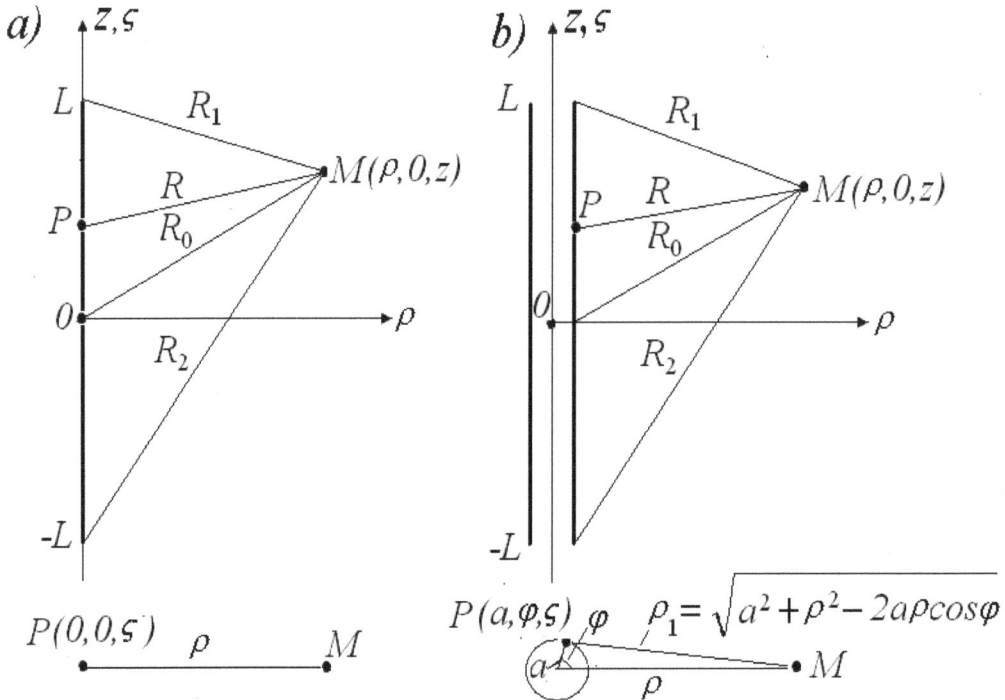

Figure 2: The dipole models shaped as a straight filament (*a*) and a thin-wall circular cylinder (*b*).

Here the sinusoidal distribution is implied to be a sinusoid with any phase. In particular, the current at the dipole or monopole end, farther from the generator, is equal to zero. In this case there is an obvious sinusoid – see (1.8). The current distribution along a folded radiator has an antinode of standing wave at the end, *i.e.,* the current follows the cosine distribution law.

For a perfectly conducting filament used as a model of a symmetrical radiator, in accordance with (1.28)

$$E_z = \frac{1}{4\pi j\omega\varepsilon} \int_0^L J(\varsigma)\left(k^2+\frac{\partial^2}{\partial z^2}\right)\left[\frac{\exp(-jkR)}{R}+\frac{\exp(-jkR_+)}{R_+}\right]d\varsigma. \tag{1.29}$$

This expression takes into account the current symmetry in radiator arms, accordingly the substitution of variable ς for $-\varsigma$ is performed at the lower arm, and designation is used: $R_+ = \sqrt{(z+\varsigma)^2 + \rho^2}$ is used. Since

$$\frac{\partial R}{\partial \varsigma} = -\frac{\partial R}{\partial z}, \quad \frac{\partial R_+}{\partial \varsigma} = \frac{\partial R_+}{\partial z}, \quad \frac{\partial^2 R}{\partial \varsigma^2} = \frac{\partial^2 R}{\partial z^2}, \quad \frac{\partial^2 R_+}{\partial \varsigma^2} = \frac{\partial^2 R_+}{\partial z^2},$$

then

$$E_z = \frac{1}{4\pi j\omega\varepsilon}\int_0^L J(\varsigma)\frac{\partial^2}{\partial \varsigma^2}\left[\frac{\exp(-jkR)}{R} + \frac{\exp(-jkR_+)}{R_+}\right]d\varsigma + \frac{k^2}{4\pi j\omega\varepsilon}\int_0^L J(\varsigma)\left[\frac{\exp(-jkR)}{R} + \frac{\exp(-jkR_+)}{R_+}\right]d\varsigma.$$

Integrating the first term of the expression by parts twice, we find

$$E_z = \frac{1}{4\pi j\omega\varepsilon}\left\{\int_0^L\left[\frac{d^2 J(\varsigma)}{d\varsigma^2} + k^2 J(\varsigma)\right]\left[\frac{\exp(-jkR)}{R} + \frac{\exp(-jkR_+)}{R_+}\right]d\varsigma + \right.$$

$$\left. + J(\varsigma)\frac{\partial}{\partial \varsigma}\left[\frac{\exp(-jkR)}{R} + \frac{\exp(-jkR_+)}{R_+}\right]\Big|_0^L - \frac{dJ(\varsigma)}{d\varsigma}\left[\frac{\exp(-jkR)}{R} + \frac{\exp(-jkR_+)}{R_+}\right]\Big|_0^L\right\}.\textbf{(1.30)}$$

If the current along the radiator is distributed in accord with (1.8), the first factor in the integrand and hence the first term of the expression are zero. It is easily verified that the second summand is zero too, since the first factor becomes zero at $\varsigma = L$, and the second factor vanishes at $\varsigma = 0$. Taking a derivative of the current distribution' function

$$\frac{dJ(\varsigma)}{d\varsigma} = -kJ(0)\frac{\cos k(L-|\varsigma|)}{\sin kL}\, signz$$

(here *signz* means sign of z), we obtain, with the account of $k/(4\pi\omega\varepsilon_0) = 30$:

$$E_z = -j\frac{30J(0)}{\varepsilon_r \sin kL}\left[\frac{\exp(-jkR_1)}{R_1} + \frac{\exp(-jkR_2)}{R_2} - 2\cos kL\frac{\exp(-jkR_0)}{R_0}\right], \qquad \textbf{(1.31)}$$

where $R_1 = \sqrt{(z-L)^2 + \rho^2}$, $R_2 = \sqrt{(z+L)^2 + \rho^2}$, $R_0 = \sqrt{z^2 + \rho^2}$ are the distances from observation point M to the upper end, to the lower end and to the middle of the radiator, respectively (see Fig. **2a**).

Let us present without proof two components of the electromagnetic field for a straight filament:

$$E_\rho = j \frac{30 J(0)}{\varepsilon_r \rho \sin kL} \left[\frac{(z-L)\exp(-jkR_1)}{R_1} + \frac{(z+L)\exp(-jkR_2)}{R_2} - 2z\cos kL \frac{\exp(-jkR_0)}{R_0} \right],$$

$$H_\phi = j \frac{J(0)}{4\pi \varepsilon_r \rho \sin kL} \left[\exp(-jkR_1) + \exp(-jkR_2) - 2\cos kL \exp(-jkR_0) \right]. \qquad (1.32)$$

The rest of components are zero, see (1.25).

If the model of a symmetrical radiator shaped as a straight circular cylinder is used, it is necessary, when calculating the field, to proceed from expression (1.28). We obtain instead of (1.31):

$$E_z = -j \frac{30 J(0)}{2\pi \varepsilon_r \sin kL} \int_0^{2\pi} \left[\frac{\exp(-jkR_1)}{R_1} + \frac{\exp(-jkR_2)}{R_2} - 2\cos kL \frac{\exp(-jkR_0)}{R_0} \right] d\phi, \quad (1.33)$$

where

$$R_1 = \sqrt{(z-L)^2 + \rho^2 + a^2 - 2a\rho\cos\phi'}, \ R_2 = \sqrt{(z+L)^2 + \rho^2 + a^2 - 2a\rho\cos\phi'},$$

$$R_0 = \sqrt{z^2 + \rho^2 + a^2 - 2a\rho\cos\phi'}.$$

Such great attention is paid to the sinusoidal distribution of the current along the radiator because the induced emf method is in particular based on this distribution.

1.3. INDUCED EMF METHOD

The induced emf method was proposed in 1922 by Rojansky and Brillouin simultaneously. Klazkin was the first to use it for radiator calculation. Later on, Pistolkors, Tatarinov, Carter, Brown *et al.* contributed to its development. Reference list in the eBook [7], which generalizes the literature results of applying the method, consists of 96 items.

The induced emf method allows determining both the active and reactive components of the antenna input impedance. Since the active component can be calculated with a similar accuracy by a much simple Poynting's vector method (see Section 1.1), the induced emf method, as emphasized in [8], is actually for practical purposes one of those for determining input reactance of antenna.

The induced emf method is constructed in the following manner. A radiator is surrounded with closed surface, and a complex power passing through this surface is found. The calculation is performed proceeding from the sinusoidal distribution of the current along the radiator. Further the power is assumed to be equal to the complex output power of an emf source (generator).

Let the closed surface be a cylinder of height *2H* and radius *b*, with a symmetrical radiator – dipole (Fig. **3**) located along its axis. The power flux density, outgoing from the volume, bounded with the closed surface, is determined by projecting Poynting's vector onto the normal to the corresponding part of the surface: $S_n = [\vec{E}, \vec{H}]_n$ - see (1.4). These projections for the cylinder side surface and top are given by

$$S_\rho = -E_z H_\phi^*, \ S_z = E_\rho H_\phi^*. \tag{1.34}$$

The integral of the normal component of Poynting's vector over the cylinder surface is the power radiated by the antenna. Here, vectors $\vec{E}$ and $\vec{H}$ in contrast to the far region, are not in phase. This is easily verified by analyzing the expressions (1.31) and (1.32). For this reason the power is a complex quantity, *i.e.,* it has both active and reactive components.

Let us combine the close surface of the cylinder with the radiator surface, *i.e.,* we shall assume that $H = L$, $b = a$. At a small radius of the radiator, the power fluxes passing through the cylinder top and bottom are small. If the radiator is a thin filament of radius tending to zero, the integrals over its upper and lower bases vanish. If the radiator is shaped as a circular cylinder of finite radius, the selected integrals tend to zero too, since the longitudinal current into the cylinder attenuates quickly, *i.e.,* the tangential component H_ϕ of magnetic field on the both ends at $\rho < a$ is close to zero.

Therefore, the power passing through the closed surface is determined by integrating over the side cylinder surface only:

$$P = \int\limits_{-L}^{L} dz \int\limits_{0}^{2\pi} S_\rho a d\phi ,$$ (1.35)

where quantity S_ρ is found from (1.34) and is independent of coordinate ϕ, since the electromagnetic field components depend on ϕ neither. As $H_\phi^* = J^*(z)/(2\pi a)$, so

$$P = -\int\limits_{-L}^{L} E_z J^*(z) dz .$$ (1.36)

Current $J(z)$ is excited by a sole generator located at the radiator center. The power of the generator is

$$P = |J(0)|^2 Z_A ,$$ (1.37)

where Z_A is the complex input impedance of the antenna. Equating the power of the emf source to the power passing through the closed surface, we find

$$Z_A = -\frac{1}{|J(0)|^2} \int\limits_{-L}^{L} E_z J^*(z) dz .$$ (1.38)

The expression (1.38) reveals the essence of the induced emf method. Another variant of the expression derivation is described in [9] and [10]. It proceeds from the equality to zero of the tangential component of the electrical field on the perfectly conducting surface of radiators. Both variants are based on the same two theses. The first thesis assumes the sinusoidal character of the current distribution along the radiator. The second thesis signifies the equality of the source complex power and the complex power passing through the closed surface.

Indeed, the expression (1.38) is useful only in the case of current distribution $J(z)$ along a radiator being known beforehand. If the current behaves sinusoidally, according to (1.8), one can find field magnitude E_z from (1.31) or (1.33) depending on the selected model of antenna and then perform the quadrature (1.38). The selection of a distribution law of the current along a filament or a thin-wall cylinder may be based on a solution of integral equations for the current, *i.e.,* on a rigorous solution of the problem. No physical base for the selection of another distribution law exists. Hence, there is no sense in speaking about the accuracy of the induced emf method proper, excluding artificially the error caused by the inexact current definition. The accuracy of the induced emf method is the combined accuracy of the formulas (1.8) and (1.38).

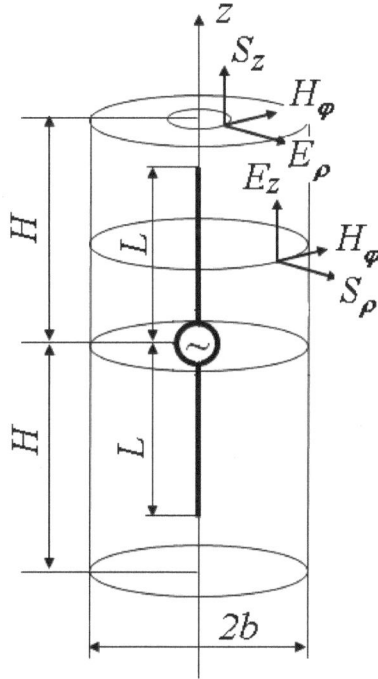

Figure 3: To calculation of power, radiated by the antenna.

As calculations show, one generator cannot create the sinusoidal distribution. Therefore, the applicability of expression (1.38) depends on how close the sinusoidal distribution is to the true current distribution along the radiator. The experience in calculations shows that (1.8) give quite an acceptable approximation, if $L/\lambda \leq 0.4$. Therefore, the first thesis is questionable, because it is of an approximate nature.

As to the second thesis, its inapplicability is obvious, since the reactive power has no physical sense, and the input reactance of antenna is determined as a result of equating two quantities, which have no physical sense. Equating two such quantities cannot be justified.

Nevertheless the induced emf method entered the antenna engineering in triumph and allowed solving many problems. Experiments confirmed constantly the accuracy of these solutions. As a rule, rejection of the sinusoidal distribution in solving new problems, where, *e.g.*, wide plates of an irregular shape are used, did not agree with the results of calculation and experimental testing. Calculations show that such distributions very often were in agreement with the physical contents of problems.

In spite of the triumphal procession and wide employment of the method, evidence of its inadequacy grew gradually. The rigorous solution of problems with help of integral equations gave results close to those using the induced emf method. They were close not only numerically, but in the explicit form, *i.e.*, in the form of an aggregate of familiar (tabulated) functions. In particular, the solutions of the equation of Hallen [11] and Leontovich-Levin [12] for the thin linear radiator give results similar to those of the induced emf method. The results are similar, but not completely identical.

Kontorovich's article [13] of 1951 quoted the expression for the input impedance of the same radiator obtained in accordance with the reciprocity theorem. A similar procedure based on the reciprocity theorem is described in [14]. In the expression, the complex conjugate quantities are absent:

$$Z_A = -\frac{1}{J^2(0)} \int_{-L}^{L} E_z J(z) \, dz \,. \tag{1.39}$$

This called "the second formulation of the induced emf method". The name "the first formulation" was given to the formula (1.38).

Unexpected results are obtained in application of the first formulation. Substituting $k = k_1 + jk_2$ for quantity k in the solution, where $k_2 \langle\langle k_1$, in order to calculate losses in metal wires, we obtain a negative resistance. A similar constitution used in the second formulation gave a positive value of the resistance loss and confirmed correctness of the second formulation. Comparing stability of both formulations against small changes of parameters, one can show that the second formulation is stable, while the first one is not.

The oscillating power theorem [15] proved by Vainshtein allowed explaining the difference between two formulations.

Two formulations of the induced emf method are the result of symbolical method use. In the electromagnetic theory, great attention is paid to signals, varying with time according to the sinusoidal law (monochromatic signals). When analyzing such signals, the instantaneous value of a quantity in form of cosine function is replaced by exponential one in accordance with expression

$$a(t) = A\cos(\omega t + \phi_a) = \mathrm{Re}\left[\dot{A}\exp(j\omega t)\right], \tag{1.40}$$

where A and ϕ_a are the amplitude and phase of cosine function, $\dot{A} = A\exp(j\phi)$ is the complex amplitude. Proceeding from (1.40), quantity $a(t)$ in the equation is replaced with quantity $\dot{A}$, with the correctness of such replacement depending on equation linearity and requiring a check in the case of product of two instantaneous values.

As known, energy characteristics are defined by a product of two physical quantities: see, for example, expression (1.4) for Poynting's vector. Strictly speaking, it is a product of two instantaneous values. Such product consists of a constant quantity equal to the average value for the oscillation period and an oscillating quantity, time-varying with double frequency. On the side surface of a vertical cylinder,

$$s_\rho(t) = \overline{S}_\rho + \tilde{S}_\rho. \tag{1.41}$$

The average value is equal to the real part of a product of two complex amplitudes: the complex amplitude of one quantity and the complex conjugate amplitude of another

$$\overline{S}_\rho = \mathrm{Re}\left[-\dot{E}_z\dot{H}_\phi^*\right]. \tag{1.42}$$

The oscillating quantity amplitude is equal to the absolute value of product of two quantities. These are the complex amplitudes of two physical quantities:

$$\tilde{S}_\rho = \left|\dot{E}_z\dot{H}_\phi\right|\exp(2j\omega t). \tag{1.43}$$

Recall that the complex amplitude corresponds to the effective value of a sinusoidal quantity, *i.e.*, the absolute value of complex amplitude is smaller than the amplitude by a factor of $\sqrt{2}$.

It is from here that the physical sense of the first and second summand in the right part of (1.41) follows. The first summand is the active part of the power flux equal to its average value. The second summand is the oscillating power flux. Reactive part $Jm\left[-\dot{E}_z \dot{H}_\phi^*\right]$ of power flux has no physical sense.

In accordance with the energy conservation law, the active output power of the source, *i.e.*, its average value over an oscillation period is equal to the active power passing through the closed surface. It is natural to consider that the power equality is valid at any time instant, *i.e.*, the oscillating part P_K of the source output power is equal to the oscillating part of the power passing through the closed surface. But such assumption for the reactive power cannot be considered well founded.

On the one hand, the complex amplitude of the oscillating power passing through the side cylinder surface of symmetrical dipole (Fig. **4a**) by analogy to (1.36) is

$$P_K = -\int_{-L}^{L} E_z(J)J(z)dz. \tag{1.44}$$

On the other hand, the complex amplitude of oscillating power produced by one generator by analogy to (1.37) is

$$P_K = eJ(0) = J^2(0)Z_A, \tag{1.45}$$

where e is the emf of the generator. By equating right parts of the expressions, we come to (1.39), *i.e.*, we obtain the second formulation of the induced emf method. As it was shown here, this is derived easily from the energy relationships, if one uses the notion "oscillating power". It is worth emphasizing that the second formulation is derived rigorously, while for derivation of the first one requires using the notion "reactive power", that has no a physical sense.

Nevertheless, it should be admitted that both formulations are valid, and upon substitution of the sinusoidal current distribution, yield the same results for a linear metal radiator situated in a lossless medium. For the same radiator in loss medium, when the propagation constant of the wave along a radiator is a complex quantity, the first formulation is not valid. A similar situation takes place in calculations of losses in the antenna wires, which are due to skin-effect, and in calculations of losses in a magnetodielectric sheath.

One can obtain the integral expression (1.39) by means of the modified method of solving Leontovich-Levin integral equation. The modified method of solving the equation allows obtaining integral formula for the radiator input impedance. As it was shown by author, the expression (1.39) is valid, if the feed point $z = h$ does not coincide with a node of the current:

$$J(h) \neq 0. \tag{1.46}$$

If the condition does not hold, then the input impedance is

$$Z_A = \frac{e}{2J(h) + \dfrac{1}{e} \displaystyle\int_{-l}^{L} E_z(J)J(z)\,dz}. \tag{1.47}$$

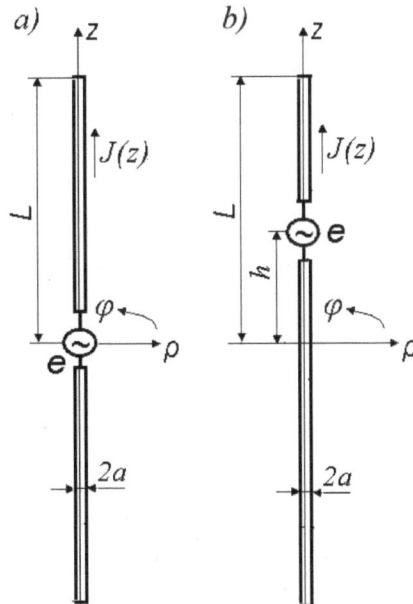

Figure 4: Symmetrical (*a*) and asymmetrical (*b*) dipoles.

The last expression permits to arrive at finite values of the input impedance at the points of parallel resonance and close to them. The induced emf method fails at the problem.

1.4. APPLICATION OF THE INDUCED EMF METHOD TO COMPLICATED ANTENNAS AND TO ANTENNA SYSTEMS

One can use the induced emf method to analyze more complicated radiators. For a radiator with a feed point displaced from the radiator middle to point $z = h$ (Fig. **4b**), the oscillating power flux through the side cylinder surface by analogy to (1.44) is

$$P_K = -\int_{-L}^{L} E_z(J) J(z) dz. \tag{1.48}$$

The oscillating power produced by one generator by analogy to (1.45) is

$$P_K = eJ(h) = J^2(h) Z_A. \tag{1.49}$$

By equating the right-hand parts of the expressions, we come to

$$Z_A = -\frac{1}{J^2(h)} \int_{-L}^{L} E_z(J) J(z) dz. \tag{1.50}$$

Note that the expression differs from (1.39) not only in the denominator, but also in current $J(z)$ distribution along a radiator and in field $E_z(J)$ of their current.

In the case of a radiator with nonzero surface impedance, one should substitute difference $E_z(J) - H_\phi Z(z)$ for $E_z(J)$ in (1.39). Here $Z(z)$ is the surface impedance, *i.e.*, the impedance of the square surface section with a side one centimeter long. Actually, in accordance with the boundary condition on the radiator surface, it is necessary to take into account that a voltage drop across the radiator proper makes no contribution to its radiation. Then for an antenna with constant surface impedance Z (Fig. **5a**) shaped as a straight circular cylinder with radius a we find

$$Z_A = -\frac{1}{J^2(h)} \int_{-L}^{L} \left[E_z(J) - ZJ(z)/(2\pi a) \right] J(z) \, dz \,. \tag{1.51}$$

If $h = 0$, current distribution $J(z)$ coincides in the first approximation with the current distribution along an open-end impedance long line:

$$J(z) = J(0) \sin k_1 (L - |z|) / \sin k_1 L \,. \tag{1.52}$$

Here $k_1 = \sqrt{k^2 - j 2 k \chi Z / (a Z_0)}$ is the propagation constant of a wave along the line, χ is a small parameter of the thin antenna theory, used in [12] (it is equal to $\chi \cong 1/\Omega$, where Ω is the parameter, used by Hallen [16]).

For a symmetrical radiator with piecewise constant surface impedance (Fig. **5b**) one can write

$$Z_A = -1/\left(J_N^2 \right) \sum_{m=1}^{2N} \int_{b_{m+1}}^{b_m} \left[E_z(J_m) - Z_m J_m(z)/(2\pi a) \right] J_m(z) \, dz \,, \tag{1.53}$$

where m is the segment number, $2N$ is the total number of segments, Z_m is the surface impedance on the mth segment. Current distribution $J_m(z)$ on the each mth segment is sinusoidal. Current distribution $J(z)$ along the radiator coincides in the first approximation, if $h = 0$, with the current distribution along an open-end stepped impedance long line:

$$J_m(z) = I_m \sin(k_m z_m + \phi_m), \qquad b_{m+1} \le z \le b_m \,, \tag{1.54}$$

where

$$I_m = A_m J(0), \qquad A_m = \prod_{p=m+1}^{N} \sin \phi_p \Bigg/ \prod_{p=m}^{N} \sin(k_p l_p + \phi_p),$$

$$\phi_m = \tan^{-1} \left\{ \frac{k_m}{k_{m-1}} \tan \left[k_{m-1} l_{m-1} + \tan^{-1} \left\langle \frac{k_{m-1}}{k_{m-2}} \tan \left[k_{m-2} l_{m-2} + \ldots + \tan^{-1} \left(\frac{k_2}{k_1} \tan k_1 l_1 \right) \ldots \right] \right\rangle \right] \right\}.$$

In these expressions $z_m = b_m - z$ is the coordinate along the mth segment, k_m is the wave propagation constant along the segment, and l_m is its length. The expressions are valid for the Nth segment too, if one considers that the product $\prod\limits_{p=N+1}^{N}$ is equal to 1.

In the case of a radiator with one lumped load Z_1 located at point $z = z_1$ (Fig. **6a**) the power is firstly radiated by the antenna:

$$P_{K1} = -\int_{-L}^{L} E_z(J)J(z)\,dz,$$
(1.55)

and, secondly, the power is wasted in the complex load:

$$P_{K2} = J^2(z_1)Z_1.$$
(1.56)

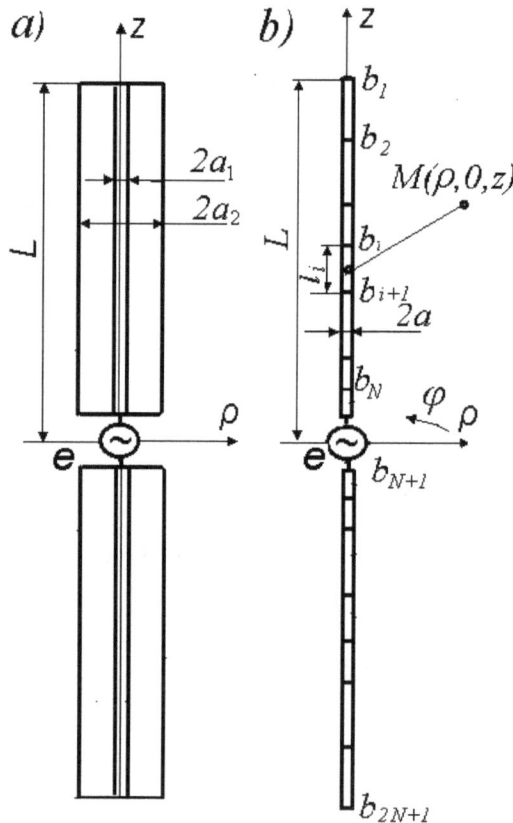

Figure 5: Antennas with constant (*a*) and piecewise constant (*b*) surface impedances.

The oscillating power produced by the generator is equal to the sum of these powers:

$$P_K = J^2(h)Z_A = P_{K1} + P_{K2},$$ **(1.57)**

that is,

$$Z_A = -\frac{1}{J^2(h)} \left\{ \int_{-L}^{L} E_z(J)J(z)\,dz - Z_1 J^2(z_1) \right\}.$$ **(1.58)**

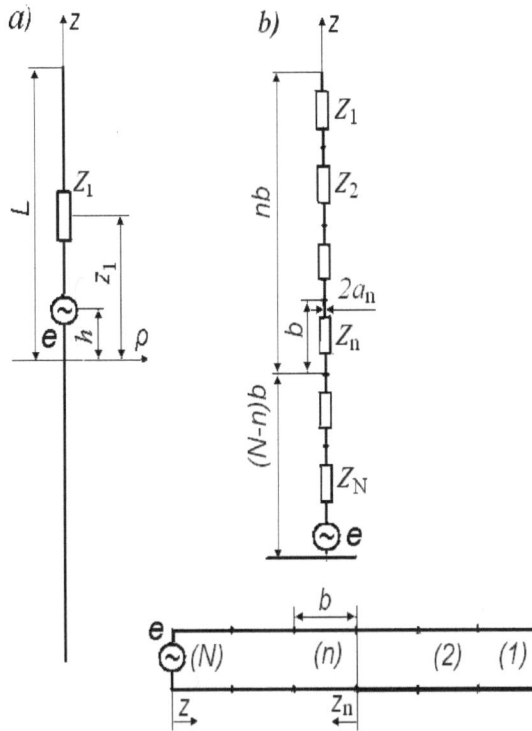

Figure 6: Antennas with one (*a*) and several (*b*) lumped impedances.

For several loads Z_n located at points $z = z_n$ of the asymmetrical radiator (Fig. **6b**)

$$Z_A = -\frac{1}{J^2(0)} \left\{ \int_{0}^{L} E_z(J)J(z)\,dz - \sum_{n=1}^{N} Z_n J^2(z_n) \right\}.$$ **(1.59)**

Free terms in (1.58) and (1.59) are proportional to the current squared and the magnitude of the lumped load. It is worth emphasizing that the connection of loads changes the current distribution along the radiator and the field of the current.

For a folded radiator (Fig. **7a**), which is an example of an antenna consisting of several parallel wires, we obtain

$$Z_A = -\frac{1}{J_g^2} \int_{-L}^{L} E_z(J) J(z) dz .$$

(1.60)

Here J_g is the generator current, $J(z)$ is the total current of an antenna. The current distribution along the antenna wires coincides in the first approximation with that along the wires of an equivalent long line and is found with the help of the theory of electrically coupled lines. Generator current J_g of a closed folded radiator is not always equal to total current $J(z)$ at $z=0$. If the radiator is opened at point A, $J_g = J(0)$. When calculating the field, it is necessary to use the total current, *i.e.*, the sum of the currents of both wires.

Multi-radiator antenna, which is shown on Fig. **7b**, is an example of a radiator consisting of wires with different length. The antenna contains the central radiator with complex load Z_1 and side radiators situated around it and connected with it at the base. In this case, one can find the antenna input impedance from (1.58). The current distribution along the antenna wires is found with the help of the theory of electrically coupled lines. The equivalent line (Fig. **7c**) consists of three wires. The first wire is equivalent to the central radiator, the second wire is equivalent to the system of the same side radiators, and the third wire is the ground.

Since the wires of the equivalent line have the different length and the complex load is connected in the central radiator, the line should be divided into three segments. The segment numbers m are shown on Fig. **7c**. Using the boundary conditions at the segment ends, one can find the current of each wire and the total currents along the segments. Function $J(z)$ is continuous in the entire interval $0 \le z \le l_1$ and behaves sinusoidally on each segment. But its derivative $dJ(z)/dz$ has a jump on the boundaries of the segments. With allowance for the derivative jump we obtain instead (1.31):

$$E_z(J) = -j\frac{15}{k}\left\{\frac{2\exp(-jkR_0)}{R_0}\frac{dJ(0)}{dz} - \left[\frac{\exp(-jkR_{11})}{R_{11}} + \frac{\exp(-jkR_{12})}{R_{12}}\right]\frac{dJ(l_1)}{dz} + \right.$$

$$\left. +\sum_{m=2}^{3}\left[\frac{\exp(-jkR_{m1})}{R_{m1}} + \frac{\exp(-jkR_{m2})}{R_{m2}}\right]\left[\frac{dJ(l_m+0)}{dz} - \frac{dJ(l_m-0)}{dz}\right]\right\},$$

(1.61)

where $R_0 = \sqrt{a^2+z^2}$, $R_{m1} = \sqrt{a^2+(l_m-z)^2}$, $R_{m2} = \sqrt{a^2+(l_m+z)^2}$; a is the radiator radius at point z, and $dJ(l_m+0)/dz$ и $dJ(l_m-0)/dz$ are the derivative values on the right and on the left of point $z = l_m$.

As noted in Section 1.3, the induced emf method does not permit to arrive at finite values of the input impedance at the points of parallel resonance, where $J(h) = 0$, and near these points. A modified method for solving the Leontovich-Levin integral equation allows using the expression (1.47) in this case. One can obtain similar expressions for more complicated radiators as well. For example, for a radiator with N lumped loads and with the feed point displaced to point $z = h$, the input impedance is

$$Z_A = \frac{e}{2J(h) + \frac{1}{e}\left[\int_{-L}^{L}E_z(J)J(z)dz - \sum_{n=1}^{N}Z_nJ^2(z_n)\right]}.$$

(1.62)

These formulas expand essentially the scope of the method. The comparison of a commonly used expression with the integral equation solution confirms as a rule this result.

So far, the subject of discussion was using the induced emf method for calculation of the radiator input impedance. Yet, this method is applied widely also to solving another problem –the estimation of the mutual influence of radiators by means of calculating their reciprocal impedances.

The analysis of two-radiator system is based on the fact that the current of one radiator establishes the field, which has the electrical component tangential to the surface of the second radiator. This component induces the field $E_\varsigma(J_1)d\varsigma$ on element $d\varsigma$ of the second radiator. For the boundary condition $E_\varsigma = 0$ to hold on this surface, the self-field of the second radiator on its surface must be equal to $-E_\varsigma(J_1)d\varsigma$. Here, the generator of the second radiator must expend the power $dP = -E_\varsigma(J_1)J_2(\varsigma)d\varsigma$ on the element $d\varsigma$ and, accordingly, the power

$$P = -\int_{-L_2}^{L_2} E_\varsigma(J_1)J_2(\varsigma)d\varsigma$$ on the entire radiator. Power P is equal to that induced

by the first radiator in the second one, and its ratio to the squared effective value of the generator current determines the impedance magnitude, which the first radiator induced in the second one:

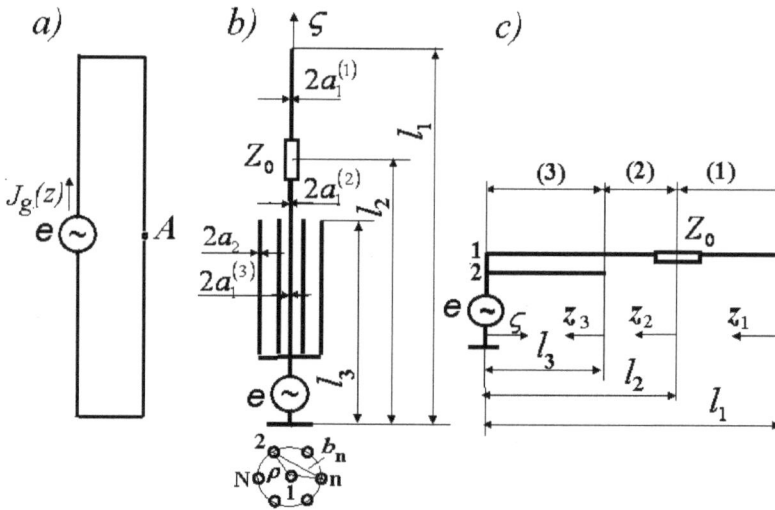

Figure 7: Folded antenna (*a*), multi-radiator antenna with the complex load (*b*), and the long line equivalent to a multi-radiator antenna (*c*).

$$Z_{21ind} = -\frac{1}{J_2^2(0)}\int_{-L_2}^{L_2} E_\varsigma(J_1)J_2(\varsigma)d\varsigma .$$ (1.63)

The Kirchhoff equation for the second radiator takes the form

$$e_2 = J_2(0)[Z_{22}+Z_{21ind}] = J_2(0)Z_{22}+J_1(0)Z_{21},$$ (1.64)

Where $Z_{21} = -\int^{L_2} E_\varsigma(f_1)f_2(\varsigma)d\varsigma$ is the mutual impedance of the first and the second radiators, $f_1(z) = J_1(z)/J_1(0)$, $f_2(\varsigma) = J_2(\varsigma)/J_2(0)$. One can write a similar expression for the first radiator. In the case of Q radiators, it has the form

$$e_p = J_1(0)Z_{p1} + \sum_{q=2}^{Q} J_q(0)Z_{pq}.$$

(1.65)

The corresponding circuit for the *p*th radiator is given on Fig. **8**.

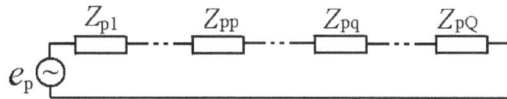

Figure 8: The circuit of the *p*th radiator with series connection of elements.

The expressions presented in this section are given in the accordance with the second formulation of the induced emf method.

The formulas, which allow calculating the mutual impedances of linear radiators for the different variants of their relative replacement, are collected in [7].

1.5. FOLDED RADIATOR WITH CAPACITOR

The last section of the chapter is dedicated to the analysis of a folded radiator with the load. On the one hand, its electrical performance is not known practically. On the other side, it is interesting as an antenna promising for the application of the compensation method.

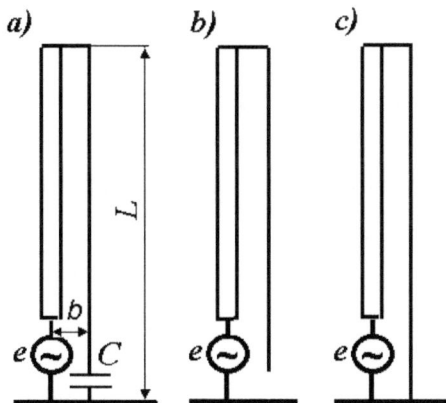

Figure 9: Folded radiators: (*a*) – with capacitor, (*b*) – open, (*c*) – shorted.

The circuit of an asymmetrical folded radiator with a capacitor is presented in Fig. **9a**. The exciting emf is connected in the base of one wire (of the left wire in Figure).

The capacitor is connected in the base of the right-hand (unexcited) wire. In particular case, this circuit is converted into two familiar ones. Such folded radiator is called an open one, if the gap of the wire occurs at the point of the capacitor placement (Fig. **9b**). The folded radiator without a capacitor and gap in the second wire (Fig. **9c**) is called a shorted one. The variant with a capacitor is the general one with respect to two others. It permits to analyze an influence of the magnitude and the sign of the reactive load on the folded radiator characteristics.

As is common in the folded radiator analysis, we divide a radiator with a load into two radiators (Fig. **10**): a circuit with in-phase currents (monopole) and a circuit with anti-phased currents (long line). For that purpose, we replace reactive load Z_1, the voltage drop across it being

$$u = \left(J_l - J_{m2}\right)Z_1 \tag{1.66}$$

(here, J_l is the current in the long line, J_{m2} is the in-phase current in the second wire base), with oppositely directed equivalent emf e_1, the magnitude of which is u. Directions of emf are shown in Fig. **10** with arrows.

We divide the input emf e of the radiator to two emf's with magnitudes ae and $(1-a)e$. The emf's are of the same direction. Also, we divide emf e_1 into two emf's of the same direction and magnitudes a_1e_1 and $(1-a_1)e_1$. In accordance with the superposition principle, the current at point A is the sum of the currents created by all generators. Therefore, as shown on Fig. **10**, one can divide the circuit in question into two circuits, with two generators in each, and then calculate and summarize the currents at point A created in each circuit. Coefficients a and a_1 are selected so that the first circuit can be a monopole (for this, emf's at points A and B must be equal): $ae = a_1e_1$, i.e.

$$a_1 = ae/e_1 . \tag{1.67}$$

The second circuit must be a long line (for that, the ratio of emf V_{l1} between point A of one wire and the ground to input emf V_l of the line must be equal to the share of the in-phase monopole current in the other wire):

$$V_{l1}/V_l = J_{m2}/J_m = m \, .$$

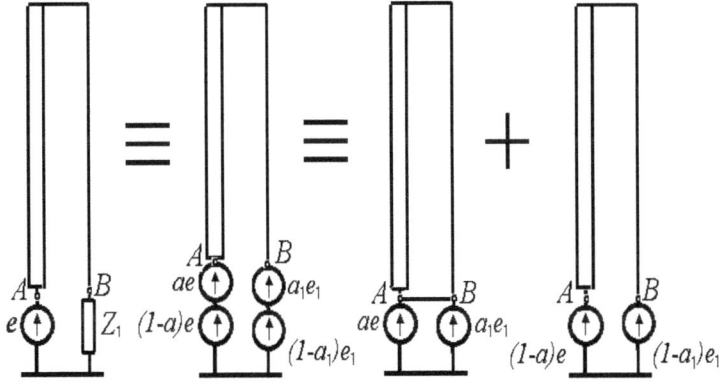

Figure 10: To calculation of a folded radiator with capacitor.

Here J_m is the in-phase current in the radiator base, which is equal to the monopole current; m is the part of the monopole current in the right-hand wire (the share of the monopole current in the left wire is equal to $p = 1 - m$). Since $V_{l1} = (1-a)e$, and $V_l = (1-a)e - (1-a_1)e_1 = e - e_1$, so

$$a = 1 - m(1 - e_1/e) \, . \tag{1.68}$$

The current in the long line is $J_l = V_l/Z_l = e(1 - e_1/e)/Z_l$. Since the monopole current is $J_m = ae/Z_m = (pe + me_1)/Z_m$, the current in the right-hand wire of the monopole is $J_{m2} = me(p + me_1/e)/Z_m$. Substituting J_l and J_{m2} in (1.66) and taking into account that $e_1 = u$, we find

$$\frac{e_1}{e} = Z_1 \frac{1/Z_l - pm/Z_m}{1 + Z_1(1/Z_l + m^2/Z_m)} \, . \tag{1.69}$$

The current in the left-hand wire of the monopole is

$$J_{m1} = pJ_m = pe(p + me_1/e)/Z_m \, .$$

The magnitude of the current in the left-hand wire of the second circuit (the magnitude of the long line current) was given earlier. The input admittance of the radiator is

$$Y_A = \frac{J_{m1} + J_l}{e} = p\left(p + m\frac{e_1}{e}\right)\Big/Z_m + \left(1 - \frac{e_1}{e}\right)\Big/Z_l = \frac{p^2}{Z_m} + \frac{1}{Z_l} + \left(\frac{pm}{Z_m} - \frac{1}{Z_l}\right)\frac{e_1}{e}.$$

Substituting ratio e_1/e from (1.69) into this expression, we obtain:

$$Y_A = \frac{Z_m + p^2 Z_l + Z_1}{Z_m Z_l \left[1 + Z_1\left(\dfrac{1}{Z_l} + \dfrac{m^2}{Z_m}\right)\right]}. \tag{1.70}$$

Verification shows that expression (1.70) reduces in specific cases to known formulas. If $Z_1 = \infty$, then

$$Z_A = 1/Y_A = Z_m Z_l\left(1/Z_l + m^2/Z_m\right) = Z_m + m^2 Z_l,$$

which coincides with the input impedance of an open folded radiator. Here, according to (1.69)

$$e_1/e = \frac{1/Z_l - pm/Z_m}{1/Z_l + m^2/Z_m} = \left(Z_m - pmZ_l\right)/Z_A,$$

i.e., $J_m = e\left(p + me_1/e\right)/Z_m = e/Z_A$, $J_{m1} = pJ_m = pe/Z_A$, $J_{m2} = mJ_m = me/Z_A$, $J_l = e\left(1 - e_1/e\right)/Z_l = me/Z_A$.

If $Z_1 = 0$, then $e_1 = 0$, $Y_A = \left(Z_m + p^2 Z_l\right)/\left(Z_m Z_l\right) = 1/Z_l + p^2/Z_m$, which coincides with the input admittance of a shorted folded radiator. Here, $J_m = pe/Z_m$, $J_{m1} = p^2 e/Z_m$, $J_{m2} = pme/Z_m$, $J_l = e/Z_l$.

Connection of the reactive impedance between the second wire and the ground was proposed in order to improve the matching of a folder radiator with a cable or a generator. The absence of positive results in this case does not rule out the possibility of using this circuit in the future.

Send Orders for Reprints to reprints@benthamscience.net

CHAPTER 2

Integral Equation Method

Abstract: Integral equations for currents in one and two straight metal radiators with exact and approximate kernel and methods of solving these equations with the help of an iterative process, perturbation method and Moment Method are considered. Results are generalized to the case of radiators with constant and piecewise-constant impedance and with lumped loads. The Moment Method with piecewise-sinusoidal basic and weighting functions is shown to correspond to the physical content of a problem and be equivalent to division of the radiator into isolated dipoles, the self- and mutual impedances of which are calculated by the method of induced emf.

Keywords: Approximate kernel, Basic functions, Boundary condition, Complicated structures, Constant impedance, Entire-domain functions, Equation for a system of radiators, Equation for two radiators, Exact kernel, Generalized method of induced emf, Hallen's equation, Integral equation for the current, King-Middleton's iterative procedure, Leontovich-Levin equation, Logarithmic singularity, Lumped loads, Metal rod with a magnetodielectric coat, Moment method, Perturbation method, Piecewise constant impedance, Piecewise-sinusoidal functions, Pocklington's equation, Radiators systems of straight wire segments, Slowing-down, Straight metal radiator, Subdomain functions, Weighting functions.

2.1. INTEGRAL EQUATION FOR A LINEAR METAL RADIATOR

As shown in Chapter 1, knowledge of the current distribution along a linear radiator allows calculating the electromagnetic field and all electrical characteristics of the radiator. For this reason the main problem of the antenna theory is finding the current distribution.

Current $J(z)$ of a dipole creates electromagnetic field $E_z(J)$ satisfying the boundary condition

$$E_z(a,z)\big|_{-L\leq z\leq L}+K(z)=0. \tag{2.1}$$

The cylindrical coordinate system is used here, a and L are the radius and the arm length of a dipole, respectively, $K(z)$ is an extraneous (impressed) emf. There is no current at the radiator ends:

$$J(\pm L) = 0. \tag{2.2}$$

Expression (2.1) is the mathematical record of the fact that the total field, which is a sum of the extraneous field and the current field, is zero on the surface of a perfectly conducting radiator. The extraneous field is specified usually as a product of potentials difference e between the gap edges with δ-function. Quantity $K_1(z) = e\delta(z)$ corresponds to the generator connection in the radiator middle, at point $z = 0$, and $K_2(z) = e\delta(z-h)$ corresponds to its displacement, *i.e.,* to the generator connection at point $z = h$.

Equality (2.1) contains as in embryo all integral equations of the thin antenna theory. The appearance of the equations is determined mostly by the selection of function $E_z(J)$. For example, using (1.27), we obtain Hallen's integral equation for the current along a filament

$$\int_{-L}^{L} J(\varsigma) G_1 d\varsigma = -\frac{j}{Z_0}\left(C \cos kz + \frac{e}{2}\sin k|z|\right), \tag{2.3}$$

where $G_1 = \exp(-jkR_1)/(4\pi R_1)$, $R_1 = |z-\varsigma|$. Using (1.28), we obtain Hallen's integral equation for the current along a straight thin-wall cylinder (the equation with exact kernel)

$$\frac{1}{2\pi}\int_{-L}^{L} J(\varsigma)\int_{0}^{2\pi} G_2 d\phi d\varsigma = -\frac{j}{Z_0}\left(C \cos kz + \frac{e}{2}\sin k|z|\right). \tag{2.4}$$

Here $G_2 = \exp(-jkR_2)/(4\pi R_2)$, $R_2 = \sqrt{(z-\varsigma)^2 + 4a^2 \sin^2 \phi/2}$. The integral equation for the current along a filament of a finite radius (the equation with approximate kernel) became a frequent practice:

$$\int_{-L}^{L} J(\varsigma) G_3 d\varsigma = -\frac{j}{Z_0}\left(C \cos kz + \frac{e}{2}\sin k|z|\right), \tag{2.5}$$

where $G_3 = \exp(-jkR_3)/(4\pi R_3)$, $R_3 = \sqrt{(z-\varsigma)^2 + a^2}$. Constant C in each equation is found from condition (2.2).

If we substitute filament field $E_z(J)$ into (2.1) in accord with (1.27) and replace R_1 with R_3, we shall obtain Pocklington's equation [17]

$$\int_{-L}^{L} J(\varsigma)\left(k^2 G_3 + \frac{\partial^2 G_3}{\partial z^2}\right)d\varsigma = -j\omega\varepsilon K(z), \tag{2.6}$$

which is the integral equation for the current along a filament of a finite radius as well.

The first solution of Hallen's equation with approximate kernel was found by Hallen himself and is described in detail in [11]. The solution uses magnitude $\Omega = 2\ln(2L/a) = 1/\chi$ as the parameter, in inverse powers of which function $J(z)$ is expanded into a series. It is obtained with the help of a successive approximation method (iterative procedure)

$$J(z) = j\frac{e}{60\Omega} \cdot \frac{\sin k(L-|z|) + N_1(z)/\Omega + N_2(z)/\Omega^2 + \ldots}{\cos kL + B_1(L)/\Omega + B_2(L)/\Omega^2 + \ldots} = J_{0H}(z)/\Omega + J_{1H}(z)/\Omega^2 + \ldots, \tag{2.7}$$

where, e.g., $J_{0H}(z) = j\dfrac{e}{60\cos kL}\sin k(L-|z|)$, $J_{1H}(z) = j\dfrac{e}{60}\left[\dfrac{N_1(z)}{\cos kL} - \dfrac{B_1(L)}{\sin k(L-|z|)}\right]$.

Functions $N_i(z)$ and $B_i(z)$ are integrals, which can be expressed in terms of integral sines and cosines.

The iterative procedure proposed by King and Middleton [18] yields more accurate result. The common expression for the current in it is similar to (2.7), but expansion parameter Ω is replaced with Ψ. For example, zero approximation instead of $J_{0H}(z)/\Omega$ is given by

$$J_{0KM}(z)/\Psi = j\frac{e}{60\Psi\cos kL}\sin k(L-|z|).$$

To find expansion parameter Ψ, magnitude $\psi(z)$ is used. This is calculated as

$$\psi(z) = \int_{-L}^{L} \frac{J_{0KM}(\varsigma)}{J_{0KM}(z)} \cdot \frac{\exp(-jkR)}{R}d\varsigma.$$

For Ψ, the value of $\psi(z)$ at point $z = z_m$, where the current is at the maximum or close to it, was taken, *i.e.,*

$$\Psi = \begin{cases} \psi(0), & kL \leq \pi/2, \\ \psi(L-\lambda/4), & kL > \pi/2. \end{cases}$$

Such selection of the expansion parameter is caused by the fact that function $\psi(z)$ is proportional to the ratio of vector potential $A_z(z)$ at point z on the antenna surface to current $J(z)$ in the same cross-section. For that reason, function $\psi(z)$ varies slowly along the antenna, or rather, it is almost constant except for free ends.

Leontovich-Levin equation [12] played an important part in the progress of the thin antenna theory. In accordance with (1.28), vector potential $A_z(\rho,z)$ of the field produced by the current along a straight circular thin-wall cylinder is

$$A_z(\rho,z) = \frac{\mu}{8\pi^2} \int_0^{2\pi} T(z,\phi) d\phi, \tag{2.8}$$

where

$$T(z,\phi) = \int_{-L}^{L} \frac{J(\varsigma)\exp(-jkR)}{R} d\varsigma, \quad R = \sqrt{(z-\varsigma)^2 + \rho_1^2}, \quad \rho_1 = \sqrt{\rho^2 + a^2 - 2a\rho\cos\phi}.$$

Integrating $T(z,\phi)$ by parts and using successively the circumstance that the radiator radius is small in comparison with its length and the wavelength, *i.e.,* neglecting the terms of order of ka and a/L and keeping the terms proportional to the logarithm of the quantities, we obtain:

$$T(z,\phi) = -2J(z)\ln p\rho_1 - \int_{-L}^{L} \exp(-jk|\varsigma-z|)sign(\varsigma-z)\ln 2p|\varsigma-z| \left[\frac{dJ(\varsigma)}{d\varsigma} - jkJ(\varsigma)sign(\varsigma-z)\right]d\varsigma,$$

where p is a constant having the dimension of inverse length. Since at $\rho > a$

$$\int_0^{2\pi} \ln p\rho_1 d\phi \equiv \int_0^{2\pi} \ln(p\sqrt{\rho^2 + a^2 - 2a\rho\cos\phi})d\phi = 2\pi\ln p\rho,$$

so, in accordance with (2.8)

$$A_z(\rho,z)=\frac{\mu}{4\pi}\left[-2J(z)\ln p\rho+V(J,z)\right],$$

where

$$V(J,z)=\int_{-L}^{L}\exp\left(-jk|\xi-z|\right)\ln 2p|z-\varsigma|\left[jkJ(\varsigma)+sign(z-\varsigma)dJ(\varsigma)/d\varsigma\right]d\varsigma.$$

Substituting $A_z(\rho,z)$ into the first equation of the system (1.25) and setting ρ equal to a, we find:

$$E_z(a,z)=\frac{1}{4\pi j\omega\varepsilon}\left[\chi^{-1}\left(\frac{d^2J}{dz^2}+k^2J\right)+\frac{d^2V}{dz^2}+k^2V\right]. \tag{2.9}$$

Here $\chi=-1/(2\ln pa)$ is a small parameter of the thin antenna theory. For constant $1/p$, one should choose the distance to the nearest inhomogeneity, *i.e.*, the smallest of three quantities: wavelength λ, antenna length $2L$ and the radius R_c of its curvature. In case of a straight radiator, the length of which does not exceed the wavelength, one can consider that $1/p = 2L$, *i.e.*,

$$\chi=1/\left[2\ln(2L/a)\right].$$

From (2.1) and (2/9), we obtain the wanted equation

$$\frac{d^2J}{dz^2}+k^2J=-\chi\left[4\pi j\omega\varepsilon K(z)+\frac{d^2V}{dz^2}+k^2V\right]. \tag{2.10}$$

The equation has along with the terms, which contain the exciting emf, the current and the current derivative, the integral item, depending on the current distribution along the conductor. It is the additional emf taking the radiation into account.

The meaning of these manipulations consists, firstly, in the separation of a logarithmic singularity: integral $V(J,z)$ is a continuous function everywhere unlike the original integral (2.8). Secondly, argument ϕ is absent in both parts of the equation (2.10) in contrast to the equation (2.4), since the integration with respect to ϕ has been accomplished here. Nevertheless, the equation was derived for the current along a straight thin-wall cylinder and is equivalent to the Hallen's equation with exact kernel.

Solving the equation (2.10) in [12] uses the perturbation method, *i.e.,* the solution is sought as expansion into a series in powers of the small parameter χ:

$$J(z)=J_0(z)+\chi J_1(z)+\chi^2 J_2(z)+...\tag{2.11}$$

Substituting this series into the equation (2.10) and equating coefficients of equal powers of χ, we come, in the case of an untuned radiator (when $J_0(z)=0$), to the set of equations:

$$\frac{d^2 J_1(z)}{dz^2}+k^2 J_1(z)=-4\pi j\omega\varepsilon K(z),\quad J_1(\pm L)=0,$$

$$\frac{d^2 J_n(z)}{dz^2}+k^2 J_n(z)=-\left[\frac{d^2 V(J_{n-1},z)}{dz^2}+k^2 V(J_{n-1},z)\right],J_n(\pm L)=0,n>1.\tag{2.12}$$

In [12], the equations (2.12) are solved in the second approximation with respect to χ, and the expression for the antenna current is given at point $z=0$. If one uses this procedure and to calculates the current at an arbitrary point of the radiator, then at $\chi=1/[2\ln(2L/a)]$ (another small parameter was used in [12]: $\chi_1=1/(2\ln ka)$), one can obtain [19]:

$$J_1(z)=j\frac{e}{60\cos\alpha}\sin(\alpha-|t|),\quad J_2(z)=\frac{e\Theta_2(t,\alpha)}{120\sin 2\alpha\cos\alpha}.\tag{2.13}$$

Here, $t=kz, \alpha=kL,$

$$\Theta_2(t,\alpha)=$$
$$=\sin(\alpha+t)\left\{\begin{array}{l}C+\ln 2(\alpha-t)-Ei[-j2(\alpha-t)]+e^{j2\alpha}\\ \langle Ei(-j4\alpha)-Ei[-j2(\alpha+t)]\rangle+e^{-j2\alpha}\ln[(\alpha+t)/(2\alpha)]\end{array}\right\}+$$

$$+\sin(\alpha-t)\begin{Bmatrix} C+\ln 2(\alpha+t)-Ei\left[-j2(\alpha+t)\right]+e^{j2\alpha}\left\langle Ei(-j4\alpha)-Ei\left[-j2(\alpha-t)\right]\right\rangle+ \\ e^{-j2\alpha}\ln\left[(\alpha-t)/(2\alpha)\right] \end{Bmatrix}-$$

$$-2\cos\alpha\sin(\alpha+|t|)\left\{e^{j\alpha}\left[Ei(-j2\alpha)-Ei(-j2|t|)\right]+e^{-j\alpha}\ln(t/\alpha)\right\}-$$

$$-2\cos\alpha\sin(\alpha-|t|)\left\{e^{j\alpha}\left[Ei(-j2\alpha)-\ln(2\gamma\alpha^2/t)\right]+e^{-j\alpha}\left[\ln 2\gamma\alpha-Ei(-j2|t|)\right]\right\},$$

and $\ln\gamma = C = 0.5772...$ is the Euler's constant. As easily verified, the current in the first approximation with respect to χ follows the sinusoidal law (1.8) and coincides with the zero approximation obtained as a result of the solving Hallen's equation:

$$J_1(z) \equiv J_{0H}(z).$$

If one uses the formulas presented in [11] for the summands of the function $J_{1H}(z)$ and the equation (2.13), one can show that

$$J_2(z) \equiv J_{1H}(z).$$

Therefore, the approximation number is a magnitude, conventional to some extent.

The input impedance of the antenna in the first approximation in accordance with (2.13) is

$$Z_{A1} = -j60\chi^{-1}\cot kL. \tag{2.14}$$

It has the reactive component only, and quantity $60/\chi$ is equal to the wave impedance of the equivalent long line.

2.2. INTEGRAL EQUATIONS FOR TWO RADIATORS

Generalizing the Leontovich-Levin equation, one can write similar equations for the currents in the system of several radiators, *i.e.,* in an antenna array. Consider two parallel symmetrical radiators of different lengths, displaced axially relative to each other (Fig. **1**). Starting from (1.25), if electrical currents $J_I(\sigma)$ and $J_{II}(\varsigma)$ of circular frequency ω run along the radiators, they create the field

$$E_z = -j\frac{\omega}{k^2}\left(k^2 A_z + \frac{\partial^2 A_z}{\partial z^2}\right).$$ **(2.15)**

In accordance with the superposition principle

$$A_z = A_{z1} + A_{z2}\ldots$$

Each radiator is modeled as a straight thin-wall circular cylinder with radius a_1 and a_2, respectively. The vector potential of the field created by the current of the cylinder is calculated with the help of (2.8), and distances R_1 and R_2 from the observation point with coordinates (ρ_1, ϕ_0, z) to an integration points (a_1, ϕ, σ) and (a_2, ψ, ς) are calculated in accord with the explication to this expression.

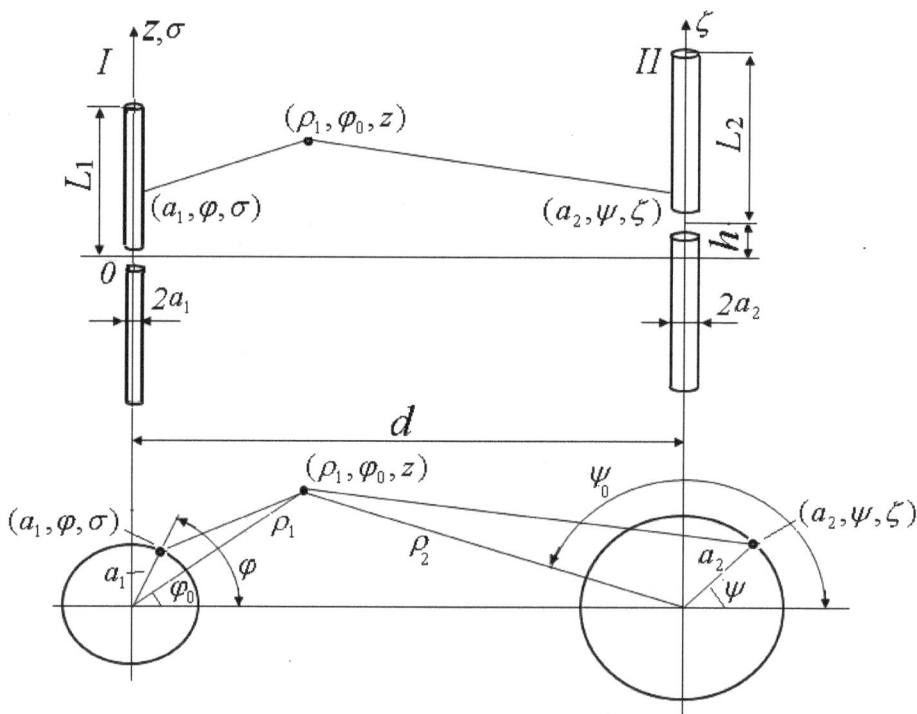

Figure 1: Systems of two parallel radiators.

If the observation point is situated near the first radiator surface, then at

$$a_1, a_2 \langle\langle d, \qquad\qquad\qquad\qquad \textbf{(2.16)}$$

where d is the distance between radiators, one can set:

$$A_{z2} = \frac{\mu}{4\pi} \int_{h-L_2}^{h+L_2} J_{II}(\varsigma) \exp(-jkR_2)/R_2 \, d\varsigma , \quad R_2 = \sqrt{(z-\varsigma)^2 + d^2} . \tag{2.17}$$

Vector potential A_{z1}, as in the case of an isolated radiator has a logarithmic singularity at small ρ_1, the separation of which gives

$$A_{z1}(a_1, z) = \frac{\mu}{4\pi} \Big[\chi_1^{-1} J_I(z) + V(J_I, z) \Big], \tag{2.18}$$

where $\chi_1 = 1/\big[2\ln(2L_1/a_1) \big]$ is a small parameter, and $V(J_I, z)$ is the integral, expression for which is presented in Section 2.1. Vector potential A_{z2} has no such singularity, since, if the assumption (2.16) is valid, quantity R_2 is not small at any ς: $R_2 \geq d - \rho_1 - a_2$. Accordingly the tangential component of the electric field created by current J_I contains a large quantity on order of χ_1^{-1}:

$$E_z(J_I, a_1, z) = \frac{1}{4\pi j\omega\varepsilon} \left\{ \chi_1^{-1} \left[\frac{d^2 J_I(z)}{dz^2} + k^2 J_I(z) \right] + \frac{d^2 V(J_I, z)}{dz^2} + k^2 V(J_I, z) \right\}, \tag{2.19}$$

and field $E_z(J_{II}, a_1, z)$ created by current J_{II} of the second radiator on the surface of the first one contains no large summand for reasons given above.

A boundary condition, similar to (2.1), must be met on the surface of the first radiator,

$$E_z(J_I, a_1, z) + E_z(J_{II}, a_1, z)\big|_{-L_1 \leq Z \leq L_1} + K_I(z) = 0, \tag{2.20}$$

where $K_I(z)$ is an extraneous (impressed) emf. Substituting (2.19) into (2.20), we obtain the equation for $J_I(z)$:

$$\frac{d^2 J_I(z)}{dz^2} + k^2 J_I(z) = -4\pi j\omega\varepsilon\chi_1 \Big[K_I(z) + W(J_I, z) + E_z(J_{II}, a_1, z) \Big], \tag{2.21}$$

where $4\pi j\omega\varepsilon W(J_I, z) = d^2 V(J_I, z)/dz^2 + k^2 V(J_I, z)$, and $J_I(\pm L_1) = 0$.

The right part of this expression contains three summands in square brackets: the first one is the exciting emf, the second one is emf taking the radiation into account, the third one is emf due to the influence of the second radiator.

While solving the equation (2.21), we represented the currents $J_I(z)$ and $J_{II}(\varsigma)$ in the form of series in powers of small parameters χ_1 and χ_2, respectively. Since functionals $W(J_I, z)$ and $E_z(J_{II}, a_1, z)$ are linear, they can also be represented in the form of a similar series. If χ_1 and χ_2 have the same order of smallness:

$$\chi_1 \sim \chi_2, \tag{2.22}$$

then the equation (2.21) reduces for untuned radiators to the set of equations, which is a generalization of the set (2.12), written for a single radiator:

$$\frac{d^2 J_{I1}(z)}{dz^2} + k^2 J_{I1}(z) = -4\pi j\omega\varepsilon K_I(z), \qquad J_{I1}(\pm L_1) = 0,$$

$$\frac{d^2 J_{In}(z)}{dz^2} + k^2 J_{In}(z) = -4\pi j\omega\varepsilon \left[W(J_{I,n-1}) + \left(\frac{\chi_2}{\chi_1}\right)^{n-1} E_z(J_{II,n-1}) \right], J_{In}(\pm L_1) = 0, n > 1. \tag{2.23}$$

As it follows from the first equation of the system (2.23), with the radiator excited by concentrated emf $K_I(z) = e_I \delta(z)$, the current in the presence of the second radiator also is of a sinusoidal nature in the first approximation:

$$\chi_1 J_{I1}(z) = j\frac{\chi_1 e_I}{60\cos kL_1}\sin k(L_1 - |z|). \tag{2.24}$$

If $n > 1$, then, in accordance with (2.23), using the method of variation of constants and considering that quantities $W(J_{I,n-1})$ and $E_{II}(J_{II,n-1})$ are known, we obtain

$$\chi_1^n J_{In}(z) = j\frac{\chi_1}{30\sin 2kL_1}\left\{ \sin k(L_1+z)\int_z^{L_1}\left[W(\chi_1^{n-1}J_{I,n-1}) + E_\sigma(\chi_2^{n-1}J_{II,n-1})\right]\sin k(L_1-\sigma)d\sigma + \right.$$

$$\left. + \sin k(L_1-z)\int_{-L_1}^z\left[W(\chi_1^{n-1}J_{I,n-1}) + E_\sigma(\chi_2^{n-1}J_{II,n-1})\right]\sin k(L_1+\sigma)d\sigma \right\}. \tag{2.25}$$

Find quantities $W\left(\chi_1^{n-1}J_{I,n-1}\right)$, by substituting the first terms of the series for current $J_I(z)$ into the expression (2.19):

$$E_z\left(\chi_1 J_{I1}\right) = -K_I(z) + W\left(\chi_1 J_{I1}\right),$$

$$E_z\left(\chi_1^n J_{In}\right) = -W\left(\chi_1^{n-1}J_{I,n-1}\right) - E_z\left(\chi_2^{n-1}J_{II,n-1}\right) + W\left(\chi_1^n J_{In}\right), n > 1,$$

i.e.,

$$E_z\left(\sum_{m=1}^{n}\chi_1^m J_{I,m}\right) = -K_I(z) - E_z\left(\sum_{m=1}^{n-1}\chi_2^m J_{II,m}\right) + W\left(\chi_1^n J_{In}\right). \tag{2.26}$$

Replacing n in (2.26) with $(n-1)$, calculating quantity $W\left(\chi_1^{n-1}J_{I,n-1}\right)$ with the help of the resulting expression, and substituting it in (2.25), we find the nth term of the current series. In particular, if $n = 2$,

$$\chi_1^2 J_{I2}(z) = j\frac{\chi_1}{30\sin 2kL_1}\left\{\sin k(L_1+z)\int_z^{L_1}\left[K_I(\sigma)+E_\sigma\left(\chi_1 J_{I1}\right)+E_\sigma\left(\chi_2 J_{II1}\right)\right]\sin k(L_1-\sigma)d\sigma+\right. \tag{2.27}$$

$$\left.+\sin k(L_1-z)\int_{-L_1}^{z}\left[K_I(\sigma)+E_\sigma\left(\chi_1 J_{I1}\right)+E_\sigma\left(\chi_2 J_{II1}\right)\right]\sin k(L_1+\sigma)d\sigma\right\}.$$

The equation (2.27) allows finding the second term of the series for the current at any point of the first radiator. For this purpose, as it follows from (2.27), it is necessary to calculate the fields of the currents, found in the first approximation. From (2.27), one can see also that, as a matter of course, the magnitude of the second term of the series depends on the geometric dimensions of the second radiator and on the relative position of radiators. In the general case, the expression (2.25), after substituting quantity $W\left(\chi_1^{n-1}J_{I,n-1}\right)$ in it, permits to find the nth summand, if the currents of both radiators are known in $(n-1)$th approximation.

From the set of equations (2.23), it follows that, when calculating the second and higher terms of the series, one can consider that the current of the first radiator is concentrates on the axis. Also, according to (2.17), one can consider that the current of the second radiator is concentrated on the axis. And the accuracy level accepted in derivation of the equation (2.21) (accuracy on order of a_1/L_1) is retained. The circumstance simplifies essentially calculations by recurrence formula, connecting nth and $(n-1)$th summands, since it allows calculating the field of a series term as the filament field. As a result, the calculating the second summand of the series for the current in the single radiator, based on using expression (2.27), is simplified, since one can use the expression (2.13) as the first summand.

Compare the results of solving the Leontovich-Levin equations for one and two radiators with those obtained by the induced emf method. The input impedance of the first radiator is

$$Z_{AI} = e_I / J_I(0) = e_I \left/ \left[\sum_{n=1}^{\infty} \chi_1^n J_{In}(0) \right] \right., \tag{2.28}$$

where $\chi_1^n J_{In}(0) = \dfrac{1}{e_I} \int\limits_{-L_1}^{L_1} \left[K_I(\sigma) + E_\sigma\left(J_I^{(n-1)}\right) + E_\sigma\left(J_{II}^{(n-1)}\right) \right] J_I^{(1)}(\sigma) d\sigma$. Here, to

simplify, the following designation is used for the first radiator current in the nth approximation: $J_I^{(n)}(z) = \sum\limits_{m=1}^{n} \chi_1^m J_{Im}$.

The input impedance in the nth approximation is equal to

$$Z_{AI}^{(n)} = \frac{e_I}{J_I^{(n-1)}(0) + \chi_1^n J_{In}(0)}. \tag{2.29}$$

Let us write the first summand of the denominator in the form

$$J_I^{(n-1)}(0) = \frac{1}{e_I} \int\limits_{-L_1}^{L_1} K_I(\sigma) J_I^{(n-1)}(\sigma) d\sigma .$$

Factor $J_I^{(1)}(\sigma)$ in the integrand of the second summand of the denominator can be replaced with $J_I^{(n-1)}(\sigma)$, *i.e.,* one can add terms of higher order of smallness to a quantity of the first order. Since the polynomial in square brackets of the integrand is a quantity of $(n-1)$th order of smallness, as is easily seen from

(2.26), the addition of terms of higher order does not change the accepted accuracy level. Hence,

$$\chi_1^n J_{In}(0) = J_I^{(n-1)}(0) + \frac{1}{e_I} \int_{-L_1}^{L_1} \left[E_\sigma \left(J_I^{(n-1)} \right) + E_\sigma \left(J_{II}^{(n-1)} \right) \right] J_I^{(n-1)}(\sigma) d\sigma .$$

As a result, we obtain

$$Z_{AI}^{(n)} = e_I \left/ \left\{ 2J_I^{(n-1)}(0) + \frac{1}{e_I} \int_{-L}^{L} \left[E_\sigma \left(J_I^{(n-1)} \right) + E_\sigma \left(J_{II}^{(n-1)} \right) \right] J_I^{(n-1)}(\sigma) d\sigma \right\} \right. . \qquad (2.30)$$

One can rewrite the expression (2.29) as

$$Z_{AI}^{(n)} = \frac{e_I}{J_I^{(n-1)}(0)} \left/ \left[1 + \frac{\chi_1^n J_{1n}(0)}{J_I^{(n-1)}(0)} \right] \approx \frac{e_I}{J_I^{(n-1)}(0)} \left[1 - \frac{\chi_1^n J_{In}(0)}{J_I^{(n-1)}(0)} \right] \right. .$$

If

$$J_I^{(n-1)}(0) \neq 0 , \qquad (2.31)$$

then

$$Z_{AI}^{(n)} = e_I \left[J_I^{(n-1)}(0) - \chi_1^n J_{In}(0) \right] \left/ \left[J_I^{(n-1)}(0) \right]^2 \right. ,$$

i.e.,

$$Z_{AI}^{(n)} \approx - \frac{1}{\left[J_I^{(n-1)}(0) \right]^2} \int_{-L_1}^{L_1} \left[E_\sigma \left(J_I^{(n-1)} \right) + E_\sigma \left(J_{II}^{(n-1)} \right) \right] J_I^{(n-1)}(\sigma) d\sigma . \qquad (2.32)$$

The expression generalizes expression (1.39), which was presented in Section 1.3 and is called "the second formulation of the induced emf method". In (1.39), the sinusoidal distribution of the current along the radiator is used to calculate the input impedance in the second approximation with respect to χ. In (2.32), the $(n-1)$th approximation for the current permits to calculate the input impedance in the nth approximation with respect to χ. The equation (1.39) is applicable only to a single radiator, whereas equation (2.32) is valid in the presence of the second radiator as well. The expression (2.30) allows writing the equation (1.47).

Comparison of the obtained results with those by the induced emf method permits to draw the following conclusion.

The integral formula of the induced emf method for the radiator input impedance, if the condition (2.31) holds, coincides completely with the integral formula obtained from the solution of the integral equation. The identity of these formulas explains the known fact that the input impedances calculated by both methods and expressed in terms of tabulated functions coincide in the second approximation with respect to χ.

Actually, if one takes the expressions (2.13) as a basis for the first term of the series (2.11) and performs the transition from the input current to the radiator input impedance, which is similar to the transition from (2.28) to (2.32), the result proves to be identical with that of calculation performed by the induced emf method. Since the condition (2.31) at the point of a parallel resonance for the sinusoidal current distribution is not met, the emf method yields incorrect results near that point, and both calculated resistance and reactance increase without bound, while the measured values of the input impedances remain finite.

The derivation of (2.32) uses conditions (2.16) and (2.22). The fulfillment of the conditions is necessary to avoid possible mistakes.

The first formulation of the induced emf method can be reduced to a form similar to expression (2.32):

$$Z_{AI}^{(n)} \approx \frac{1}{\left|J_I^{(n-1)}(0)\right|^2} \int_{-L_1}^{L_1} \left[E_\sigma\left(J_I^{(n-1)}\right) + E_\sigma\left(J_{II}^{(n-1)}\right)\right] J_I^{(n-1)*}(\sigma)\, d\sigma. \qquad (2.33)$$

It is shown in [19] that the expression is obtainable by the direct transition from (2.32). But for that, the equality $J_1^{(1)}(\sigma) = -J_1^{(1)*}(\sigma)$ needs to hold. In accordance with this equality, the current should be purely reactive, *i.e.,* there should be no losses either in the radiator, or in the environment.

To conclude the section, it is necessary to say a few words about an antenna with a feed displaced from the radiator center to point $z = h$ (see Fig. **4b**). The expression (2.32) was derived for the extraneous emf without defining $K(z)$ concretely. For this reason, it is applicable to the radiator with $h \neq 0$, if current $J_I(0)$ and all terms of its series are replaced with current $J_I(h)$ and the terms of a new series. For the single radiator, we obtain the expression, which generalizes the expression (1.50) written in accordance with the induced emf method:

$$Z_{AI}^{(n)} \approx -\frac{1}{\left[J_I^{(n-1)}(h)\right]^2} \int_{-L_1}^{L_1} E_\sigma\left(J_I^{(n-1)}\right) J_I^{(n-1)}(\sigma) d\sigma.$$

2.3. INTEGRAL EQUATIONS FOR COMPLICATED STRUCTURES

Two previous sections are dedicated to integral equations for the currents in straight metal radiators. Radiators with distributed and concentrated loads are more complicated variants. An antenna in the form of a metal rod coated with a magnetodielectric layer (Fig. **5a**) of Chapter 1 is an example of a radiator with distributed load. In contrast to (2.1), the boundary condition on the surface of a dipole with distributed load is given by

$$\frac{E_z(a,z) + K(z)}{H_\phi(a,z)}\bigg|_{-L \leq Z \leq l} = Z(z), \tag{2.34}$$

where $E_z(a, z)$ and $H_\phi(a, z)$ are the tangential component of the electric field and the azimuthal component of the magnetic field, respectively, and $Z(z)$ is the surface impedance, which is in the general case dependent on coordinate z. The boundary condition of such kind is valid, if the structure of the field in one of the media (inside an antenna, *e.g.,* in a magnetodielectric sheath) is known and independent of the field structure in the other medium (ambient space).

The radiators with boundary conditions (2.34) met on their surface and with the surface impedance that substantially changes the distribution of current along the antenna already in the first approximation are called the impedance ones.

In accordance with the equivalence theorem, one can, when calculating the field, replace the radiator with the field on its boundary, and afterwards use only the field. However, for clearness and simplicity, it is expedient to metallize the antenna surface. Surface density $\vec{j}_S$ of the electric current is related to magnetic field strength $\vec{H}$ as $\vec{j}_S = [\vec{e}_\rho, \vec{H}]$, where $\vec{e}_\rho$ is the unit vector in the ρ direction. Then

$$H_\phi(a,z) = j_z(z) = J(z)/(2\pi a),\qquad(2.35)$$

where $J(z)$ is the linear current along a metalized antenna (it is equal to the total radiator current).

The tangential component of the field is determined by expression (2.19). Substituting (2.19) and (2.35) into (2.34), we obtain the equation for the current along an impedance radiator:

$$\frac{d^2 J(z)}{dz^2} + k^2 J(z) = -4\pi j\omega\varepsilon\chi\left[K(z) + W(J,z) - \frac{J(z)Z(z)}{2\pi a}\right],\qquad(2.36)$$

which should satisfy the condition (2.2). Three summands in the right-hand part of the equation take into account the exciting emf, the radiation, and the presence of the distribution load, respectively.

As before, we shall seek the solution as a series in powers of small parameter χ, presenting the surface impedance as $2jkZ(z)/(aZ_0) = \chi^{-1}U$. That allows arriving at the set of equations for the untuned radiator:

$$\frac{d^2 J_1(z)}{dz^2} + k_1^2 J_1(z) = -4\pi j\omega\varepsilon K(z),\qquad J_1(\pm L) = 0,$$

$$\frac{d^2 J_n(z)}{dz^2} + k_1^2 J_n(z) = -4\pi j\omega\varepsilon W(J_{n-1},z),\qquad J_n(\pm L) = 0, n > 1.$$

$$(2.37)$$

Here $k_1^2 = k^2 - U$. If both summands are of the same order of smallness, the surface impedance substantially affects the distribution of current, and one must attach the meaning of a new wave propagation constant along an antenna to the quantity $k_1 = \sqrt{k^2 - j2k\chi Z(z)/(aZ_0)}$. From the first equation of system (2.37) it follows that the current distribution along the antenna has in the first approximation sinusoidal nature

$$\chi J_1(z) = j\frac{k\chi e}{60k_1 \cos k_1 L} \sin k_1 (L - |z|).$$
(2.38)

Ratio k_1/k is usually referred to as the slowing-down.

Solving the equation for $J_2(z)$ in the system (2.37) allows finding the current in the second approximation, calculating the active component of input impedance and defining the value of reactive component more precisely. If one uses the modified method of solving the equation for the current described in the previous section, the additional summand $Z/(2\pi a)\sum_{m=1}^{N} \chi^m J_I^{(m)}(z)$ will appear in the right-hand part of expression (2.26).

If the condition (2.31) holds, then, by analogy to (2.32), we find for the single radiator

$$Z_A^{(n)} \approx -\frac{1}{\left[J^{(n-1)}(0)\right]^2} \int_{-L_1}^{L_1} \left[E_z\left(J^{(n-1)}\right) - \frac{Z}{2\pi a}J^{(n-1)}\right] J_I^{(n-1)}(z)\,dz .$$
(2.39)

This generalizes the expression (1.49) written in accordance with the induced emf method.

A radiator with constant surface impedance is a particular case of a radiator with impedance varying along the antenna. Let, for example, the radiator consist of $2N$ sections of length l_m, surface impedance Z_m being constant on each of them (Fig. **5b**) of Chapter 1. We consider that the radiator to be symmetrical, and the emf to be connected at its center. The equation for current $J_m(z)$ along the mth section of a radiator takes the form

$$\frac{d^2 J_m(z)}{dz^2} + k^2 J_m(z) = -4\pi j\omega\varepsilon\chi \left[K(z) + \sum_{i=1}^{2N} W(J_i, z) - \frac{J(z)Z_m}{2\pi a} \right], \quad b_{m+1} \leq z \leq b_m, \quad (2.40)$$

Considering that the impedance affects essentially the current distribution in the first approximation, we introduce propagation constant $k_m = \sqrt{k^2 - j2k\chi Z_m/(aZ_0)}$ on the each section and write the current as a series in powers of small parameter χ to obtain the set of equations:

$$\frac{d^2 J_{m1}(z)}{dz^2} + k_m^2 J_{m1}(z) = -4\pi j\omega\varepsilon K(z),$$

$$\frac{d^2 J_{mn}(z)}{dz^2} + k_m^2 J_{mn}(z) = -4\pi j\omega\varepsilon \sum_{i=1}^{2N} W(J_{i,n-1}, z), \quad b_{m+1} \leq z \leq b_m, \quad n > 1. \qquad (2.41)$$

The current and the terms of the current series are continuous along the radiator and absent at its ends. From the first equation, it follows that the current distribution along the each antenna segment has in the first approximation sinusoidal character. In order to find the law of the current distribution along the entire radiator, it is necessary to complement the conditions of current continuity on the segment boundaries with those of charge continuity, *i.e.,* the equality of derivations on the left and the right at each boundary. The conditions mean continuity of voltage along entire radiator, except for the point of the generator placement.

The above-mentioned conditions permit expressing the amplitude and phase of the current at any section through those of the preceding segment current, and, therefore, through segment parameters and one of the currents. The current distribution along entire radiator coincides in the first approximation with that in a stepped long line open at the end.

For the symmetrical radiator excited at the center, the current distribution is determined by the expressions (1.54). If the condition (2.31) holds, the expression for the input impedance in the nth approximation with respect to χ takes the form

$$Z_A^{(n)} \approx -\frac{1}{\left[J_N^{(n-1)}(0) \right]^2} \sum_{m=1}^{2N} \int_{b_{m+1}}^{b_m} \left\{ E_z \left[J_m^{(n-1)} \right] - \frac{Z_m J_m^{(n-1)}}{2\pi a} \right\} J_m^{(n-1)}(z)\, dz, \qquad (2.42)$$

i.e., the expression (1.53), obtained by the induced emf method, is generalized.

In the course of searching for the radiator with the impedance varying along its length, the issue of rational changing the surface impedance along the antenna, which allows improving the antenna-cable matching, arose invariably. The analysis of the problem shows that, at unchanged frequency of the first resonance, the surface impedance must be concentrated at a small antenna segment near the generator. A typical wire antenna with a lengthening coil in the base meets this requirement.

An example of a radiator with concentrated load is given in Fig. **6a** of Chapter 1. The integral equation for the current in such antenna is easily derived from the equation for the current in a metal dipole. The connection in a wire (at point $z = z_n$) of concentrated complex impedance Z_n is equivalent to connection of additional concentrated emf $e_n = -J(z_n)Z_n$, which produces the impressed field

$$E_n = -J(z_n)Z_n\delta(z-z_n).\tag{2.43}$$

The boundary condition for the electric field on the radiator surface with N loads will take the form

$$E_z(a,z)\big|_{-L\le Z\le L}+K(z)-\sum_{n=1}^{N}J(z_n)Z_n\delta(z-z_n)=0,\tag{2.44}$$

i.e.,

$$\frac{d^2J(z)}{dz^2}+k^2J(z)=-4\pi j\omega\varepsilon\chi\left[K(z)+W(J,z)-\sum_{n=1}^{N}J(z_n)Z_n\delta(z-z_n)\right].\tag{2.45}$$

If the radiator is symmetric and loads Z_n in its both arms are identical and placed at identical distances z_n from the coordinate origin, it follows from (2.44) that

$$E_z(a,z)\big|_{-L\le Z\le L}+K(z)-\sum_{n=1}^{N/2}J(z_n)Z_n\left[\delta(z-z_n)+\delta(z+z_n)\right]=0.\tag{2.46}$$

For example, Hallen's equation (1.68) for the current along a filament yields

$$\int_{-L}^{L} J(\varsigma)G_1 d\varsigma = -\frac{j}{Z_0}\left\{ C\cos kz + \frac{e}{2}\sin k|z| - \frac{1}{2}\sum_{n=1}^{N/2} J(z_n)Z_n\left[\sin k|z-z_n| + \sin k|z+z_n|\right] \right\}.$$

This equation was used in paper [20]. If the radiator has only one load Z_1 connected in the wire (at point $z = z_n$), it follows from (2.45) that

$$\frac{d^2 J(z)}{dz^2} + k^2 J(z) = -4\pi j\omega\varepsilon\chi\left[K(z) + W(J) - J(z_1)Z_1\delta(z-z_1)\right]. \tag{2.47}$$

Three summands in the right-hand part of the expression take into account the exciting emf, the radiation and the presence of the load. We seek the solution as a series in powers of small parameter χ, which allows arriving at the set of equations

$$\frac{d^2 J_1(z)}{dz^2} + k^2 J_1(z) = -4\pi j\omega\varepsilon\left[K(z) - \chi J_1(Z_1)Z_1\delta(z-z_1)\right], \quad J_1(\pm L) = 0,$$

$$\frac{d^2 J_n(z)}{dz^2} + k^2 J_n(z) = -4\pi j\omega\varepsilon\left[W(J_{n-1}) - \chi J_n(z_1)Z_1\delta(z-z_1)\right], \quad J_n(\pm L) = 0, \quad n > 1. \tag{2.48}$$

The equations are written provided that Z_1 has the magnitude on order of $1/\chi$, i.e., it is comparable with the antenna wave impedance. The solution of the first equation for the particular case when the antenna feed point is displaced from its center, i.e., $K(z) = e\delta(z-h)$, takes the form

$$\chi J_1(z) = j\frac{\chi e}{30\sin 2kL}\sin k(L+\gamma_2 h)\sin k(L-\gamma_2 z) +$$

$$+\frac{\chi^2 e}{900\sin^2 2kL}\frac{Z_1 Z_2}{Z_1 + Z_2}\sin k(L+\gamma_1 z_1)\sin k(L+\gamma_3 z_1)\sin k(L-\gamma_3 h)\sin k(L-\gamma_1 z), \tag{2.49}$$

where

$$Z_2 = -j\frac{30\sin 2kL}{\chi\sin k(L+z_1)\sin k(L-z_1)}, \quad \gamma_1 = \begin{cases} +1, z_1 \le z, \\ -1, z_1 > z, \end{cases} \quad \gamma_2 = \begin{cases} +1, h \le z, \\ -1, h > z, \end{cases} \quad \gamma_3 = \begin{cases} +1, h \ge z_1, \\ -1, h < z_1. \end{cases}$$

As to be expected, the current along the radiator with one concentrated load contains two sinusoidal terms: one of them is caused by the generator, and the other one is caused by the load.

The solution of the equations for $J_n(z)$ at $n>1$ is found by replacing quantity $K(\varsigma)$ in equation for $J_1(z)$ with $W(J_{n-1})$. If one takes into account (2.49), we get at the excitation point:

$$\chi^n J_n(h) = \frac{1}{e} \int_{-L}^{L} W\left(\chi^{n-1} J_{n-1}\right) \chi J_1(\varsigma) d\varsigma$$

Let us use an equality of the type (2.26), in which the additional summand in the form $\sum_{m=1}^{n} \chi^m J_m(z_1) Z_1 \delta(z-z_1)$ appears in accordance with (2.47) because of the concentrated load Z_1. From this equality we find quantity $W\left(\chi^{n-1} J_{n-1}\right)$ and substitute it in $\chi^n J_n(h)$. If the condition (2.31) holds, then, by analogy to (2.32),

$$Z_A^{(n)} \approx -\frac{1}{\left[J^{(n-1)}(h)\right]^2} \left\{ \int_{-L}^{L} E_\sigma \left(J^{(n-1)}\right) J^{(n-1)}(\sigma) d\sigma - Z_1 \left[J^{(n-1)}(z_1)\right]^2 \right\}. \tag{2.50}$$

This expression at $Z_1 = 0, h = 0$ coincides with (2.32) in the absence of the second radiator. As seen from (2.50), if a concentrated load is connected in the antenna wire, then an absolute term, proportional to the impedance magnitude and the squared current at the point of connection, will appear in the formula for Z_A along with the integral. The addition of such term does not contradict the logic of the induced emf method. The expression (2.50) generalizes the expression (1.57), obtained by this method. In the case of several (*N*) loads with magnitudes Z_n, located at points $z = z_n$ (see Fig. **6b**) of Chapter 1, we come to the expression generalizing the formula (1.58).

Therefore, the solution of the integral equation for currents in antennas of different types confirms and defines more precisely the results determined by the method of induced emf when its second formulation is used. The conclusion is valid also for radiators made of several parallel wires. They are considered in Section 3.2. The results of using the theory of the impedance antennas and the antennas with concentrated loads are considered in Chapter 6.

2.4. INTEGRAL EQUATIONS FOR A SYSTEM OF RADIATORS

In Section 2.2, the system consisting of two radiators was analyzed with the help of the integral equation. The expression (2.32) of the section shows clearly that an input impedance of a radiator in the system is equal to the sum of self-radiator impedance $Z_{II}^{(n)}$ in the nth approximation with respect to χ and the additional (coupled) impedance, equal to a product of mutual impedance $Z_{III}^{(n)}$ of radiators and the ratio of currents at radiator centers. By analogy, in the case of several (Q) radiators, the strength of the electric field on the pth radiator surface is

$$E_p = \sum_{q=1}^{Q} E_p\left(J_q\right),$$
(2.51)

where $E_p\left(J_q\right)$ is the field along the pth radiator, created by current $J_q = J_q(0) f_q(\sigma)$ of the qth radiator, $f_q(\sigma)$ is the current distribution law in the qth radiator. The oscillating power, produced by the pth radiator with current $J_p = J_p(0) f_p(\sigma)$ to all radiators, is

$$P_p = -\sum_{q=1}^{Q} \int_{-L_p}^{L_p} E_p\left(J_q\right) J_p(0) f_p(\sigma) d\sigma,$$

whence the input impedance of the pth radiator is

$$Z_p = \frac{P_p}{\left[J_p(0)\right]^2} = \sum_{q=1}^{Q} J_q(0) \frac{Z_{qp}}{J_p(0)},$$
(2.52)

where $Z_{qp} = -\int_{-L_p}^{L_p} E_p\left(f_q\right) f_p(\sigma) d\sigma$ is the mutual impedance of the qth and pth radiators.

In the notation system adopted in Section 2.2, where the order of quantity smallness is taken into account, the expression is of the form:

$$Z_p^{(n)} = -\frac{1}{J_p^{(n-1)}(0)} \sum_{q=1}^{Q} J_q^{(n-1)}(0) \int_{-L_p}^{L_p} E_\sigma\left(f_q^{(n-1)}\right) f_p^{(n-1)}(\sigma) d\sigma.$$

Multiplying the radiator input impedance by the source current, we obtain the magnitude of emf, connected at the pth radiator center:

$$e_p = J_p(0)Z_p = \sum_{q=1}^{Q} J_q(0)Z_{qp}, p = 1, 2, \dots Q. \tag{2.53}$$

This is the Kirchhoff equation for a closed circuit. According to the equation, the emf in the circuit is the sum of the voltage drops across the elements. Since the equality is valid for each radiator, then the set of equations is written in fact with the help of (2.53). The expression (2.53) is identical to the equation (1.65) written in accordance with the logic of the induced emf method.

Note that (2.53) corresponds to the series connection of circuit elements (see Fig. **8**) of Chapter 1. The series circuit is employed widely in the analysis of radiator systems. The input impedance of each radiator is calculated usually in accordance with expression of the type (1.39). For this reason, the series circuit is valid for the radiator system with the arm length smaller than 0.4λ.

At higher frequencies, using a parallel circuit is expedient. Here, the input impedance is calculated in accordance with an expression of the type (1.47).

In spite of seeming diversity of described methods, they have an essential disadvantage. They are developed for specific radiators and possess no flexibility and freedom for the analysis of arbitrarily constructed radiators. The method, which allows treating a wire structure consisting of straight wire segments located arbitrarily and connected partially with each other, offers in this context much greater possibilities (Fig. **2a**). Consider the currents running here along thin perfectly conducting filaments. Two segments of a filament are shown in Fig. **2b**. The distance from point O_p of the pth segment to element ds of the sth segment is

$$R = \left| \vec{r}_p + p\vec{e}_p - \vec{r}_s - s\vec{e}_s \right|_{p=0}, \tag{2.54}$$

where $\vec{r}_p$ and $\vec{r}_s$ are radii-vectors from the origin to points O_p and O_s of the corresponding segments, p and s are coordinates measured along the segments, $\vec{e}_p$ and $\vec{e}_s$ are the unit vectors, their directions coincident with wire axes.

Let us write for the current along the sth segment: $\vec{j}_s = j_s(s)\vec{e}_s$. According to

(1.21), $\vec{A}_s(\vec{j}_s) = A_s(j_s)\vec{e}_s$. Here, $A_s(j_s) = \mu \int_{S_1}^{S_2} J(s')G_3 ds'$, where S_1 and S_2 are

the coordinates of beginning and end of the sth segment on the s-axis. In order to find the vector potential of the total field, one has to sum the vector potentials of all segment fields:

$$\vec{A} = \sum_{n=1}^{N} \vec{A}_{sn}(\vec{j}_s) = \sum_{n=1}^{N} A_{sn}(j_{sn})\vec{e}_{sn},$$ (2.55)

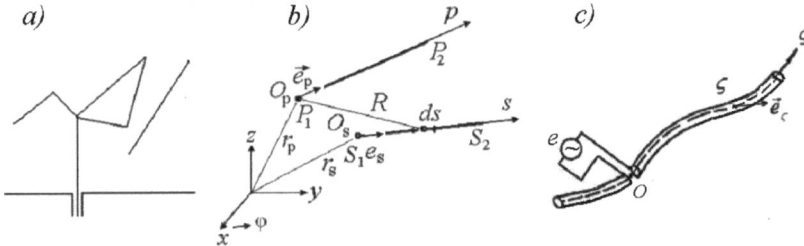

Figure 2: Wire antenna of straight segments (*a*), its two segments (*b*) and curvilinear variant (*c*).

where n is the segment number, N is the number of segments and

$$A_{sn}(j_{sn}) = \mu \int_{Sn1}^{Sn2} J(s'_n)G_3 ds'_n.$$

In accordance with (1.20) and (1.21), the field of the sth segment at point O_p is

$$\vec{E}_s(O_p) = \frac{1}{j\omega\varepsilon} \int_{S_1}^{S_2} J(s)\left[k^2 G_3\vec{e}_s + graddiv(G_3\vec{e}_s)\right]ds.$$

Here, the differentiation in the last term is performed with respect to the coordinates of the observation point. Since in the rectangular coordinate system $R = \sqrt{(x_p-x_s)^2 + (y_p-y_s)^2 + (z_p-z_s)^2}$, where x_p, y_p, z_p are the coordinates of the observation point, and x_s, y_s, z_s are the coordinates of the integration point, then $grad_p G = -grad_s G$ (indices p and s show, with respect to which coordinate the differentiation is performed). Take into account that in accordance with the gradient definition $\vec{e}_s grad_s G = \partial G/\partial s$. Using these relationships and the mathematical identity $div(G\vec{e}_s) = \vec{e}_s\,gradG$, we find: $div_p(G\vec{e}_s) = -\partial G/\partial s$, whence it follows

$$\vec{E}_s(O_p) = \frac{1}{j\omega\varepsilon}\int_{S_1}^{S_2} J(s)\left[k^2 G_3\vec{e}_s - grad_p\left(\frac{\partial G_3}{\partial s}\right)\right]ds.$$

The projection of the sth wire field onto direction p is calculated as a product of quantity $\vec{E}_s(O_p)$ and $\vec{e}_p$:

$$E_{ps} = \vec{E}_s(O_p)\vec{e}_p = \frac{1}{j\omega\varepsilon}\int_{S_1}^{S_2} J(s)\left[k^2 G_3\vec{e}_s\vec{e}_p - \frac{\partial^2 G_3}{\partial p\partial s}\right]ds,\qquad (2.56)$$

and the projection of the total field is the sum of the fields projections of all segments:

$$E_p = \sum_{n=1}^{N} E_{pn} = \frac{1}{j\omega\varepsilon}\sum_{n=1}^{N}\int_{S_{n1}}^{S_{n2}} J_n(s_n)\left[k^2 G_3\vec{e}_n\vec{e}_p - \frac{\partial^2 G_3}{\partial p\partial s_n}\right]ds_n.$$

Substituting this field in (2.1), we get the equation generalizing the Pocklington's equation:

$$\sum_{n=1}^{N}\int_{S_{n1}}^{S_{n2}} J_n(s_n)\left[k^2 G_3\vec{e}_n\vec{e}_p - \frac{\partial^2 G_3}{\partial p\partial s_n}\right]ds_n = -j\omega\varepsilon K_p(p).\qquad (2.57)$$

If $N=1$, equation (2.57) converts to (2.6),

$$\int_{-L}^{L} J(\varsigma)\left[k^2 G_3\vec{e}_\varsigma\vec{e}_z - \frac{\partial^2 G_3}{\partial z\partial\varsigma}\right]d\varsigma = -j\omega\varepsilon K(z).\qquad (2.58)$$

Here, the replacement of variables is performed: $p \to z$, $s_n \to \varsigma$. Furthermore, one should take into account the Green's function symmetry relative to the coordinates of the observation and integration points: $\partial G_3/\partial z = -\partial G_3/\partial\varsigma$.

Let the wire antenna is a polygonal (broken) line, along which combined coordinate ς is measured, and the lengths of straight segments tend to zero. Then from (2.57), we obtain the Pocklington's integral equation for the current in a curvilinear conductor (Fig. **2c**):

$$\int_{(L)} J(\varsigma)\left[k^2 G_3 \vec{e}_\varsigma \vec{e}_z - \frac{\partial^2 G_3}{\partial z \partial \varsigma}\right] d\varsigma = -j\omega\varepsilon K(z). \tag{2.59}$$

Here, $\vec{e}_z$ and $\vec{e}_\varsigma$ are tangent unit vectors at the observation and integration points. If the curvilinear conductor is symmetrical relative to some middle point, this equation completely coincides with (2.58) in form.

Other derivations of this equation are given in [11] and [21].

2.5. GENERALIZED INDUCED EMF METHOD. IMPEDANCE LONG LINE

An analytical solution of an antenna radiation problem has been obtained for a small number of the simplest variants of radiators. As a rule, small-scale radiators situated in free space are considered. This is due to the difficulty of the problem. In this connection, digital methods allowing reducing the problem to solution of a set of linear algebraic equations became frequent practice in solving integral equations for the antenna current. These methods permit to find characteristics of complex antennas with size, great in comparison with a wavelength, and take into account the influence of nearby antennas and metal bodies.

Reducing integral equation to a set of algebraic equations is accomplished with the help of the Moment Method.

In the general case the integral equation for the current in a wire antenna has the form

$$\int_{(l)} J(\varsigma) K(z,\varsigma) d\varsigma = F(z), \tag{2.60}$$

where $J(\varsigma)$ is the sought function (the current distribution along a wire), $K(z,\varsigma)$ is the equation kernel, dependent on coordinate z of the observation point and on coordinate ς of the integration point, $F(z)$ is a known function determined by extraneous sources of the field. The terms proportional to the current may enter into function $F(z)$; here, this is of no great importance. The integral is taken over an all wire length. It is easy to verify that the equations considered earlier are particular cases of the equation (2.60).

Unknown current $J(\varsigma)$ is expressed in the form of a sum of linearly independent functions $f_n(\varsigma)$, which are called the basis ones:

$$J(\varsigma) = \sum_{n=1}^{N} I_n f_n(\varsigma),\tag{2.61}$$

where I_n are unknown coefficients, which are complex in the general case. Substituting (2.61) into (2.60), we obtain:

$$\sum_{n=1}^{N} I_n \int_{(l)} f_n(\varsigma) K(z,\varsigma) d\varsigma = F(z).\tag{2.62}$$

Introduce the second system of linearly independent functions $\phi_p(z)$, they are called the weight ones. If we multiply both parts of equation (2.62) by $\phi_p(z)$ and integrate over entire wire length, and then repeat the operation at different p, we shall obtain the set of equations:

$$\sum_{n=1}^{N} I_n \int_{(l)} \phi_p(z) \int_{(l)} f_n(\varsigma) K(z,\varsigma) d\varsigma dz = \int_{(l)} \phi_p(z) F(z) dz, \quad p=1,2...N.\tag{2.63}$$

Obviously number N of equations (2.63) must coincide with number N of unknown quantities. The integration result of each expression is its moment; hence the method's name.

If the system of weight functions coincides with that of the basis functions such variant of the Moment Method is known as Galerkin's method. Then

$$\sum_{n=1}^{N} I_n \int_{(l)} f_p(z) \int_{(l)} f_n(\varsigma) K(z,\varsigma) d\varsigma dz = \int_{(l)} f_p(z) F(z) dz, \quad p=1,2...N.\tag{2.64}$$

One can rewrite this set of equations as

$$\sum_{n=1}^{N} I_n Z_{np} = U_p, \qquad p = 1, 2 \ldots N, \tag{2.65}$$

where $Z_{np} = \int_{(l)} f_p(z) \int_{(l)} f_n(\varsigma) K(z,\varsigma) d\varsigma dz$, $U_p = \int_{(l)} f_p(z) F(z) dz$. Equation (2.65) is valid also for set of equations (2.63), if one replaces $f_p(z)$ with $\phi_p(z)$ in formulas for Z_{np} and U_p.

Expression (2.65) is the set of linearly independent algebraic equations with N unknown I_n, having the dimensionality of the current. Coefficients Z_{np} and U_p have the dimensions of the impedance and voltage; they can be calculated, *e.g.*, by digital integration. Accordingly, one can interpret the expression (2.65) as Kirchhoff equation for the pth contour with current I_p and emf U_p, which enters into the system of N coupled contours, and Z_{pp} is the self-impedance of the contour element, and Z_{np} is the mutual impedance of the nth and pth contours.

Set of equations (2.65) can be solved on the computer with the help of standard software. If one writes down the set in the matrix form:

$$[I][Z] = [U], \tag{2.66}$$

where $[Z]$ is the impedance matrix, $[I]$ and $[U]$ are the current and a voltage vectors, then one can say that the solution is obtained with the help of the standard method of matrix inversion:

$$[I] = [Z]^{-1}[U]. \tag{2.67}$$

Substitution of obtained values of I_n into (2.61) permits to calculate current distribution $J(\varsigma)$, and afterwards all electrical characteristics of the radiator.

The calculation of matrix elements Z_{np} in practice may prove to be difficult, since it is connected with the digital double integration. To alleviate the difficulties, one can use δ - functions as weights: $\phi_p(z) = \delta(z - z_p)$. Then, the integral in the calculation of Z_{np} becomes a single one, the calculation of U_p requires no integration, and the expression (2.63) takes the form

$$\sum_{n=1}^{N} I_n \int_{(l)} f_n(\varsigma) K(z_p, \varsigma) d\varsigma = F(z_p), \, p=1, 2...N.$$

One can obtain the equation directly from (2.60) and (2.61), if the left- and right-hand parts of the equation (2.60) are equated to each other at isolated points, their number N corresponding to that of obtained equations. For this reason, the variant of the Moment Method is known as the point-matching technique or the collocation method (see, *e.g.*, [20]).

The collocation method ensures an exact equality of functions found in the left- and right-hand parts of the equation (2.60), at least at N points. In the intervals between the points, the difference between the two parts of the equation may increase sharply. When using the Moment Method with weight functions of another type, the equality may not take place on the whole at any point of the interval, where the value of z is changed. But equating both function moments (integration with some weight) means decreasing to the minimum of difference between the left- and right-hand parts at all points of interval, where z changes. This property in the final analysis may prove to be more important than the exact equality at isolated points. Therefore, Galerkin's method allows providing, as a rule, an essentially more accurate solution than the collocation method. Yet, the collocation method is useful sometimes, too.

The choice of basis functions is of a great importance for using the Method Moments, since the successful selection of the system permits to decrease the amount of calculation at given accuracy or to increase the accuracy within the same calculation time. For that end, as a rule, the basis functions must suit the physical sense of a problem, *i.e.,* coincide, in the first approximation, with the actual distribution of the current along a radiator or its elements.

Basis functions are subdivided into two types: entire-domain functions, which are other than zero along the entire radiator length, and subdomain functions, which are other than zero along a wire segment. As basis functions of the first type, one can use, *e.g.*, for example, terms of Fourier series and Tchebyscheff or Legendre polynomials. Their application area is limited mainly to isolated radiators of a simple shape. For a wire antenna of an intricate shape, subdomain basis functions are employed typically. In particularly, such approach is expedient, if the antenna consists of arbitrarily situated segments of straight wires partially connected with each other. A straight radiator also may consist from physically isolated wire segments, if lumped loads are connected in its arms at given distances from each other. One can exemplify subdomain basis functions with piecewise-constant (impulse) functions (Fig. **3a**), piecewise-linear functions (Fig. **3b**), and piecewise-parabolic functions (Fig. **3c**). These basis sets are particular cases of a wider class of basis functions – of polynomials. A simpler variant of the current approximation with a polynomial is proposed in [22]:

$$J(\varsigma) = \sum_{m=0}^{M_n} I_{nm} (\varsigma - \varsigma_n)^m , \ \varsigma_n < \varsigma < \varsigma_{n+1} .$$

Here, M_n is the selected degree of the polynomial on the *n*th segment, and I_{nm} are unknown coefficients. Comparing the expression with (2.61), we obtain:

$$I_n = I_{n0}, \ f_n(\varsigma) = \sum_{m=0}^{M_n} \frac{I_{nm}}{I_{n0}} (\varsigma - \varsigma_n)^m \text{ at } \varsigma_n \leq \varsigma \leq \varsigma_{n+1} \text{ and } 0 \text{ elsewhere.}$$

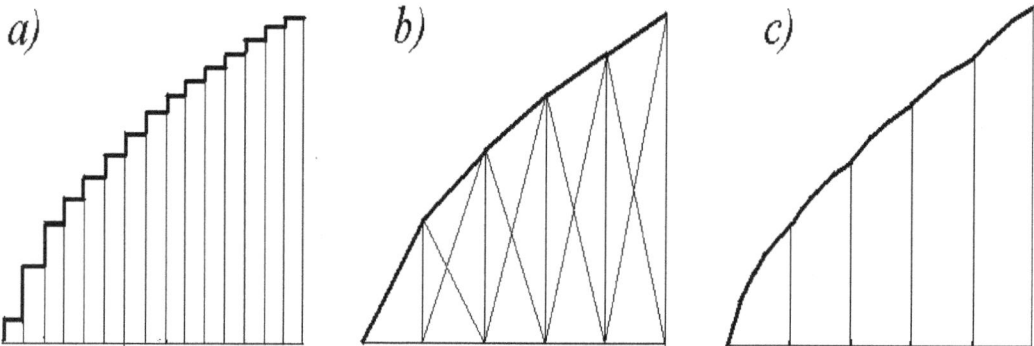

Figure 3: The current as the sum of impulse (*a*), piecewise-linear (*b*) and piecewise-parabolic (*c*) basis subdomain functions.

By analogy with the entire-domain basis, one can use the Fourier expansion as subdomain basis functions. A special case of such expansion is piecewise-sinusoidal functions:

$$J(\varsigma) = \frac{I_p \sin k(\varsigma_{p+1} - \varsigma) + I_{p+1} \sin k(\varsigma - \varsigma_p)}{\sin k(\varsigma_{p+1} - \varsigma_p)}, \quad \varsigma_{p-1} \leq \varsigma \leq \varsigma_{p+1}. \tag{2.68}$$

Comparing this expression with (2.61), one can write:

$$f_p(\varsigma) = \begin{cases} \dfrac{\sin k(\varsigma - \varsigma_{p-1})}{\sin k(\varsigma_p - \varsigma_{p-1})}, & \varsigma_{p-1} \leq \varsigma \leq \varsigma_p, \\[3mm] \dfrac{\sin k(\varsigma_{p+1} - \varsigma)}{\sin k(\varsigma_{p+1} - \varsigma_p)}, & \varsigma_p \leq \varsigma \leq \varsigma_{p+1} \\[3mm] 0 & elsewhere. \end{cases} \tag{2.69}$$

Using representation (2.61) with the basis functions of the form (2.69) is equivalent to dividing the wire into short dipoles with overlapped arms and with centers at points ς_p, with I_p being the current at the center of the pth dipole. In this sense, equations (2.61) and (2.69) are the generalization of expression (1.8). Since weighting become piecewise-linear as short dipole lengths decrease then Fig. **3b** permits to visualize how the collection of the subdomain basis functions produces the total current distribution along an antenna.

Work [23] proposed to use the functions in the form (2.69) as the basis and weighs. Such variant of the Moment Method has two advantages. Firstly, a rapid convergence of results is ensured, *i.e.,* the matrix [Z] dimension is small in comparison with the matrix dimensions when using other basis and weight functions. This means that application of piecewise-sinusoidal functions as the basis and weights ones corresponds to the physical content of the problem. Secondly, closed expressions containing integral sines and cosines can be used to calculate many matrix elements.

Substitute the current distribution (2.61) with weight functions (2.69) into the equation (2.57) for the complex wire radiator. In accordance with Galerkin's method, we multiply both parts of the equation by weight function $f_s(z)$ and integrate along the entire wire length. Repeating the operation at different s, we obtain a set of P equations of (2.65) type with P unknown quantities I_p, with the coefficients in them being

$$Z_{ps} = -\frac{1}{j\omega\varepsilon} \int_{z_{s-1}}^{z_{s+1}} f_s(z) \int_{\varsigma_{p-1}}^{\varsigma_{p+1}} f_p(\varsigma) \left[k^2 G_3 \vec{e}_\varsigma \vec{e}_z - \frac{\partial^2 G_3}{\partial z \partial \varsigma} \right] d\varsigma \, dz \quad,$$

$$U_s = \int_{z_{s-1}}^{z_{s+1}} f_s(z) K_s(z) \, dz \quad .$$

(2.70)

Comparing (2.70) with expression (1.39), where quantity E_{ps} is substituted from (2.56), it is easy to verify that the formula for Z_{ps} corresponds to the mutual impedance between pth and nth dipoles, calculated by the induced emf method. As seen from (2.70), the dipoles are considered as isolated ones, *i.e.*, the current of each one follows the sinusoidal law. Substituting extraneous field $K_s(z)$ in (2.70) shows that quantity U_s is the emf of the generator connected at the center of the sth dipole. Therefore, the set of equations (2.65) with coefficients Z_{ps} and U_s is the set of Kirchhoff equations for the collection of dipoles constituting the wire antenna.

Thus, the variant of Galerkin's method that was proposed by Richmond and used to calculate the current distribution in a complex wire radiator is equivalent to division of the radiator into isolated dipoles, their self- and mutual impedances being calculated by the induced emf method. For this reason, one can refer to Richmond's method as the generalized induced emf.

If the lumped loads are connected in the antenna, it is expedient to divide the wire into short dipoles so that the load is placed at the dipoles centers. Then, in accordance with (2.43) one can generalize the equations set (2.65) and write it in the form:

$$\sum_{p=1}^{P} I_p Z_{ps} = U_s - I_s Z_s \ , \ \text{s=1, 2...N,} \tag{2.71}$$

or in matrix form

$$[I][Z] = [U - IZ]. \tag{2.72}$$

The accuracy of the induced emf method for the dipole calculation is known to decrease as the dipole length increases. The accuracy of calculation is acceptable at dipole arm length $L \leq 0.4\lambda$. The advantage of the generalized induced emf method consists in the fact that one can divide the long dipole into several short dipoles, *e.g.*, with the arm length no greater than 0.2λ. That allows ensuring the required exactness.

Calculation of the coefficients Z_{ps} requires the digital double integration. But the problem is simplified essentially if the method described in [7] is used to calculate the mutual impedance of two arbitrarily situated dipoles. Here, the double integrals are reduced to single ones, and each integral is a sum of alternating series. The series terms are calculated by recurrence formulas, almost as quickly as the terms of the power series. This manner of calculating Z_{ps} is useful in general-purpose software.

From all the above it follows that the induced emf method is a constant companion and satellite of the integral equation method. Also it is inseparable from the concept of an equivalent open-end long line with the sinusoidal current distribution coinciding with the one along the symmetrical dipole. In the case of a usual line of the metal wires, the propagation constant of a wave along the line is equal to that of a wave in the air.

An impedance long line is the same kind of a counterpart of an impedance dipole, as a usual line is that of a metal dipole. In contrast to a metal antenna, tangential component E_z of the electric field on the surface of an impedance dipole is other than zero, which results in an additional voltage drop across each its element: $dU = E_z dz$. Using boundary condition (2.34) and taking (2.35) into account, we get

$$dU = \frac{ZJ(z)}{2\pi a} dz \, . \tag{2.73}$$

Thus, infinitesimal element dz of line, equivalent to a symmetric dipole, contains, besides inductance $d\Lambda_1 = \Lambda_1 dz$ and capacity $dC = C_1 dz$, additional impedance $Zdz/(\pi a)$ (factor 2 takes into account that the dipole consists of two wires). Here C_1 and Λ_1 are the capacity and inductance per unit length. The long line, equivalent to an impedance dipole, is shown in Fig. **4**.

The telegraph equations for such line are

$$-\frac{dU(z)}{dz} = J(z)\left(j\omega\Lambda_1 + \frac{Z}{\pi a} \right), \quad -\frac{dJ(z)}{dz} = j\omega C_1 U(z), \tag{2.74}$$

hence

$$\frac{d^2 J}{dz^2} + k_1^2 J(z) = 0, \quad \frac{d^2 U}{dz^2} + k_1^2 U(z) = 0, \tag{2.75}$$

where $k_1^2 = k^2 - jZ\omega C_1 / (\pi a)$.

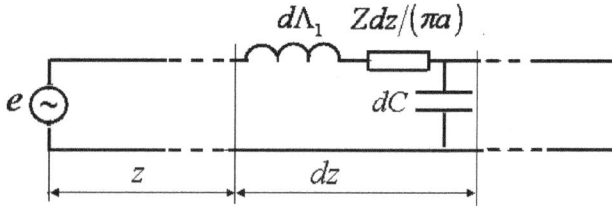

Figure 4: Equivalent impedance long line.

We shall assume capacity C_d per unit length of a dipole to be equal to the self-capacitance of an infinitely long wire of radius a. Since radius a is much less than length $2L$ of the antenna, the surface of zero potential can be placed at distance $2L$ from the antenna. Then the capacity per unit length of a line compared with a symmetric dipole is

$$C_1 = C_d/2 = \pi\varepsilon/\ln(2L/a)$$

and

$$k_1^2 = k^2 - j\,\omega\varepsilon Z \big/ \big[a\ln(2L/a)\big],$$

which coincides with the quantity obtained in Section 1.4.

Solving (2.74) in the ordinary way, we find the current and input impedance of an open- ended impedance line:

$$J(z) = J(0)\sin k_1 (L-z)\big/\sin k_1 L, \quad Z_l = -jW_l \cot k_1 L, \tag{2.76}$$

where $J(0)$ is the generator current, and W_l is the wave impedance of the line,

$$W_l = \frac{\omega\Lambda_1 - j(Z/\pi a)}{k_1} = \frac{k}{\omega C_1} = 120\frac{k_1}{k}\ln\frac{2L}{a}. \tag{2.77}$$

Thus, a dipole with constant surface impedance can be considered in a first approximation as an equivalent long line differing from a usual line by the presence of impedance $Z/(\pi a)$ per unit length. Here, the current in the dipole follows the sinusoidal law with propagation constant k_1 differing from propagation constant k in a metal antenna, and the wave impedance of the dipole is greater than that of a metal antenna by a factor of k_1/k. A calculation of the equivalent long line yields results coinciding with those of solving the integral equation in a first approximation.

The equivalent long line permits to analyze antenna structures.

CHAPTER 3

New Methods of Analysis

Abstract: Additional methods of analysis are considered. It is shown that reducing three-dimensional conic problem to the two-dimensional one and using the complex potential method enables one to calculate the capacitance per unit length and the wave impedance for a dipole with inclined arms and also the same for an infinite long line and a metal radiator of two convergent filaments or conic shells. The theory of electrically coupled lines permits analyzing multiple-wire structures of antennas and cables. The mathematical programming method allows selecting the loads to create an antenna with the characteristics as close to the given one as possible. The compensation method is proposed to protect living organisms and electronic devices from strong electromagnetic fields in the near region of an antenna.

Keywords: Additional radiator, Complex potential method, Dipole with inclined arms, Efficiency, Electrically coupled lines, Electrodynamic wave impedance, Electrostatic wave impedance, In-phase current distribution, Inverse problem of the radiators theory, Long line, Metal radiator of two convergent shells, Mathematical programming method, Objective function, Pattern factor, Phase step in a signal reradiating, Protecting devices against irradiation, Protecting living organisms against irradiation, Required current distribution, Required electrical characteristics, Selection of loads, Three-dimensional problem, Transformation of variables, Transition from a cone to a cylinder, Travelling wave ratio, Two convergent charged shells.

3.1. RELATIONSHIP BETWEEN CONIC AND CYLINDRICAL PROBLEMS

The first two chapters of the eBook are dedicated mostly to rigorous methods of calculating thin straight radiators and also systems of widespread parallel radiators of such type. Those are traditional, widely applied and sufficiently well tried methods. The third chapter gives brief description of comparatively new methods that extend potential feasibility of antenna electrical characteristics computation.

Strictly speaking, the calculation of a linear radiator or a system of parallel linear radiators is a one-dimensional or a two-dimensional problem. Three-dimensional problems are considerably more complicated. The difference between them is clear by exemplified by the electrostatic problems. The calculation of the electrical field of charged bodies is simplified essentially if all quantities characterizing the field depend only on two coordinates. Such field is called a plane parallel one. The two-dimensional (*i.e.,* planar in fact) problem was treated widely for a different number of wires with diverse shape of their cross-section while a three-dimensional problem is solved only for a few particular cases.

In this connection of interest is the attempt of using the results obtained in solving the two-dimensional (plane) problem to calculate the electrostatic field in the three-dimensional problem when a relative position of metal bodies resembles a two-dimensional variant.

Calculating the field of two infinitely long charged filaments, converging to a single point (Fig. **1a**) is an example of such three-dimensional problem. The linear densities of both filaments charges are the same in magnitude and opposite in sign:

$$q_1 = -q_2 = q .$$ (3.1)

The two-dimensional problem of two parallel filaments (Fig. **1b**) is an analog of this three-dimensional problem. It is solved by the complex potential method [24]. The scalar potential of the two parallel filaments field is

$$U(x, y) = \frac{q}{2\pi\varepsilon} \ln(\rho_2/\rho_1),$$ (3.2)

where ε is the medium permittivity and ρ_1 and ρ_2 are the distances from the observation point M to the filaments axes, with

$$\rho_1^2 = (b+x)^2 + y^2, \quad \rho_2^2 = (b-x)^2 + y^2 .$$

Here b is half a distance between the filaments axes.

Lines of equal potential (equipotential lines) U=const in the plane problem are circumferences with centers situated on the axis of abscissas. It follows from here, in particular, that the field of two parallel metal cylinders is of the same nature, since one can always place the equivalent line axes so that two surface of equal potential in their field may coincide with surfaces of metal cylinders (Fig. **2a**). Lines of field strength (lines of force) V=const are circumferences with centers situated on the axis of ordinates.

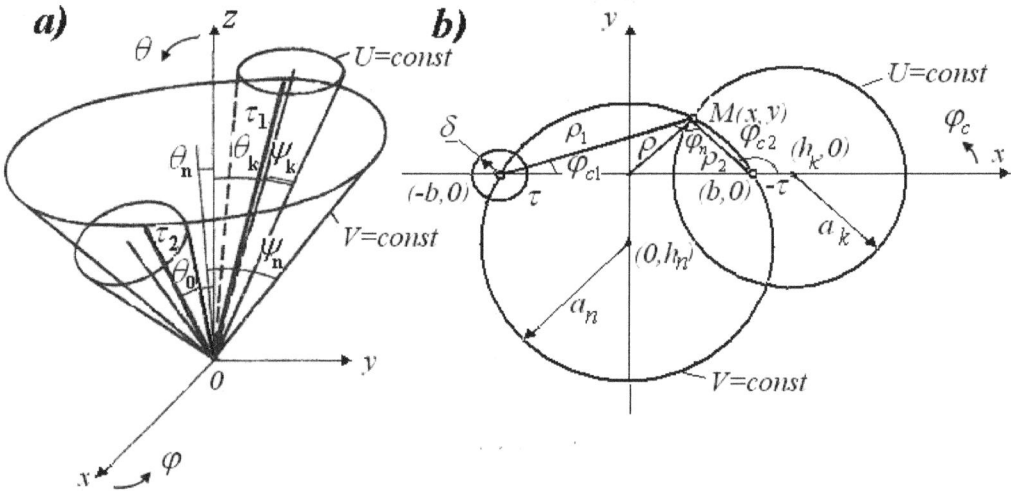

Figure 1: Three-dimensional (*a*) and two-dimensional (*b*) problems for two infinitely long filaments.

In accordance with the uniqueness theorem, an electrostatic problem solution is to satisfy Laplace equation, and the surfaces of conducting bodies are to coincide with those of equal potential. The three-dimensional problem of two convergent charged filaments (see Fig. **1a**) is a particular case of the conic problem, where conducting bodies are shaped as a cone with the vertex at the coordinate origin (Fig. **2b**). Conic and cylindrical problems are compared in [25] where the Laplace equation is shown to hold true in transition from one problem to the other, if the substituted variables are related by equations:

$$\rho = \tan(\theta/2), \quad \phi_c = \phi. \tag{3.3}$$

Here ρ and ϕ_c are the cylindrical coordinates, and θ and ϕ are the spherical ones.

The result of such transformation of variables is the mapping of the spherical surface of arbitrary radius R onto the plane (ρ, ϕ_c). The intersection line of the spherical surface with any circular cone (with a vertex at the coordinate origin) transforms into a circumference.

It follows from the above that the three-dimensional conic problem can be reduced to the two-dimensional one, where the coordinates of the conducting bodies are related with those in the original problem by equations (3.3). So the case of two convergent charged filaments situated at angle $2\theta_0$ to each other in plane $x0z$ (see Fig. **1a**) corresponds to two parallel filaments situated at distance $2b = 2\tan(\theta_0/2)$ from each other (see Fig. **1b**). The case of two metal cones with an angle 2ψ at the vertex of each cone and with angle $2\theta_1$ between cones axes (see Fig. **2b**) corresponds to two cylinders of radius $a = (c-d)/2$, the distance between their axes being $2h=c+d$ (see Fig. **2a**). Since, according to (3.3),

$$c = \tan\left[(\theta_1 + \psi)/2\right], \quad d = \tan\left[(\theta_1 - \psi)/2\right], \tag{3.4}$$

it is easy to verified that

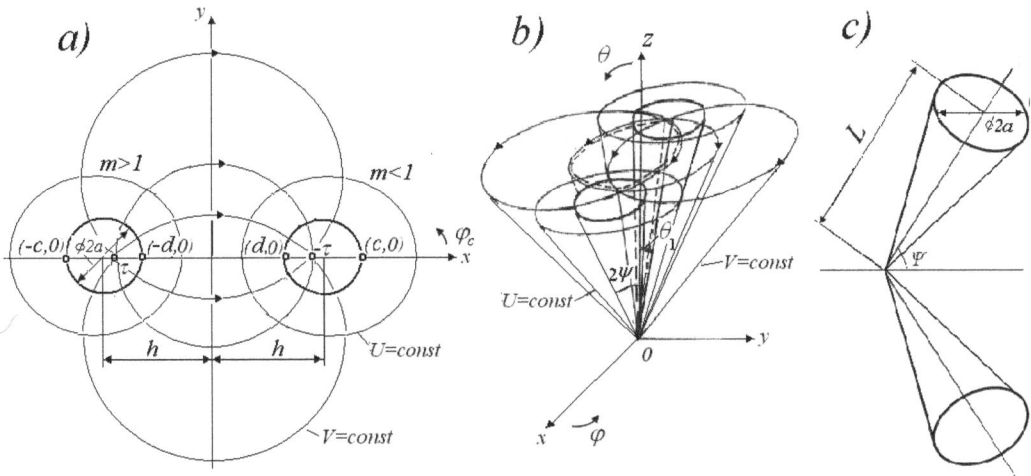

Figure 2: Two-dimensional problem for cylinders (*a*), three-dimensional problem for cones (*b*) and a dipole with inclined arms (*c*) as the particular case.

$$a = \frac{\sin\psi}{\cos\theta_1 + \cos\psi}, \quad h = \frac{\sin\theta_1}{\cos\theta_1 + \cos\psi}. \tag{3.5}$$

It is worth emphasizing that when passing from a cone to a cylinder the cylinder axis does not coincide with the cone axis.

The scalar potential of the electric field for two convergent charged filaments by analogy with (3.2) is

$$U(\theta,\psi) = \frac{q}{2\pi\varepsilon} \ln(\rho_2/\rho_1), \qquad (3.6)$$

where

$$\rho_1^2 = \left(\tan\frac{\theta_0}{2} + \cos\phi\tan\frac{\theta}{2}\right)^2 + \sin^2\phi\tan^2\frac{\theta}{2}, \quad \rho_2^2 = \left(\tan\frac{\theta_0}{2} - \cos\phi\tan\frac{\theta}{2}\right)^2 + \sin^2\phi\tan^2\frac{\theta}{2}.$$

Surfaces of equal potential U=const in given case are the circular cones, their axial lines lying in plane $x0z$. Surfaces of field strength V=const, where the lines of force are situated, also are the circular cones. Their axial lines lie in plane $y0z$.

The reduction of the conic problem to the cylindrical one allows calculating the capacitance per unit length and wave impedance of a long line excited at the cone vertex formed by two convergent filaments or cones. It is known, *e.g.*, that capacitance C per unit length of a line consisting of two wires with a radius a spaced at distance $2h$ and its wave impedance W are respectively

$$C = \frac{\pi\varepsilon}{\ln\left[h/a + \sqrt{(h/a)^2 - 1}\right]} = \frac{\pi\varepsilon}{\cosh^{-1}(h/a)}, \quad W = \frac{1}{cC} = 120\cosh^{-1}(h/a). \qquad (3.7)$$

Here c is the light velocity.

For two convergent cones, we obtain, according to (3.5)

$$C_1 = \frac{\pi\varepsilon}{\cosh^{-1}(\sin\theta_1/\sin\psi)}, \quad W_1 = 120\cosh^{-1}(\sin\theta_1/\sin\psi). \qquad (3.8)$$

In the particular case of two convergent cones with an angle 2ψ at the vertex of each cone and with angle 2Ψ between cone axes situated symmetrically with respect to the horizontal plane, we find

$$C_2 = \frac{\pi\varepsilon}{\cosh^{-1}\left(\sin\Psi/\sin\psi\right)}, \quad W_2 = 120\cosh^{-1}\left(\sin\Psi/\sin\psi\right). \tag{3.9}$$

As can be seen from Fig. **2c**, expressions (3.9) are written for a dipole with inclined conic arms. If the dipole arms are shaped as thin cylindrical wires of length L and radius a, with $a \langle\langle L$, one can, in the first approximation, replace the cylinder with a cone and consider that $\sin\psi \approx a/L$. From here,

$$C_2 \approx \frac{\pi\varepsilon}{\cosh^{-1}\left(\dfrac{L\sin\Psi}{a}\right)}, \quad W_2 \approx 120\cosh^{-1}\left(\frac{L\sin\Psi}{a}\right). \tag{3.10}$$

The case of two convergent charged shells with angular width 2α situated along the surface of a circular cone with angle $2\theta_0$ at the vertex (Fig. **3a**) is of specifical interest. The line of two coaxial cylindrical shells (Fig. **3b**) of radius $a = \tan\left(\theta_0/2\right)$ with the same angular width corresponds to it as the plane problem.

Summing the fields from pairs of symmetrically located parallel filaments 1-1', 2-2', 3-3', *etc.* (see Fig. **3b**), their charges being the same in the magnitude and the contrary in sign, one can understand the nature of the electrostatic field of two cylindrical shells. Lines of equal potential for each pair are circumferences with centers situated on the straight line passing through the filaments of the pair. The envelope of the circumferences with the same value of constant m is the line of equal potential for the field of the shells. It is a curve line of a complicated shape, extended along both sides of each shell and slightly bent toward the ends. One of the equipotential lines is also the axis of structure symmetry, *i.e.,* the y-axis.

Lines of the field strength for each pair of the filaments are circumferences centered on the symmetry axis. Two lines coincide with the circumference, where the shells are situated, *i.e.,* they close gaps between the latter. Lines of the field strength inside the circumference connect the symmetrically placed filaments with each other and cross the lines of equal potential at right angles.

The field structure in the case of two convergent shells (see Fig. **3a**) is of a similar nature, the only difference being that the surfaces of equal potentials U=const and the surfaces of field strength *V*=const coincide with conic rather than cylindrical surfaces.

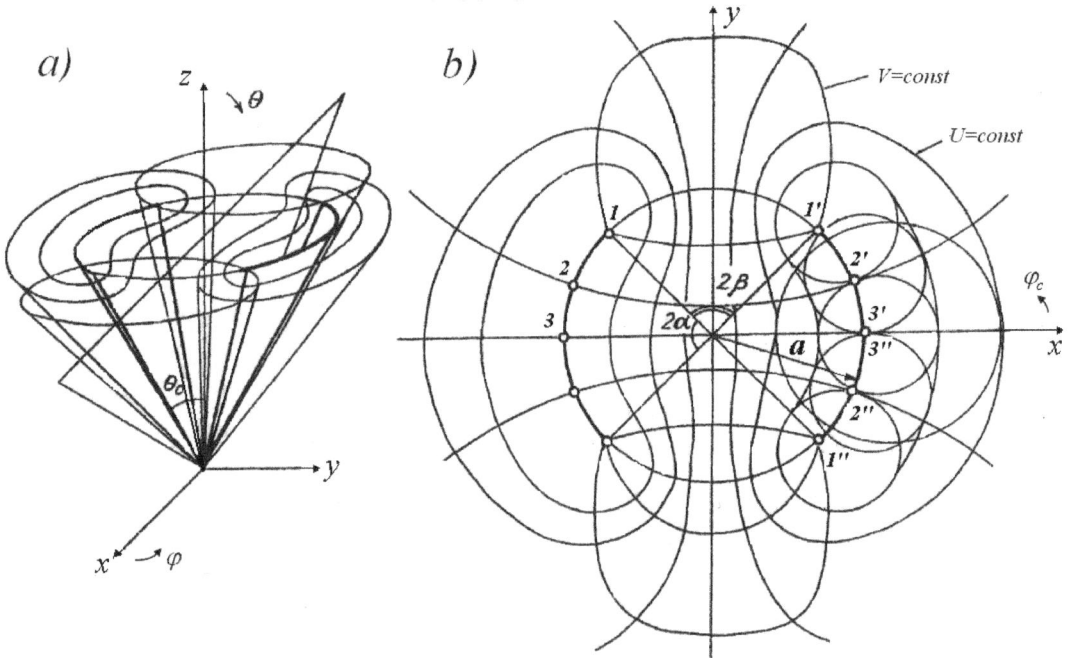

Figure 3: Three-dimensional (*a*) and two-dimensional (*b*) problems for two shells.

The capacitance per unit length and the wave impedance of the line formed by two cylindrical shells [26, 27] are equal to

$$C = \frac{\varepsilon K\left(\sqrt{1-k^2}\right)}{K(k)}, \quad W = 120\pi \frac{K(k)}{K\left(\sqrt{1-k^2}\right)}. \tag{3.11}$$

Here $K(k)$ is the complete elliptic integral of the first kind of argument k, where $k = \tan^2(\beta/2)$, and $\beta = (\pi - 2\alpha)/2$ is a half of an angular slot width, *i.e.*, C and W depend only on slot width 2β and hence angular width 2α of the metal shell. As it follows from (3.11), C and W are independent of cylinder radius a. It means that both expressions are valid for conic shells as well.

Quantities C and W are constant along the conic line, *i.e.*, two-wire line of the convergent shells is a uniform one. Input impedance Z_l of a uniform two-wire line is known to tend to its wave impedance W as its length increases. Therefore, the input impedance of an infinitely long line excited by a generator situated near the cone vertex is

$$Z_l(k) = 120\pi \frac{K(k)}{K\left(\sqrt{1-k^2}\right)}. \qquad (3.12)$$

The structure in question can be treated on the one hand as a two-wire line and on the other hand as an antenna. The antenna is a symmetrical V-radiator, with the arms shaped as two convergent metal shells situated along the surface of the circular cone. One can even think of the structure as a slot cut through a conic screen. If $Z_R(2\alpha)$ is the input impedance of a metal radiator with angular arm width 2α, and $Z_S(2\beta)$ is the impedance of a slot antenna with width 2β, then, as it follows from the above, if the structure length is great, we find

$$Z_R(2\alpha) = Z_S(2\beta) = Z_l(k) = 120\pi K(k) \big/ K\left(\sqrt{1-k^2}\right). \qquad (3.13)$$

Two variants of slots cut through a metal cone are shown in the Fig. **4**. In the first variant (see Fig. **4a**) the slot edge coincides with the cone generatrix. The variant was considered earlier. In the second variant (see Fig. **4b**), the slot edge is a helix, and the metal cone with a slot cut in it forms a bifilar helix, excited near the vertex. The angular slot width is considered to be constant.

Strictly speaking, the expressions (3.11-3.13) are valid only for the variant of the slot antenna shown in Fig. **4a**. But, if the angular width of the metal shell in both variants is the same, one can believe with a high probability that the capacitances per unit length and the wave impedances and consequently the input impedances of radiators are the same in both cases.

Compare the input impedances of the metal and the slot radiators with the same width. If, for example, the metal shell width is $2\alpha = 2\pi/3$, then $\beta/2 = \pi/12$, $k^2 = 0.00515$, $K\left(\sqrt{1-k^2}\right)\big/K(k) = 2.56$, *i.e.,* $Z_R(2\pi/3) = 120\pi/2.56$. In fact, the slot impedance is the input impedance of the metal radiator situated next to it. If the slot width is $2\beta = 2\pi/3$, then $\beta/2 = \pi/6$, $k^2 = 0.111$,

$K\left(\sqrt{1-k^2}\right)\!\Big/K(k)=1.56.$ Accordingly, $Z_S\left(2\pi/3\right)=120\pi/1.56.$ Therefore, the impedances of the metal and the slot radiators of the same width $2\pi/3$ are related to each other by $Z_R\left(2\pi/3\right)Z_S\left(2\pi/3\right)=\left(120\pi\right)^2\!\Big/\left(2.56\cdot1.56\right)$, whence it follows

a)   b)

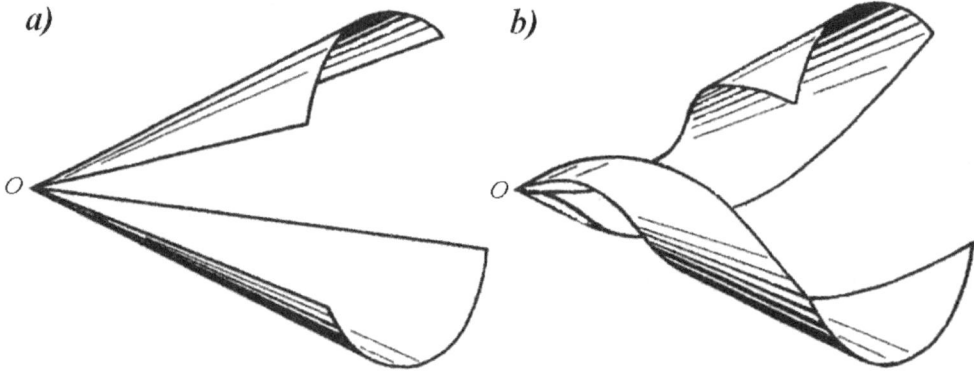

Figure 4: Slot with the straight (*a*) and helical (*b*) edges on the cone.

$$Z_S=\left(60\pi\right)^2\!\Big/Z_R\,. \tag{3.14}$$

Here Z_R is the input impedance of a metal radiator identical to the slot in the shape and dimensions. It is easily verified that (3.14) holds for radiators of any width.

Radiators with the same width of the metal shell and the slot are of particular interest. Setting $2\alpha=2\beta=\pi/2$, we obtain $k^2=0.0294$, $K\left(\sqrt{1-k^2}\right)\!\Big/K(k)=2.0$, *i.e.,*

$$Z_R\left(\pi/2\right)=Z_S\left(\pi/2\right)=60\pi\,. \tag{3.15}$$

As can be seen from (3.15), the infinite long radiator mounted on a cone has the constant and purely active input impedance, and hence the high level of matching with the cable in an unlimited frequency range. If the radiator is of finite size, the frequency range is limited, but remains a sufficiently wide one. One can read about that in Section 4.2.

3.2. ELECTRICALLY COUPLED LINES

The theory of electrically coupled lines [28] permits the analysis of multiple-wire structures of antennas and cables. In particular, it allows treating structures of N

closely spaced parallel wires with due account of grounding. An asymmetrical line of two wires (Fig. **5**) is the simplest example of such structure. It is equivalent to a folded radiator opened at the end (Fig. **6b**). Expressions for the current and potential of wires 1 and 2 in the line take the form:

$$-\frac{\partial u_1}{\partial z} = jX_{11}i_1 + jX_{12}i_2, \quad u_1 = j\frac{1}{k^2}\left(X_{11}\frac{\partial i_1}{\partial z} + X_{12}\frac{\partial i_2}{\partial z}\right),$$

$$-\frac{\partial u_2}{\partial z} = jX_{22}i_2 + jX_{12}i_1, \quad u_2 = j\frac{1}{k^2}\left(X_{22}\frac{\partial i_2}{\partial z} + X_{12}\frac{\partial i_1}{\partial z}\right).$$

$$(3.16)$$

Here u_i is the potential of the ith wire relative to ground, i_i is the current along the ith wire, $X_{ik} = \omega\Lambda_{ik}$ is the inductive impedance per unit length (self- or mutual).

The two left equations of the system (3.16) are based on the fact that the potential decrease at segment dz of each wire is the result of the emf influence. The emf's are induced by the self- current and by the currents of adjacent wires. The other two equations were written on the basis of the electrostatic equations relating charges and potentials with due account of the continuity equation.

The current dependence on coordinate z is adopted in the form $\exp(\gamma z)$, where γ is the propagation constant. Differentiation of equations on the right and substitution them into those

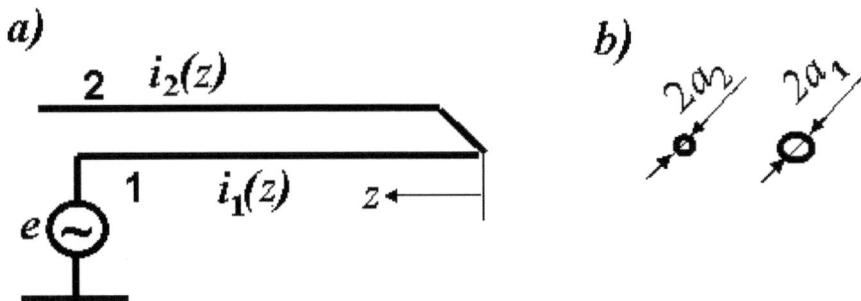

Figure 5: The asymmetrical line, which is equivalent to a folded radiator open at the end: (*a*) – circuit, (*b*) - cross section.

on the left reduces them to a set of homogeneous equations, which shows that propagation constant γ in the system of two metal wires is equal to k. We search the solution of the set of equations in the form of $u_1 = A\cos kz + jB\sin kz$ and assume $z = 0$ in order to determine constant quantities A and B. As a result, we obtain

$$i_{1(2)} = I_{1(2)}\cos kz + j\left[\frac{U_{1(2)}}{W_{11(22)}} - \frac{U_{2(1)}}{W_{12(21)}}\right]\sin kz,$$

$$u_{1(2)} = U_{1(2)}\cos kz + j\sum_{s=1}^{2}\rho_{1(2)s}I_s\sin kz, \tag{3.17}$$

where $I_{1(2)}$ and $U_{1(2)}$ are the current and potential at the beginning of wire 1 or 2 (at point $z = 0$), $W_{1(2)s}$ and $\rho_{1(2)s}$ are the electrostatic and electrodynamic wave impedances between wire 1 or 2 and wire s.

In the general case for an asymmetrical line of N wires, the expressions for the current and potential of nth wire take the form:

$$i_n = I_n\cos kz + j\left(\frac{2U_n}{W_{nn}} - \sum_{s=1}^{N}\frac{U_s}{W_{ns}}\right)\sin kz,$$

$$u_n = U_n\cos kz + j\sum_{s=1}^{N}\rho_{ns}I_s\sin kz, \tag{3.18}$$

where I_n and U_n are the current and potential at the beginning of nth wire (at point $z = 0$), respectively, W_{ns} and ρ_{ns} are the electrostatic and electrodynamic wave impedances between the nth wire and the sth wires, with

$$\rho_{ns} = \frac{p_{ns}}{c}, \quad W_{ns} = \begin{cases} 1/(c\beta_{ns}), n = s, \\ -1/(c\beta_{ns}), n \neq s. \end{cases} \tag{3.19}$$

Here, p_{ns} is the potential coefficient (taking into account a mirror image in the perfectly conducting ground surface), β_{ns} is the coefficient of electrostatic induction, c is the light velocity. The coefficients β_{ns} and p_{ns} are related as follows:

$$\beta_{ns} = \Delta_{ns}/\Delta_N,$$

where $\Delta_N = |p_{ns}|$ is the $N \times N$ determinant, and Δ_{ns} is the cofactor of the determinant Δ_N.

For an asymmetrical line of two wires, we can write

$$\frac{1}{W_{11}} = \frac{\rho_{22}}{\rho_{11}\rho_{22} - \rho_{12}^2}, \quad \frac{1}{W_{22}} = \frac{\rho_{11}}{\rho_{11}\rho_{22} - \rho_{12}^2}, \quad \frac{1}{W_{12}} = \frac{\rho_{12}}{\rho_{11}\rho_{22} - \rho_{12}^2}. \qquad (3.20)$$

The ordinary two-wire line open at the end is a useful equivalent of the monopole. It permits to find the current distribution along a monopole wire. The asymmetrical line of two wires situated above ground is a counterpart of a folded antenna. The folded antenna is a particular case of parallel wire structure. In the case, two parallel wires are situated at a small distance from each other as compared with the wavelength. The asymmetrical line shown in Fig. **5** is a counterpart of a folded antenna open at the opposite end of the wire, which is farther from the generator. Two asymmetrical variants of folded antennas – closed (*a*) and open (*b*) - are shown in Fig. **6**.

If the wires of an asymmetrical line have unequal lengths, or if the loads are connected into them, one must divide the line to segments. The expressions for the current and potential of the *n*th wire at the *m*th segment take the form:

$$u_n^{(m)} = U_n^{(m)} \cos kz_m + j \sum_{s=1}^{N} \rho_{ns}^{(m)} I_s^{(m)} \sin kz_m,$$

$$i_n^{(m)} = I_n^{(m)} \cos z_m + j \left(\frac{2U_n^{(m)}}{W_{nn}^{(m)}} - \sum_{s=1}^{N} \frac{U_s^{(m)}}{W_{ns}^{(m)}} \right) \sin kz_m, \qquad (3.21)$$

where $I_n^{(m)}$ and $U_n^{(m)}$ are the current and potential of the *n*th wire at the beginning of the *m*th segment (at point $z_m = 0$), respectively, M is the number of wires in the *m*th segment, and $W_{ns}^{(m)}$ and $\rho_{ns}^{(m)}$ are the electrostatic and electrodynamic wave impedances between wires *n* and *s* at segment *m*. The equations (3.21) generalize the expressions (3.18).

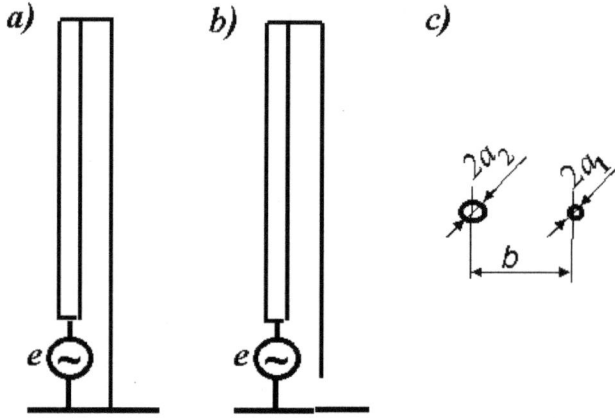

Figure 6: Asymmetrical folded radiators: (*a*) – closed, (*b*)– open at the end, (*c*)– cross section.

To solve the set of equations, the boundary conditions are used. They establish the absence of currents at free ends of the wires, the continuity of the current and potential along each wire, the abrupt changes in potential at the points of connecting loads and generator *e*. If we calculate the current magnitude $J(0)$ at the feed point, we shall find the input impedance of the asymmetrical line,

$$Z_l = e/J(0). \tag{3.22}$$

This is equal approximately to the reactive impedance of the antenna, whose equivalent is the given asymmetrical line. One can find the antenna impedance more accurately, if antenna is treated as a line radiator, the current along which being equal to the total current along the line.

When calculating the antenna input impedance, one needs, as a rule, to find field E_ς at antenna surface. And it should be kept in mind that while current function $J(\varsigma)$ is continuous along the entire length of the antenna and sinusoidal at the each segment, function $dJ/d\varsigma$ may have a jump near the segment boundaries.

The equations (3.21) use wave impedances $W_{ns}^{(m)}$ and $\rho_{ns}^{(m)}$, and the equations (3.18) use similar ones. The magnitudes of the impedances, as seen from (3.19), are determined by the potential coefficients. The coefficients are found by the method of mean potentials in accordance with the actual position of antenna wires. The simplest variant of this method is the Howe's method. The mutual potential coefficient of two parallel wires of equal lengths, their size and position being given in Fig. **7**, is easily shown to be

$$p_{ns} = p(L,l,b)/(2\pi\varepsilon),\tag{3.23}$$

where

$$p(L,l,b) = \frac{1}{2L}\left[(L+l)\,sh^{-1}\frac{L+l}{b}+(L-l)\,sh^{-1}\frac{L-l}{b}-2Lsh^{-1}\frac{l}{b}-\sqrt{(L+l)^2+b^2}\right.$$

$$\left.-\sqrt{(L-l)^2+b^2}+2\sqrt{l^2+b^2}\ \right],$$

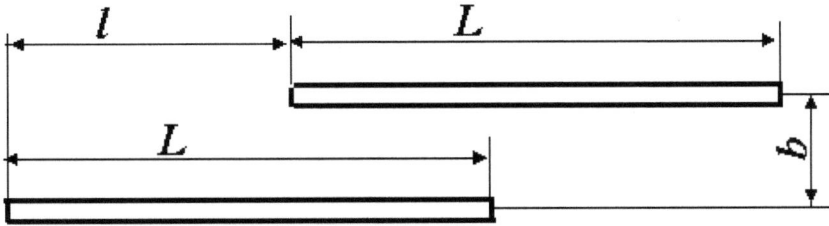

Figure 7: The mutual location of wires.

i.e., $\rho_{ns}=p_{ns}/c=p(L,l,b)/(2\pi\varepsilon_0\varepsilon_r c)=60\,p(L,l,b)/\varepsilon_r$.

Then the self-potential coefficient of the nth wire at the mth segment with the account of the mirror image is

$$p_{nn}^{(m)} = p\left[l_m-l_{m+1},0,a_n^{(m)}\right]-p\left[l_m-l_{m+1},l_m+l_{m+1},a_n^{(m)}\right],\tag{3.24}$$

where l_m and l_{m+1} are the coordinates of the mth segment boundaries, $a_n^{(m)}$ is the radius of the nth wire at the mth segment. The mutual potential coefficient between the nth and the sth wires at the mth segment is

$$p_{ns}^{(m)} = p\left[l_m-l_{m+1},0,b_{ns}^{(m)}\right]-p\left[l_m-l_{m+1},l_m+l_{m+1},b_{ns}^{(m)}\right].\tag{3.25}$$

Here, $b_{ns}^{(m)}$ is the distance between the axes of wires n and s at the mth segment. Summing up, it is important to note the general principle underlying the coupled lines theory. The theory permits to find, in the first approximation, the current distribution along each wire with the view of using the distribution later to calculate the active component and to define more precisely the reactive component of the input impedance with the help of the induced emf method. This principle is known to be used in calculations by the induced emf method of the impedance of a linear radiator, the current distribution along it coinciding in the first approximation with the current distribution along a uniform long line.

A similar approach is used in calculations of the electrical characteristics of an impedance antenna, *i.e.,* of a radiator with nonzero boundary conditions holding on the surface. One can obtain the current distribution along the radiating structure by analyzing the integral equation for the current. Here, the laws of a current distribution along radiator wires and along the wires of the equivalent line (or of system of lines) are identical. The advantage of equivalent lines is the maximum simplicity in finding the law and the efficiency of applying the obtained results to designing radiators with required characteristics.

The results of using the coupled lines theory are considered in Chapter 5.

3.3. RADIATOR SYNTHESIS

The characteristics of a linear radiator with fixed dimensions can be changed, if it is coated with a layer of magnetodielectric, or concentrated loads connected in it. Impedance antennas, with nonzero boundary conditions secured on their surface, and antennas with concentrated loads (in contrast to metal antennas without loads) give an additional degree of freedom in the development of a radiator with given characteristics. The freedom is greater, if the surface impedance varies along the radiator, or there are many loads, and they are connected along the whole radiator length.

Loads can be used to solve the inverse problem of the radiators theory, namely, to create an antenna with required electrical characteristics. Of great practical importance is the particular case of the problem, *i.e.,* creation of a radiator exhibiting in a wide frequency ranges a high matching level and the radiation maximum in the plane perpendicular to the radiator axis.

A typical linear radiator (thin, without loads) fails to meet the requirements. The reactive component of its input impedance is great everywhere, except for the vicinity of the series resonance, and this results in the antenna mismatch with a cable. If the length of the radiator arm is greater than 0.7λ, there is a decrease of the radiation in the plane, perpendicular to the radiator axis, since the current distribution along a linear radiator without loads is close to the sinusoidal one and anti-phased sections are formed on the current curve at high frequencies.

Connecting concentrated loads across the radiator length, one can, depending on their magnitudes and points of connection, obtain the current distribution other than the sinusoidal one. The experimental results presented in [29] show that a radiator with linear in-phase current distribution along its axis has good characteristics (high matching level, the pattern of necessary shape) in a wide frequency range. In particular, such distribution is created by capacitive loads, the capacitances of which decrease towards the antenna ends under the exponential law. These results confirm the known fact that the radiation maximum in the direction, perpendicular to the radiator axis, is attained, if the current is in-phase along the entire length of the antenna. Besides, with the in-phase current, a long radiator has high radiation resistance, which allows increasing the matching level.

The problem of selecting loads to create almost linear in-phase current distribution in the radiator is considered in Chapter 6. The analysis is based on the impedance antenna theory, more precisely, on the impedance long line method. It allows determining potentialities of radiators with loads. In addition, one can use the results obtained with the help of the method for the solution of the antenna optimization problem by the mathematical programming method.

The mathematical programming method [30] enables to choose the loads for creating an antenna with given characteristics, or more precisely, with the characteristics as close to the given ones as possible. The reservation is due to the fact that the variation interval of radiator parameters is bounded, *i.e.*, not every value of antenna electrical characteristic can be realized practically. Different characteristics are at the optimum at different values of parameters. Moreover, an antenna should exhibit certain properties not at a single fixed frequency, but in the entire operation range. Therefore, the selected parameters are a result of a compromise, reached with the help of the mathematical programming method.

The problem of mathematical programming in the general case is stated as follows: one has to find vector $\vec{x}$ of parameters that minimizes some objective function $F(\vec{x})$ under imposed constraints $\phi_i(\vec{x}) \geq 0$. Depending on the type of functions $F(\vec{x})$ and $\phi_i(\vec{x})$, mathematical programming is divided into linear, convex and nonlinear ones. In the case at hand, the problem is solved by nonlinear programming methods, since the type of function $F(\vec{x})$ is unknown.

The objective function (or general functional) is a sum of several partial functionals $F_j(\vec{x})$ with weighting coefficients p_j and penalty function F_{ip}:

$$F(\vec{x}) = \sum_j p_j F_j(\vec{x}) + \sum_i F_{ip} . \qquad (3.26)$$

The partial functional is an error function of some or other antenna characteristic. The weighting function allows taking account of the characteristic being important and the sensitivity of a partial functional to vector $\vec{x}$ changes. A penalty function is zero, if the parameters lie within given interval, and has a great value, if even just one of parameters falls outside the interval limits.

For an antenna with loads (see Fig. **6b**), controlled parameters x are load magnitudes, coordinates z_n of their connection points and the wave impedance W of the cable. Loads are understood as simple elements or a set of simple elements (capacitors with capacitance C_n, coils with inductance Λ_n and resistors with resistance R_n). Values z_n, W, C_n, Λ_n and R_n are to be real-valued, positive and frequency-independent, and z_n are to be smaller than antenna length L. These requirements, naturally, limit the variation interval of parameters.

Different manners of an error function construction are known. Good results are produced by quasi-Tchebyscheff criterion

$$F_j(\vec{x}) = \frac{1}{N_f} \left[\frac{f_{j0}}{f_{j\min}(\vec{x})} - 1 \right] \left\{ \sum_{n_f} \left[\frac{\left(f_{j0}/f_j(\vec{x}) \right) - 1}{\left(f_{i0}/f_{j\min}(\vec{x}) \right) - 1} \right]^s \right\}^{1/s} . \qquad (3.27)$$

Here, N_f is the number of points of the independent argument (*e.g.*, the number of frequencies in given range), n_f is the frequency number, $f_j(\vec{x})$ is one of an antenna electrical characteristics, $f_{j\min}(\vec{x})$ is its minimal value in the considered interval, f_{j0} is the hypothetical value of the characteristic that is to be approached, S is the index of power allowing to control the method sensitivity.

The choice of function $f_j(\vec{x})$ depends on the stated problem. When a wide-range radiator is created, one must use, for functions $f_j(\vec{x})$, *e.g.*, the travelling wave ratio (*TWR*) in the cable and the pattern factor (*PF*), equal to the average level of radiation at given angles:

$$TWR = \frac{2a}{a^2 + b^2 + 1 + \sqrt{(a^2 + b^2 + 1)^2 - 4a^2}}, \quad PF = \frac{1}{K}\sum_{k=1}^{K} F(\theta_k). \tag{3.28}$$

Here $a = R_A/W$, $b = X_A/W$ are, respectively, the active and reactive components of antenna impedance referred to the cable wave impedance, and $F(\theta_k)$ is the magnitude of normalized pattern in the vertical plane for angle θ_k within the limits of angular sector from θ_1 to θ_K (*e.g.*, from 90° to 60°). If resistors with resistances R_n are used as loading elements, the above functions $f_j(\vec{x})$ are supplemented with the antenna efficiency

$$\eta_A = 1 - \frac{1}{J_A^2 R_A}\sum_{n=1}^{N}|J_n|^2 R_n, \tag{3.29}$$

where N is the number of loads, J_n and J_A are the currents in the nth load and in the antenna base, respectively.

In the cases, when no analytical expression for observation function $F(\vec{x})$ is available, the function minimum is found by a numerical method based on searching by the gradient. The gradient method is an iterative procedure, whereby we move step by step from one set of parameters $\vec{x}$ to another one in the direction of the maximum decrease of the function (the steepest descent method):

$$\vec{x}_{m+1} = \vec{x}_m - \alpha_m gradF(\vec{x}_m). \tag{3.30}$$

Here m is the iteration number, α_m is the scale parameter, determined as a result of a linear search of the functional minimum in the antigradient direction.

At iteration, the minimum of the functional and the values of parameters, where it is attained, are determined. It is in essence searching for the minimum of a function of one variable, α. The method with increasing the step (*e.g.*, with doubling it) and subsequent function interpolation in the considered interval by a polynomial of the given power is the most rational. It is convenient to use the cubic interpolation, since the number of interpolation nodes is large enough (four), and the root of the derivative (the value of α, causing the derivative to vanish) is found analytically. If the first step results in an increase, rather than decrease, of the observation function, the step should be reduced by a factor of 10^p, where $p = 1, 2...$, whereupon the linear search continues again with a step doubling.

A modification of the steepest descent method is that of the conjugate gradients.

The calculation is over, when the decrease of the observation function from iteration to iteration becomes smaller than a preset value or the quantity of iterations exceeds certain limit ($m \geq M$).

The mathematical programming method (synthesis) presupposes multiple computations of the antenna electrical characteristics at different initial parameters (analysis). Performing such calculations requires including a special program into the synthesis software. It allows determining, at given loads and exciting emf's, all electrical characteristics of an antenna, *i.e.*, calculating functions $f_j(\vec{x})$ for known vector $\vec{x}$ of initial parameters.

The most laborious in the calculation is that of self- and mutual impedances between short dipoles (see Section 2.5). Therefore, in order to speed up the calculations, it is expedient to fix points of concentrated load connections, so that the location of short dipoles and their mutual impedances may not vary from iteration to iteration. If there are rather many loads, *i.e.,* the distances between them are small in comparison with the wavelength, the restriction will have no effect on the synthesis results.

As the initial values of the loads, those found by the method of an impedance long line (see Chapter 6) are taken. The results of calculations show the computational process in the case to speed up. But of most importance is the reduction of the probability of the error due to the fact that the arbitrary choice of the initial loads may cause the optimization process to come to a local, rather than true, extremism of the observation function.

The synthesis program can be applied also to determining loads with providing the required current distribution along the radiator. The point is that the problem in the synthesis of antennas with given electrical characteristics is often broken up into two stages: the distribution of current is calculated at the first stage, and antenna parameters securing such distribution are determined at the second stage. The first stage has been investigated adequately. It covers a wide class of problems (the above problem of creating a wide-range dipole is one of possible variants). The second stage of synthesis has received far less attention.

In principle, if the required current distribution along a wire antenna is known, one can split the wire into short dipoles and set currents I_s at the centers of these dipoles. For piecewise-sinusoidal basis functions, they are equal to the currents at the corresponding antenna points. Loads magnitudes Z_s, which one must connect at these points to establish currents I_s thereat, are determined from equation (2.71).

But the loading impedances calculated by the approach, include both active and reactive components, varying with frequency. The active component of load impedance may prove to be negative, which is an evidence of impossibility to create such distribution with the help of passive loads. As to the reactive component, it is still necessary to solve the problem of its implementation over given frequency range with the help of a set of simple elements.

Therefore, it is expedient to consider the problem of creating an antenna with the chosen type of loads that, in the desired range, ensures not the given current distribution, but the one as close to it as possible. The problem, as well as the treated earlier problem of creating a wide-range dipole, is solved by the mathematical programming method.

Let it is necessary to obtain current distribution $J(z)$. In the case, it is expedient to use for functions $f_j(\bar{x})$ (the electrical characteristics of an antenna) either both real and imaginary current components

$$f_1 = \operatorname{Re} J(z,f), \; f_2 = \operatorname{Im} J(z,f),$$
(3.31)

or the amplitude and phase of the current:

$$f_3 = |J(z,f)|, \; f_4 = \tan^{-1}\left[\operatorname{Im}(z,f)/\operatorname{Re}(z,f)\right].$$
(3.32)

Construct an error function for each electrical characteristic using, *e.g.*, a root-mean-square criterion

$$F_j = \frac{1}{N_f N_l} \sum_{n_f=1}^{N_f} \sum_{n_l=1}^{N_l} (f_j - f_{j0})^2 ,$$
(3.33)

where N_f and N_l are the number of frequencies and division points on the wire, n_f is the frequency number, n_l is the point number along the antenna.

Functions $f_j(\bar{x})$ for a known vector of initial parameters are calculated with a special program. As the initial parameters, the values found by the method of an impedance long line are taken. The results of using this and the mathematical programming method are considered in Chapter 6.

3.4. COMPENSATION METHOD

The compensation method was proposed to protect living organisms and devices against strong electromagnetic fields in an antenna near region. The protection of devices is necessary, since the irradiation of nearby devices can disrupt their normal operation, cause spontaneous device switching on and switching off, change operating regimes, *etc.* The protection of living organisms is required, firstly, in the vicinity of powerful transmitting centers where the electrical field strength is great, and, secondly, near mobile equipment in the transmitting mode, which is located next to the user. A cellular phone is, in particular, such a device. During a phone conversation, it is placed next to user's head, and its transmitter irradiates sensitive human organs (brain, eye, *etc.*). But the problem is not limited to the cellular phones, since a man often uses or services a low-power transmitter, which he carries or places nearby in a production area or in a vehicle.

Along with the proliferation of mobile communication systems, there has been an increasing concern about possible hazards to the user's body, especially to the head, while exposing it to the irradiation by a handset antenna. The absorbed energy in today's cellular phones can be as high as a half of all radiated energy. In order to minimize any possible health risk, it is necessary to reduce the amount of that energy. The specific absorption rate (SAR) is commonly used to evaluate this energy level [31].

The rather obvious idea of head protection by means of screening, *i.e.,* by shading effect, is not practicable. The near field has no ray structure, and hence the shadow formed behind a metal screen can only cover an area approximately equal to the screen size. For example, in order to protect the head of the cellular phone user, the screen must be much larger than the cross section of the handset housing. For similar reasons, one should discard the idea of using an absorber, *i.e.,* a dielectric shield that absorbs some part of energy. The distortion of the antenna pattern is another obvious disadvantage of using screens and absorbers.

The protective action of the compensation method proposed by M. Bank [32] is based on a different principle: the mutual suppression of fields created by various radiating elements in a certain area. The diverse variants of the principle implementation (interference of two radiators fields, folded antenna, and cavity antenna) possess a potentiality to reduce irradiation of user's organism, especially his head, without distorting the antenna pattern in the horizontal plane. In accordance with the key variant of the compensation method, as shown in Fig. **8**, main radiator 1 is supplemented with second (auxiliary) radiator 2 situated in the plane passing through the head center and the feed point of the main radiator. The second radiator is placed between the head and the main radiator and is excited approximately in anti-phase to it (not exactly in anti-phase because the phase of the field progresses along the interval between the radiators). So, the radiator fields will compensate each other at a certain point inside the head, and the point will be surrounded with a zone of a weak field - a dark spot will arise.

The dipole moment of the additional radiator must be smaller than that of the main radiator, since the field close to an electrical radiator decreases quickly. In order to get the same field strengths at the compensation point, if the currents differ substantially, it is enough to space the radiators by about a few centimeters (1-2 cm at frequency 1 GHz). Therefore, the total far region field differs slightly from the main radiator field, *i.e.,* the pattern stays close to circular. The relatively small field of the additional radiator ensures that the total field in the far region is almost unaffected by the presence of the additional radiator, and the pattern remains close to one in the no-compensation condition.

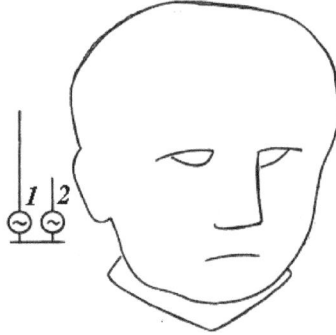

Figure 8: Two radiators next to the head.

The diagram of Fig. **9** giving the linear radiator placement near the human head is the main one in the use of compensation method. As seen from the Figure, feed point A_1 of the main radiator and compensation point A are located along the horizontal straight line passing through head center O. Onto the straight line, we place feed point A_2 of the additional radiator, at distance b from the main one. Assume that both radiators are vertical and have equal lengths and the cylindrical coordinate system origin coincides with point A_1. Inside the head at compensation point A (at distance ρ_0 from point A_2), the fields of both radiators must be equal in magnitude and opposite in sign:

$$E_{z2}\left(b+\rho_0,0,0\right)=-E_{z1}\left(b+\rho_0,0,0\right). \tag{3.34}$$

As linear radiators with finite lengths, one can employ dipoles and monopoles (with one and two feed points). The preferred variant is the monopoles with a feed

point in their base. Linear radiators create two electric field components: E_z and
E_ρ. If a straight perfectly conducting filament is used to model a vertical linear
radiator, the electric field components of a symmetrical vertical electrical dipole
of finite length in cylindrical coordinate system (ρ,ϕ,z) are given by (1.31) and
(1.32). The field components of a monopole of the same arm length are smaller by
a factor 2. Similar expressions hold for the additional radiator.

As can be seen from (1.31) and (1.32), the simultaneous compensation of both E_z
and E_ρ field components by adjustment of the additional radiator current J_{A2} is
impossible. Since E_z is greater than E_ρ, and E_ρ along the ρ-axis (at $z=0$) is
zero, we prefer to compensate the E_z components.

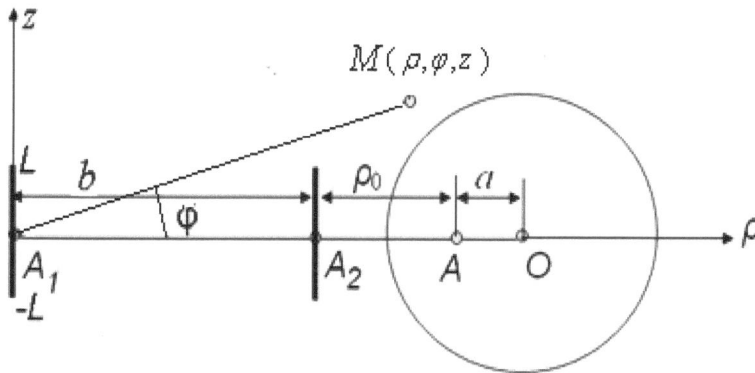

Figure 9: The diagram of phone radiator placement close to a human head.

We assume the feed point A_1 of the main radiator and the compensation point A to
lie on a horizontal straight line passing through head center O. The line is defined
as the structure axis. In this case, feed point A_2 of the additional radiator must also
lie on the structure axis (see Fig. **9**). Under these conditions, component E_z of the
main radiator in a homogeneous medium with relative permittivity ε_r is
determined by (1.31), and $\rho_1 = \rho$. For the additional radiator, a similar expression
can be written, with $\rho_2^2 = \rho_1^2 + b^2 - 2\rho_1 b\cos\phi$ (Fig. **10**). For simplicity sake, we
assume the both dipoles arm lengths to be the same.

Find the current and the input impedance of the radiator, situated in the near
region of the neighboring radiator. If the emf is connected in the input of the first
radiator, the circuit appears, as shown in Fig. **11**. The radiators are the monopoles

with finite lengths, R_1 and R_2 are the internal impedances of the generators: R_1 is the output impedance of the first generator; R_2 is the impedance at the second generator no-excitation input (or at the input of a cable leading to the generator). Generally $R_1 = R_2 = R$.

The calculation method for a system with two linear radiators is based on the folded dipole theory and on the superposition principle. Connect two voltage generators, equal in magnitude to $e_1/2$ and opposite in direction, in a base of the right radiator. We also divide the main generator into two generators, the same in magnitude and in direction. According to the superposition principle, the current at each point is the sum of the currents produced by all generators. Therefore, as shown in Fig. **12**, one can divide the circuit in question into two circuits with two generators in each one and then calculate and add the currents at points C and D, produced in each of the circuits. This procedure allows analyzing the antenna system as a superposition of two sub-systems with in-phase currents (even mode) and anti-phased currents (odd mode).

Let the wires of the first circuit carry only anti-phased currents (odd mode), that is currents, equal in magnitude and opposite in direction; the two parallel wires acting as an open-ended transmission line with the load $2R$ in the base. Let the second circuit carry only in-phase currents (even mode), *i.e.,* the potentials of both wires points situated at the same height (including the radiator bases) are identical. For that end the generator emf's in the wire bases of the second circuit must be equal, if the wire radii are equal. In this case, the parallel wires act as a monopole with resistance $R/2$ between the generator and the ground.

For the two-wire line, we can write

$$e_1 = J_l\left(Z_l + 2R\right),$$ (3.35)

where J_l is the current in the line base, $Z_l = -jW_l \cot kL$ is the input impedance of line with length L, $W_l = 120\ln\left(b/a\right)$ is the line wave impedance, b is the distance between the wires, and $2a$ is the diameter of each wire. The current at point C is $J_{Cll} = e_1Y_1$, and the current at point D is $J_{Dll} = -e_1Y_1$, where $Y_1 = 1/\left(-jW_l \cot kL + 2R\right)$.

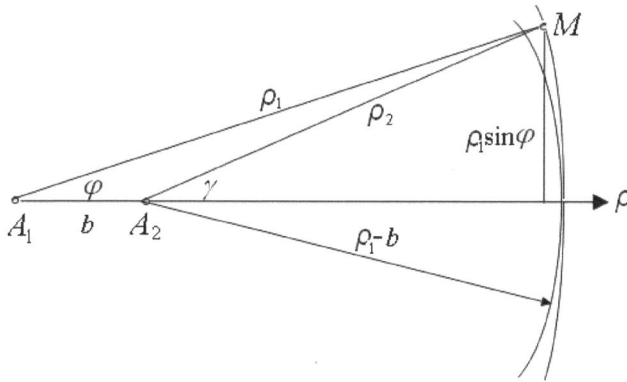

Figure 10: Two linear radiators in cylindrical coordinate system (top view).

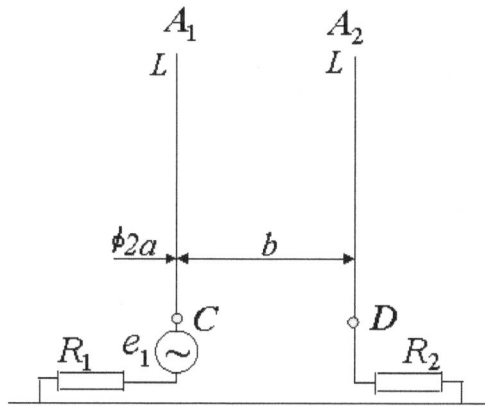

Figure 11: Two radiators in the near region of each other.

For the monopole, we can write

$$e_1/2 = J_r(Z_r + R/2), \tag{3.36}$$

where J_r is the current at the monopole base, and $Z_r = Z_m(L, a_e)$ is the input impedance of the monopole with length L and equivalent radius a_e, equal to $\sqrt{ab}$. The currents at points C and D are the same and equal

$$J_{C1r} = J_{D1r} = e_1 / (4Z_r + 2R) = e_1 Y_2, \tag{3.37}$$

$$\text{where } Y_2 = 1/\left[4Z_m(L, a_e) + 2R\right].$$

So, if emf e_1 is connected in the first radiator input, the currents in the first and the second radiator bases are

$$J_{11} = e_1(Y_1 + Y_2), \quad J_{21} = e_1(-Y_1 + Y_2).$$
(3.38)

If emf e_2 is connected in the second radiator input, so, similarly to (3.38), the currents in the first and the second radiator bases are

$$J_{12} = e_2(-Y_1 + Y_2), \quad J_{22} = e_2(Y_1 + Y_2).$$
(3.39)

According to the superposition principle, the terminal currents of the radiators are

$$J_{A2} = (e_2 - e_1)Y_1 + (e_1 + e_2)Y_2,$$

$$J_{A1} = J_{11} + J_{12} = (e_1 - e_2)Y_1 + (e_1 + e_2)Y_2.$$
(3.40)

The input admittances of the radiators are

$$Y_{A1} = J_{A1}/e_1 = Y_1 + Y_2 + e_2(Y_2 - Y_1)/e_1,$$

$$Y_{A2} = J_{A2}/e_2 = Y_1 + Y_2 + e_1(Y_2 - Y_1)/e_2.$$
(3.41)

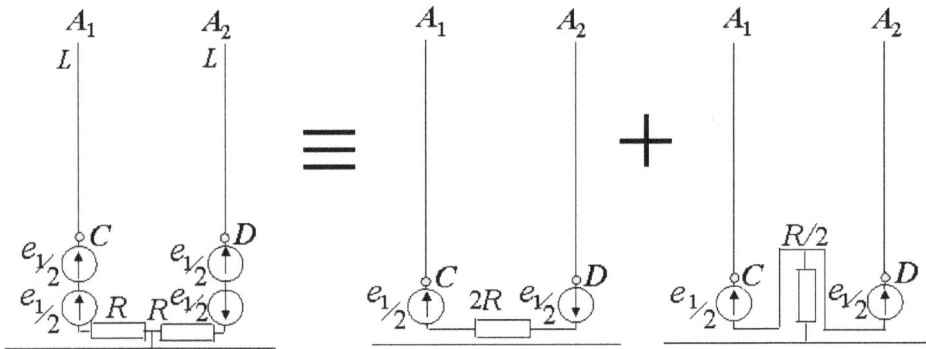

Figure 12: Division of the considered circuit into two circuits.

As is well known, the current and the input impedance of a radiator are affected significantly by the neighboring radiator currents. For a system of two radiators one can write

$$e_1 = J_{A1}Z_{11} + J_{A2}Z_{12}, \ e_2 = J_{A1}Z_{21} + J_{A2}Z_{22}. \tag{3.42}$$

Here, e_1 and e_2 are the driving emf's connected, respectively, to the terminals of the first and second radiators, Z_{11} and Z_{22} are their self-impedances, and Z_{12} and Z_{21} are their mutual impedances. Either expression in (3.42) is a Kirchhoff equation corresponding to a series connection of circuit elements.

The expression in (3.42) can be rewritten as

$$J_{A1} = \frac{e_1 Z_{22} - e_2 Z_{12}}{Z_{11}Z_{22} - Z_{12}Z_{21}}, J_{A2} = \frac{e_2 Z_{11} - e_1 Z_{21}}{Z_{11}Z_{22} - Z_{12}Z_{21}}, \tag{3.43}$$

that is, the current in each radiator is the sum of the currents produced by both its own generator, and the generator of the neighboring radiator (because of the mutual coupling between radiators). The ratio of the currents depends on the level of mutual coupling between the radiators, which is determined by their dimensions and position.

Comparing (3.38)-(3.39) with (3.43), one can determine the radiator self- and the mutual impedances (for wires of equal radii). Considering that

$$J_{11} = \frac{e_1 Z_{22}}{Z_{11}Z_{22} - Z_{12}^2} = e_1 (Y_1 + Y_2), \ J_{12} = \frac{-e_2 Z_{12}}{Z_{11}Z_{22} - Z_{12}^2} = e_2 (Y_2 - Y_1),$$

$$J_{12} = \frac{-e_2 Z_{12}}{Z_{11}Z_{22} - Z_{12}^2} = e_2 (Y_2 - Y_1), \ J_{21} = \frac{-e_1 Z_{12}}{Z_{11}Z_{22} - Z_{12}^2} = e_1 (Y_2 - Y_1),$$

we obtain:

$$Z_{11} = Z_{22}, Y_1 + Y_2 = \frac{Z_{11}}{Z_{11}^2 - Z_{12}^2}, Y_1 - Y_2 = \frac{Z_{12}}{Z_{11}^2 - Z_{12}^2}. \tag{3.44}$$

Adding and subtracting the left-hand and right-hand parts of last two expressions, we find:

$$2Y_1 = 1/(Z_{11} - Z_{12}), \ 2Y_2 = 1/(Z_{11} + Z_{12}),$$

that is,

$$Z_{11} + Z_{12} = 1/(2Y_2), \quad Z_{11} - Z_{12} = 1/(2Y_1),$$

and consequently

$$Z_{11} = Z_{22} = 0.25(1/Y_1 + 1/Y_2) = Z_m(L, a_e) - j0.25W_l \cot kL + R,$$
$$Z_{12} = 0.25(1/Y_2 - 1/Y_1) = Z_m(L, a_e) + j0.25W_l \cot kL. \tag{3.45}$$

One can see from (3.45) that the close proximity between the radiators results in an additional term of each radiator self-impedance. This is in contrast to the self-impedance of a single radiator. The additional term corresponds to the two-wire line, in spite of the fact that the second wire is not connected to a generator. Since the second radiator wire is situated near the first wire, at a small distance compared to the wavelength, its effect is similar to connecting a load in the first wire, *e.g.*, a reactance or a horizontal wire segment. That is, the second wire is a concurrently part of the first radiator. A similar approach is valid also for the second radiator and for the mutual coupling between them.

The presented analysis shows, on what input impedances of both radiators depend and to what extent they change. It confirms the feasibility of the proposed device. The results of calculating the shape and dimensions of dark spots and the data on the reduction of power dissipated in the user's head, caused by mounting a compensation device, are given in Chapter 7. The additional radiator influence on antenna patterns is checked, calculations and experiments are compared. Recommendations for special cases of dark spots positions are given. A circuit permitting to create the dark spot nearby the transmitter during its operation in a wide and continuous range rather than at one or two fixed frequencies (as in the cellular phone) is proposed.

3.5. RECIPROCITY THEOREM FOR A REFLECT ARRAY

In the theory of antenna arrays, the issue of calculating steps of field amplitude and phase in a signal reradiating has long remained open. This problem is resolved if one applies the reciprocity theorem to the analysis of a relationship between the fields incident onto an array element (single radiator) and reflected from that element.

At incidence of an electromagnetic wave on a flat perfectly conducting metal surface, it is known that the wave is reflected at an angle equal to that of incidence, the amplitudes of the incident and reflected fields are identical, and the wave phase changes in a stepwise fashion by π. However, when a metal surface is replaced with a system of radiators, *e.g.*, with a linear equispaced array, the direction and the phase of a reflected field could be essentially changed, because they depend on parameters and electric characteristics of a separate radiator.

An example of such structure is the in-phase reflective array (Fig. **13a**). It is a flat equivalent of a parabolic reflector. The structure consists of primary feed 1 (*e.g.*, a horn) and an equispaced array of secondary microstrip radiators 2, situated in one plane along surface 3. In order to sum up for the secondary radiators signals to add up in the direction, perpendicular to the array plane, their phases should be identical. Since distances r_i (*i* is the reradiator number) between the feed and an arbitrary reradiator are not identical, this results in a phase path difference, which should be compensated with a phase step in the signal reradiating.

The method for calculating the phase step of the reflected field in comparison with the incident field phase in the signal reradiating can be constructed on the basis of the reciprocity theorem.

The reciprocity theorem for two antennas (see, *e.g.*, [14]) states: if emf e_a applied to the terminals of antenna A establishes current I_b at the input of an antenna B (Fig. **14a**), then equal emf e_b applied to the terminals of an antenna B will create at the input of an antenna A the same current I_a (Fig. **14b**), *i.e.*,

$$I_b/e_a = Y_{ab} = Y_{ba} = I_a/e_b, \tag{3.46}$$

where $Y_{ab} = |Y_{ab}|\exp(j\phi_{ab})$ and $Y_{ba} = |Y_{ba}|\exp(j\phi_{ba})$ are the mutual admittances between antennas. It follows that

$$\phi_{ab} = \phi_{ba}, \tag{3.47}$$

i.e., the difference of phases between the exciting emf and the current excited in an adjacent antenna is the same in both cases.

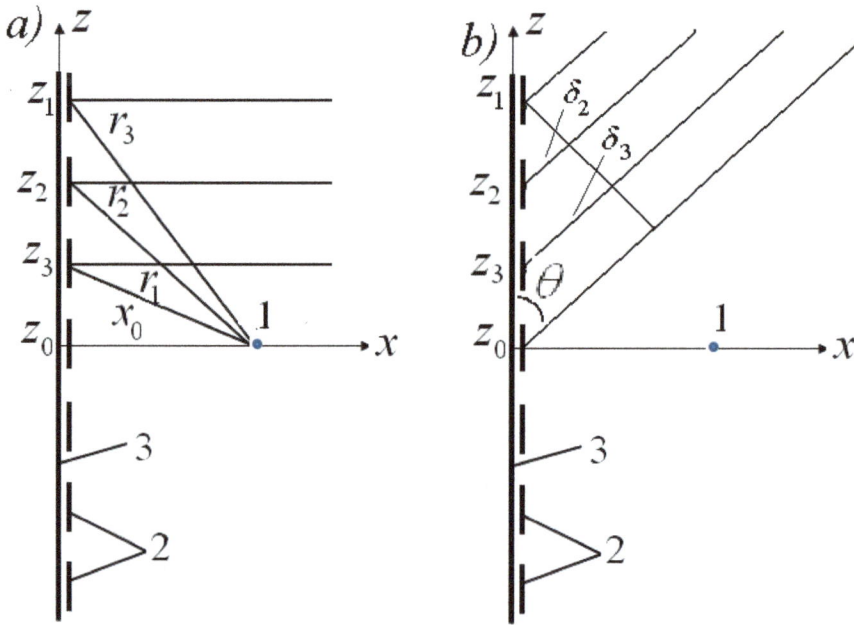

Figure 13: Reflect array with the radiation direction perpendicular to the array plane (*a*) and with the radiation in any desired direction (*b*).

In our example incident field E_a acts as a signal source instead of the antenna A with emf e_a. Introduce into the circuit a field's source – antenna A excited by the generator e_a with infinitely high output resistance. Usually, a linear antenna is thought as the aggregate of elementary dipoles with appropriate currents, and the radiation field phases outstrip the current phases by $\pi/2$. The phases difference between the current I_b and emf e_a is (see Fig. **14a**)

$$\phi_{ab} = \phi_{11} + \phi_{12} + \phi_{13} + \phi_{14},$$ (3.48)

where ϕ_{11} is the phase shift between the current in the antenna A and emf e_a (in this case it is absent), ϕ_{12} is the phase shift between the radiated field E_a and the current in the antenna A (it is $\pi/2$), ϕ_{13} is the phase shift due to the distance between antennas, ϕ_{14} is the phase difference between current I_b and field E_a, that is

$$\phi_{14} = \phi_{ab} - \pi/2 - \phi_{13}. \tag{3.49}$$

The other source of a signal is the current in antenna B (rather than emf e_b), which creates the reflected field E_b.

A current distribution along the receiving antenna differs from that along a transmitting antenna. One should think of antenna B as the aggregate of elementary dipoles, each of which is excited by its generator. The currents in the dipoles create the in-phase fields. Let the current in the middle dipole (at the antenna center) be excited by the generator e_b; it is equal to the product of emf e_b and the dipole admittance. Since the dipole impedance is capacitive in effect, the current phase outstrips the emf phase by $\pi/2$. That assertion holds true for other dipoles of the antenna B. Accordingly, the phase difference between current I_a and emf e_b is equal (see Fig. **14b**)

$$\phi_{ba} = \phi_{21} + \phi_{22} + \phi_{23} + \phi_{24}, \tag{3.50}$$

where ϕ_{21} is the phase shift between the current in antenna B and emf e_b (in this case it is $\pi/2$), ϕ_{22} is the phase difference between field E_b and the current in antenna B, ϕ_{23} is the phase shift due to the distance between antennas (it is ϕ_{13}), ϕ_{24} is the phase shift between the reflected field E_b and current I_a (it is zero, since the input impedance of antenna A is infinitely large), that is

$$\phi_{22} = \phi_{ba} - \pi/2 - \phi_{13}. \tag{3.51}$$

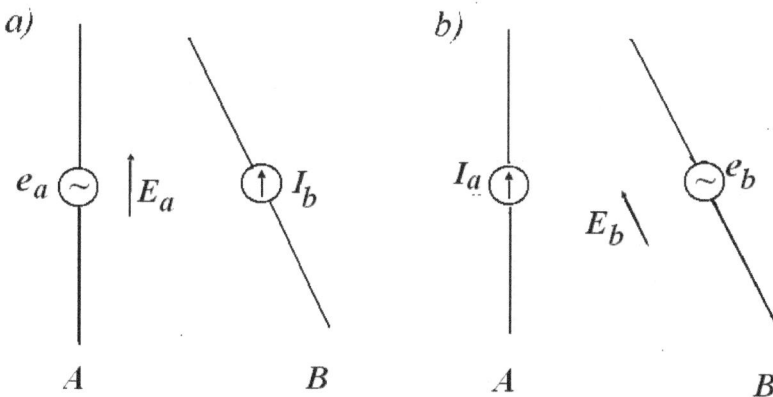

Figure 14: The reciprocity theorem: (a)–antenna A radiates, (b)- antenna B radiates.

In accordance with the equation (3.47), this implies that the increment of the phase of the receiving antenna current compared with that of the incident field and the increment of the phase of the reflected field compared with that of the receiving antenna current are identical in magnitude.

As to the amplitudes of incident $|E_1|$ and reflected $|E_2|$ fields, since the total tangential component of both fields on a perfectly conducting metal surface is zero, then at the reflection point

$$|E_2| = |E_1| \cos \gamma \cos \delta , \tag{3.52}$$

where γ is the angle of ray incidence onto an antenna, δ is angle of reflection.

The realistic variants of reflect arrays are considered in Section 8.3.

Send Orders for Reprints to reprints@benthamscience.net

CHAPTER 4

Self-Complementary Antennas

Abstract: The performance of a slot antenna situated on infinite conic metal surface with circular cross-section is shown to be similar to that of situated on the infinite flat metal surface. The proof is based on the transition from a *V*-shaped magnetic radiator to a double-sided slot cut in the mentioned surface. If, to simplify the design, metal and slot radiators are placed along the regular pyramid sides, the antenna performance changes but remains close to that of the self-complementary antenna. Flux density distribution over long line cross section is analyzed. The transition from the parabolic problem to the flat one is done.

Keywords: Complementary principle, Directivity of 3-dimensional radiator, Duality principle, Double-sided slot, Electric *V*-shaped radiator, Finite dimensions of the radiator, Flux density distribution over long line cross section, Magnetic *V*-shaped radiator, Metal sheet as a set of divergent wires, Parabolic long line, Parabolic problem, Phantom model, Phantom vessel shape as an example of a parabolic surface, Pyramid-shaped volume radiator, System of parabolic coordinates, Three-dimensional radiator, Transformation of variables, Transition from a paraboloid to a cylinder, Two slot antennas on curvilinear metal surfaces.

4.1. THREE-DIMENSIONAL RADIATORS

The structure of two convergent charged shells of angular width 2α situated along the surface of a circular cone with angle $2\theta_0$ at vertex (see Fig. **3a**) was considered in Section 3.1. It is excited by a generator placed at the cone vertex. One can treat this structure, on the one hand, as a two-wire line, and, on the other hand – as an antenna. The antenna is a symmetrical *V*-radiator, with the arms shaped as two convergent metal shells situated along the surface of the circular cone. Finally, one can interpret the structure as a slot antenna cut in the conical screen.

As it was shown, if the structure length is great, the impedances of the metal and slot radiators with the same angular width, *i.e.,* identical in the shape and dimensions, are related by (3.14). If the width of each shell and of the interval between them is equal to $\pi/2$, then expression (3.15) is valid.

Equality (3.14) coincides with expression for input impedance Z_S of an arbitrary slot antenna, which is cut in a flat perfectly conducting metal screen of infinite dimensions and infinitesimal thickness (Fig. **1**). It shall be recalled that Z_R in the expression is the input impedance of a metal radiator, with the shape and dimensions coinciding with those of a slot.

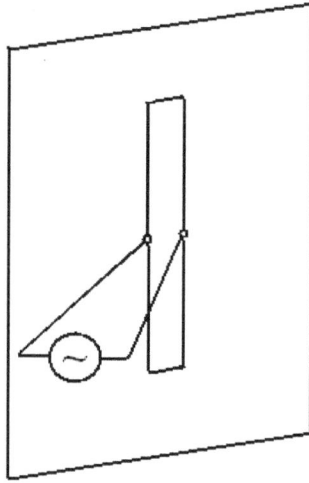

Figure 1: Flat slot antenna.

The expression for the input impedance of a slot antenna cut in a flat metal screen is determined by two methods: by means of the duality principle [33-36] and by means of complementary principle [37]. In particular, one of the creators of self-complementary antennas Prof. Mushiake describes his results in [36]. It was shown in Section 3.1 that it is valid not only for a slot cut in a flat screen, but also for a slot cut in the circular cone.

The complementary principle stems from an interrelation between scattering properties of the metal and slot radiators of the same shape and dimensions. The particular case of the interrelation occurs, when the metal radiator and identical slot radiator fill the entire plane. In this case, the structure is termed self-complementary. The variants of flat self-complementary radiators are shown in Fig. **2**.

As noted earlier, impedance Z_S of the slot (magnetic) radiator and impedance Z_R of the metal (electric) radiator, identical to the slot in shape and dimensions, are the same in the case of the self-complementary structure, *i.e.,* the equality (3.15) holds independently of the structure kind and the operation frequency. Therefore, the structure has the constant and purely active input impedance in a wide frequency range. The fact that such structure needs not be necessarily a flat one is noted in [38], namely, if the angular width of a metal strip and a slot in the case of the double-start helix are the same, one may speak of the self-complementary structure, bearing in mind the identity of surface area covered by the metal shell and free of it.

In the general case, the expression (3.14) is valid for the input impedance of a symmetrical double-sided slot antenna of an arbitrary shape and dimensions, situated on a circular metal cone of an infinite length and excited at its vertex (Fig. **3**). Let us use the duality principle for its derivation and consider a symmetrical magnetic *V*-shaped radiator (Fig. **4a**) situated in a free space. Its radiation resistance R_M is related to the radiation resistance R_E of an electrical radiator, similar in shape and dimensions, by expression

$$R_M = (120\pi)^2 / R_E .$$

(4.1)

The expression is well known. It follows from comparison of the powers radiated by both antennas. If one compares the oscillating powers of both radiators, we shall obtain, similarly to (4.1),

$$Z_M = (120\pi)^2 / Z_E ,$$

(4.2)

where Z_M and Z_E are the input impedances of the magnetic and electric radiators, respectively.

Let us pass from the magnetic *V*-shaped radiator to a slot antenna. For this end, we shall divide each arm of the magnetic radiator with a conic metal surface (of a circular cross section) passing through their axes. Since the shape of magnetic lines of force coincides with that of the surface, the radiator field undergoes no change as a consequence of the metal surface insertion. Actually, one can show that surfaces of field strength, along which the lines of force pass, are the circular cones. The axis of one cone coincides with bisector of the angle between arms of *V*- shaped radiator.

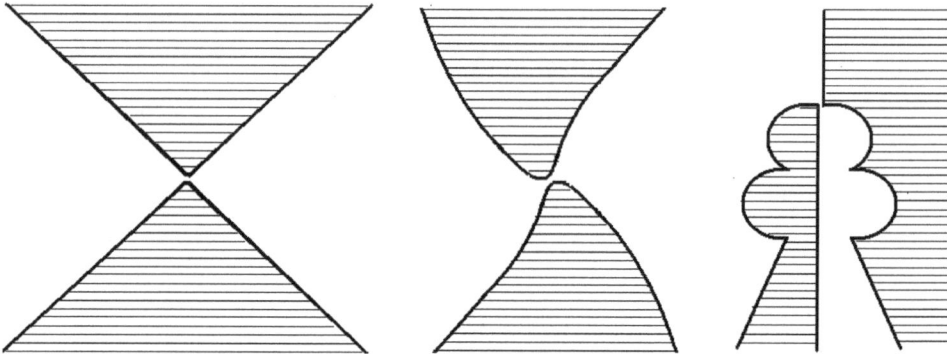

Figure 2: Flat self-complementary structures.

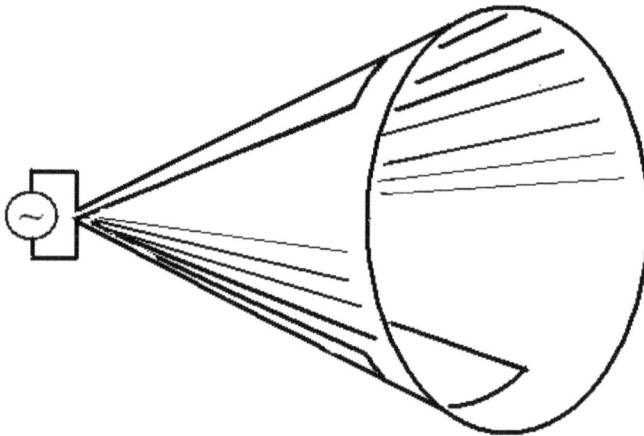

Figure 3: Symmetrical slot antenna on a circular cone.

The insertion of the metal surface divides the magnetic radiator into two ones situated on its either side (inside and outside of the conic spatial angle). Since this does not change magnetomotive force e_M exciting the radiator and the oscillating power created by the radiator

$$P = e_M J_M,$$

(4.3)

the share of magnetic current J_M in each newly formed radiator is equal to the share of the power radiated into each subspace. To be definite assume that m is the power share in the smaller subspace (inside the cone) and $1-m$ is the power share in the greater subspace. The input admittance of the magnetic radiator located inside the cone is

$$Y_1 = e_M /(mJ_M) = Y/m ,$$

(4.4)

where $Y = 1/Z_m$ is the total admittance of the original radiator.

Let the cross section of the magnetic radiator with current mJ_M be shaped as a curvilinear rectangle with the sides b and a, and $b \rangle\rangle a$ (Fig. **4b**). The greater side b is parallel to the metal surface and is arc-shaped. Such radiator is equivalent to a one-sided slot of width b. (It is necessary to note that the value of b changes along the slot axis.) The second radiator is equivalent to a similar one-sided slot with current $(1 - m)J_M$. Its input admittance is

$$Y_2 = e_M /[(1-m)J_m] = Y/(1-m) .$$

(4.5)

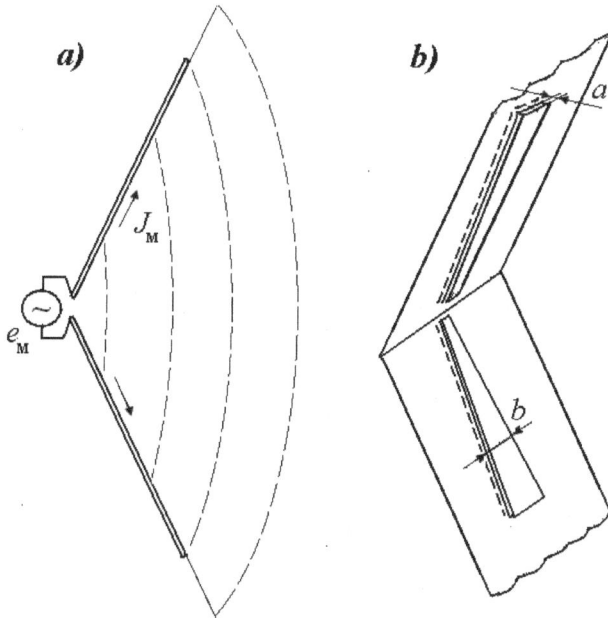

Figure 4: Magnetic *V*-shaped radiator (*a*) and double-sided slot (*b*).

If one changes both slots to a single double-sided one and takes into account that its admittance is equal to the sum of both slot admittances, then

$$Y_s = Y_1 + Y_2 = Y/[m(1-m)] .$$

(4.6)

The input impedance of the double-sided slot is

$$Z_s = 1/Y_s$$

or, taking into account (4.2) and (4.6),

$$Z_s = m(1-m)Z_M = (120\pi)^2 \, m(1-m)/Z_E \, .$$

(4.7)

In the particular case of a straight magnetic radiator, the conical surface becomes a flat one and divides the space to two equal halves. Accordingly, the current share in each of the newly formed radiators is equal to $m = 1/2$, and equality (4.7) transforms to expression (3.14).

In the case of the self-complementary structure, the metal and slot radiators are identical in the shape and dimensions. And so

$$Z_s = Z_m = 120\pi \sqrt{m(1-m)} \, ,$$

(4.8)

i.e., the volume self-complementary structure has the properties similar to those of the flat structure. It has the constant and purely resistive input impedance, which ensures a high level of its matching in a wide frequencies range.

The performed analysis shows that a choice of the surface to place the self-complementary radiation structures is not arbitrary. This surface is a circular cone, which transforms in the limiting case into the plane.

The shape of a metal and slot radiator situated at the circular cone may be different – similarly to that of flat radiators (see Fig. **2**). The simplest shape results, if the slot edge coincides with the cone generatrix.

4.2. ACTUAL ANTENNAS ENERGY FLUX THROUGH A LONG LINE CROSS-SECTION

The flat asymmetrical antenna shaped as a triangular metal sheet suspended on two grounded metal supports (Fig. **5a**) is an example of the self-complementary radiator. The triangular sheet consists of wire divergent from the lower triangle apex (from the feed point) at equal angles. The angle at the apex is equal to $\pi/2$. It is easily verified that with allowance for a mirror image, the described antenna is a symmetrical flat radiator composed of metal and slot radiators, with the same angular width of both radiators being $\pi/2$.

Symmetrical variant of such radiator (Fig. **5b**) is used as stand-alone horizontal antenna and as an element of an antenna array.

Antennas presented in Fig. **5** differ from structures in Fig. **2**. Firstly, they have finite dimensions: the metal radiator is limited from the top by the horizontal wire, and the slot radiator is limited on both sides by vertical shunts realized in the shape of grounded metal supports. Secondly, a metal sheet is not made of a solid plate, but as a set of wires divergent from the feed point. For this reason, it is difficult to accomplish the identity of the shape and dimensions of the metal radiator and the slot near the feed point. Approximate coincidence of dimensions limits the frequency range from above. From below, it is bounded by the arm height of the radiator. If the arm is 0.25λ and higher, the standing wave ratio (SWR) in the feed cable is lower than 2.0. The flat self-complementary antenna can ensure this SWR in the range with the frequency ratio greater than 20-30.

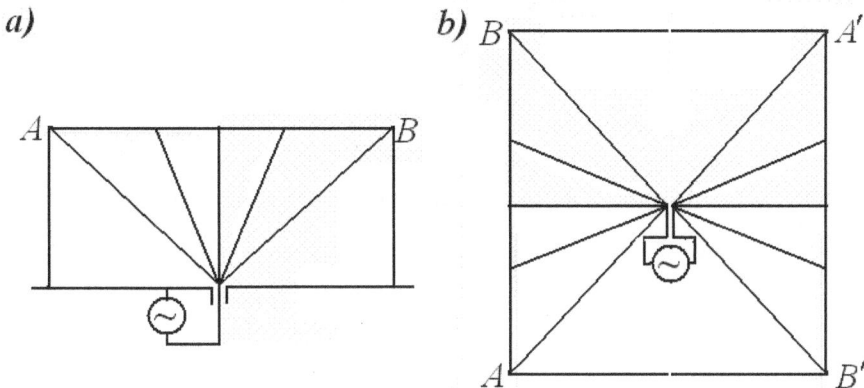

Figure 5: Asymmetrical (*a*) and symmetrical (*b*) variants of flat self-complementary antenna.

Three-dimensional antenna (volume radiator) constructed to the same principle is shown in Fig. **6a**. Here, the metal sheet is located at a sharp angle to ground along the surface of the circular cone. This permits to increase the arm length of the radiator at the same height of supports, *i.e.,* extend the range to low frequencies.

At the same time the antenna directivity increases, the antenna radiation intensifies in the direction of the sheet inclination. For the dimensions of metal and slot radiator to coincide, the distance between supports must be twice their height. Then, vertices A and B and their mirror images coincide with the vertices of regular quadrangle (square).

To simplify the design, the arm of a metal radiator can be taken in the shape of a flat triangle, with the wires located along a pyramid side (see Fig. **6b**). In this case, in contrast to the Fig. **6a** the metal and slot radiators are not located along a smooth conical surface, the structure is no self-complementary, and the expressions (3.14) and (3.15) are invalid. But in the range of decameter waves and at lower frequencies it is easier to implement a radiator located along the pyramid sides than a conic radiator.

The analysis of the structure can be performed by the method used in Section 3.1 in order to calculate the input impedances of metal and slot radiators in terms of wave impedance of the infinitely long uniform line. Consider a pyramid of the rectangular cross section, with two metal sides (shaped as flat cones) and two air sides (Fig. **7**). A flat cone differs from the triangular plate in that the base of the isosceles triangle is replaced with an arc of circumference with the center at vertex O of the pyramid.

Find input impedance Z_R in terms of the expression for the wave impedance of a line formed by two flat cones with angle 2α at the vertex, which are placed in the planes located at angle 2β to each other [38]. The expression coincides with (3.11), and

$$k = \left(\frac{1 - \sin \alpha}{1 + \sin \beta} \right)^{\pi/2\beta} ,$$

(4.9)

and angles α and β are related by the equation

$$\tan \alpha / \sin \beta = b/d ,$$

(4.10)

where b is the plate width in given cross section, and d is the distance between plates (see Fig. **7**).

The wave impedance of a two-wire long line formed by two flat cones, like that of a conical line with circular cross section, is defined by the angles at its vertex. It depends on the relation between magnitudes b and d, which is constant along the long line and independent of the absolute values of these magnitudes. This means that the line is also uniform, and that its input impedance tends to the wave impedance with increasing line length.

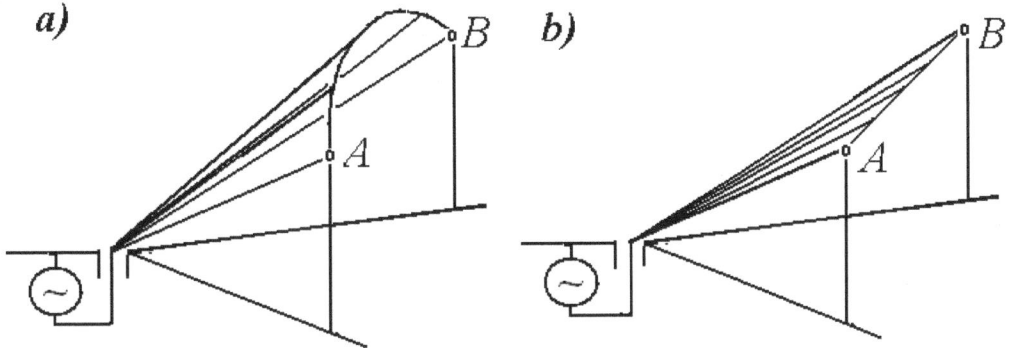

Figure 6: Antennas located on the circular cone (*a*) and on the pyramid (*b*).

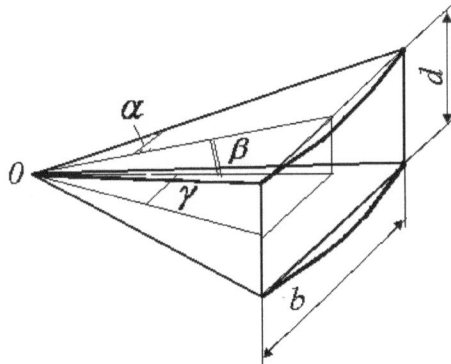

Figure 7: Placement of radiators along the pyramid sides.

As seen from the presented expressions, quantity W is a function of two arguments: the ratio b/d and angle α. If, for example, $b/d = 1$, the change of α from 15 to 30° (and, accordingly, the change of β from 15.5 to 35.3°) causes the wave impedance to increase from $42.5\,\pi$ to $45\,\pi$, i.e., $Z_m = Z_s = (42.5 - 45)\pi$.

If $b/d = 5$, $\beta = 15°$, then $Z_R = 12.7\pi$. In this case, Z_s is equal to the input impedance of a metal radiator situated side by side with the slot. For a metal radiator with $b/d = 1/5$, angle β is found from the expression $\tan\beta/\tan\gamma = d/b$, where $\gamma = 15°$ (since at the interchange of b and d, angles β and γ follow suit). It is easily verified that $\beta = 53.3°$, wherefrom $Z_s = 105\pi$, i.e., $Z_R Z_S = (36.6\pi)^2$.

The calculations have shown the product $Z_R Z_S$ to depend on the ratio b/d, but be always smaller than $(60\pi)^2$.

The radiators of infinite length situated at the cone and pyramid with the same angular width of the metal shell and the slot have the constant and purely resistive input impedance, which permits to secure a high level of matching in an unlimited frequency range. In the case of finite dimensions of such radiator the frequency range is limited, but remains rather wide.

It is necessary to note the essential drawback of the flat vertical antenna. At frequencies, where the antenna height exceeds 0.7λ, the main lobe of the vertical pattern deviates from the perpendicular to the antenna axis (deviates from the ground). This effect limits the antenna frequency range from above.

In the case of a thin radiator, one can extend the antenna frequency range, if one connects in it lumped capacitive loads permitting to create the in-phase current distribution along an antenna wire. Here, one may proceed in a similar manner. The capacitive loads in a flat triangular metal radiator can be shapes as horizontal slots. For the shape and dimensions of a triangular slot radiator to coincide with those of the metal one, it is necessary to mount vertical metal plates in the slot radiator, with each plate width having to coincide with that of the corresponding horizontal slot (Fig. **8**). For a flat structure of such kind, the expression (3.15) remains valid.

In the case of plane-parallel electrostatic field, flux function $V(x,y)$ defines the flux of vector $\vec{E}$ through the cylindrical surface of unit length located between given and zero surfaces of the field strength. For two parallel infinitely long charged filaments the flux function is

$$V = -\frac{q}{2\pi\varepsilon}\left(\varphi_{c2} - \varphi_{c1}\right)$$

(4.11)

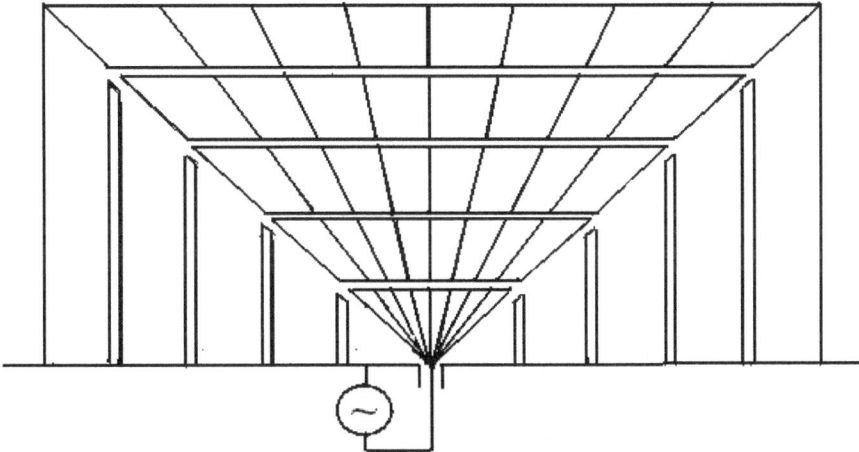

Figure 8: Self-complementary antenna with loads.

Here, (see Fig. 1b) q is linear density of positive and negative charges of each filament, ε is the medium permittivity, φ_{ci} is the cylindrical coordinate of the ith metal filament. Consider the cylindrical surface of unit length along the z-axis with the cross section shaped as a circumference of small radius $\delta \langle\langle b$ situated around the left charged filament. For any point at that surface, angle φ_{c2} is constant and equal to π, *i.e.,* the magnitude of the flux increment inside given angle $\varphi_{c2} - \varphi_{c1}$ is proportional to φ_{c1}. That means, in particular, that the share of the flux of a charged filament, which falls into the volume bounded by the cylindrical surface, passing through both filaments and having the cross section shaped as a circumference with the center at the coordinate origin, is equal to a half of the total flux.

Likewise in the case of two convergent charged filaments (see Fig. 1a) a half of the flux is directed inside a circular cone passing through both filaments, its axis coinciding with z-axis.

In accordance with the electrostatic analogy method, the share of the constant current inside the volume bounded by a circular cylinder (or cone) passing through two high-conductivity wires located in a homogeneous weakly conductive medium, will be the same, if a constant voltage is applied to the wires.

In accordance with the conformity principle the magnetic field of constant linear currents coincides with the electric field of linear charges, if the currents and the charges are equally distributed in space. This means, in particular, that a half of the energy flux propagating along a two-wire long line of parallel (or divergent) wires is concentrated inside the circular cylinder (or cone) passing through the wires.

Similar postulates are valid for the system of two wires with finite radius (see Fig. **2**), if one considers the flux inside the cylindrical (or conical) surface of circular cross section passing through the filaments creating the field, which equipotential surfaces coincide with the outer surfaces of the cylindrical (or conical) wires.

4.3. RELATIONSHIP BETWEEN PARABOLIC AND CYLINDRICAL PROBLEMS

Comparison of conic and cylindrical problem allows calculating the input impedance of symmetrical double-sided slot antenna located on a circular metal cone and excited at its vertex. The circular cone generatrix is a straight wire. If a curvilinear wire, *e.g.*, parabolic one, acts as a generatrix, the metal surface assumes the shape of a paraboloid. Of interest here is calculating the field of two infinitely long charged filaments of curvilinear shape located on the paraboloid surface and converging to its vertex (Fig. **9**).

The analysis of such structure is facilitated by the use of the system of parabolic coordinates (σ,τ,ψ) [39]. This is a system of orthogonal curvilinear coordinates, their coordinate surfaces being confocal paraboloids of rotation $(\sigma = const, \tau = const)$ with focuses at the origin of coordinates and half-planes $(\psi = const)$, passing through the axis of rotation (see Fig. **9**). Rectangular coordinates are related to parabolic coordinates by the follows:

$$x = \sigma\tau\cos\psi \ , \ y = \sigma\tau\sin\psi \ , \ z = \left(\tau^2 - \sigma^2\right)/2 \ . \tag{4.12}$$

The parabolic wires are located along the surface of a paraboloid, or more exactly along the curve of the surface intersection with the half-plane, passing through the axis of rotation ($\sigma = const$, $\psi = const$).

As in the case of convergent straight wires, it is expedient to reduce the calculation problem for the electrostatic field of two charged parabolic filaments to the flat problem for two parallel filaments (see Fig. **1b**). To this end the Laplace equation is in accordance with the uniqueness theorem to hold true at the transition from one problem to another, and the wires are to coincide with the lines of equal potential.

In the cylindrical coordinates system (ρ, φ_c, z), the Laplace equation for a potential U has the form

$$\frac{\partial}{\partial \rho}(\rho \frac{\partial U}{\partial \rho}) + \frac{1}{\rho}\frac{\partial^2 U}{\partial \varphi_c^2} = 0$$

$\qquad$ (4.13)

Here, the account is taken of $\partial U / \partial z = 0$, *i.e.*, the lines parallel to *z*-axis have the constant potential (the field is plane-parallel).

In the system of parabolic coordinates the Laplace equation has the form

$$\frac{1}{\sigma^2 + \tau^2}\left[\frac{1}{\sigma}\frac{\partial}{\partial \sigma}\left(\sigma \frac{\partial U}{\partial \sigma}\right) + \frac{1}{\tau}\frac{\partial}{\partial \tau}\left(\tau \frac{\partial U}{\partial \tau}\right) + \left(\frac{1}{\sigma^2} + \frac{1}{\tau^2}\right)\frac{\partial^2 U}{\partial \psi^2}\right] = 0.$$

$\qquad$ (4.14)

As seen from (4.14), the equation is symmetrical with respect to σ and τ, that is, the equation for each unknown quantity is valid. In particular, for σ we obtain

$$\frac{1}{\sigma^2}\left[\frac{1}{\sigma}\frac{\partial}{\partial \sigma}\left(\sigma \frac{\partial U}{\partial \sigma}\right) + \frac{1}{\sigma^2}\frac{\partial^2 U}{\partial \psi^2}\right] = 0.$$

$\qquad$ (4.15)

Here, $U(\sigma) = U(\tau)$, if other coordinates are the same. If, for example, $\frac{\partial U}{\partial z} = 0$, then in accordance with (4.12) $\sigma = \sqrt{\tau^2 - 2z}, \tau = \sqrt{\sigma^2 + 2z}$, *i.e.*

$$\frac{\partial \sigma}{\partial z} = -\frac{1}{\sigma}, \frac{\partial \tau}{\partial z} = \frac{1}{\tau},$$ (4.16)

and $\dfrac{\partial U}{\partial z} = \dfrac{\partial U}{\partial \sigma}\dfrac{\partial \sigma}{\partial z} + \dfrac{\partial U}{\partial \tau}\dfrac{\partial \tau}{\partial z} = 0$, whence

$$\frac{1}{\tau}\frac{\partial U}{\partial \tau} = \frac{1}{\sigma}\frac{\partial U}{\partial \sigma}.$$ (4.17)

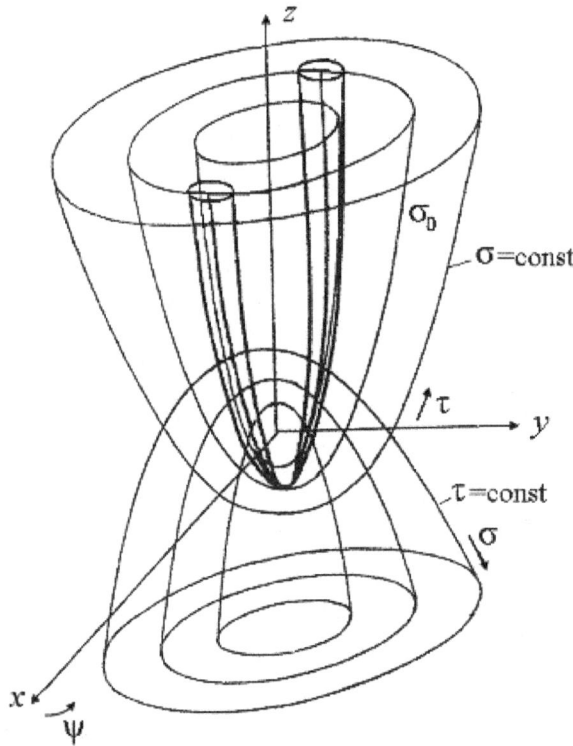

Figure 9: Surface in the shape of a circular paraboloid.

Comparison of (4.13) and (4.15) shows that these equations coincide, if the substituted variables are related by equations:

$$\rho = \sigma, \varphi_c = \psi .$$ (4.18)

Here ρ and φ_c are the cylindrical coordinates, and σ and ψ are the parabolic ones. Hence, the Laplace equation holds true in the transition from the parabolic problem to the cylindrical one, if expressions (4.18) are valid. The statement presented in [48] that this requires the condition $\tau \rangle\rangle \sigma$ to be met is incorrect.

The transformation of variables in accordance with (4.18) results in the mapping of parabolic surface $\tau = const$ (for arbitrary τ) onto the plane (ρ, φ_c). The line of the surface intersection with any paraboloid $\sigma = const$ is transformed into a circumference. Two parallel wires spaced at distance $2b = 2\sigma_0$ (see Fig. **3.1***b*) correspond to the case of two charged parabolic filaments situated along surface $\sigma = \sigma_0$ in the plane xOz (see Fig. 9). Two metal cylinders of radius $a = (c-d)/2$ (see Fig. **3.2***a*), their axes spaced at $2h = c + d$, correspond to two solid geometrical figures located between parabolic surfaces $\sigma_1 = c$ and $\sigma_2 = d$.

The scalar potential of the electric field for two parabolic charged filaments situated along surface $\sigma = \sigma_0$ (with linear charge density $\pm q_0$) is similarly to (3.2)

$$U(\sigma,\psi) = \frac{q}{2\pi\varepsilon}\ln(\rho_2/\rho_1), \tag{4.19}$$

where $\rho_2^2 = (\sigma_0 - \sigma\cos\psi)^2 + \sigma^2\sin^2\psi$, $\rho_1^2 = (\sigma_0 + \sigma\cos\psi)^2 + \sigma^2\sin^2\psi$.

The reduction of the parabolic problem to the cylindrical one permits to calculate capacitance C_3 per unit length and wave impedance W_3 of a long line consisting of two solid figures with the parabolic axes. By analogy with (3.7), we find:

$$C_3 = \frac{\pi\varepsilon}{ch^{-1}[(\sigma_1 + \sigma_2)/(\sigma_1 - \sigma_2)]}, \quad W_3 = 120ch^{-1}[(\sigma_1 + \sigma_2)/(\sigma_1 - \sigma_2)]. \tag{4.20}$$

Since the input impedance of a uniform two-wire line tends as its length increases to its wave impedance, the input impedance of the line is

$$Z_{AB} = 120ch^{-1}\frac{\sigma_1 + \sigma_2}{\sigma_1 - \sigma_2}. \tag{4.21}$$

The case of two charged converging shells of angular width 2α located along the surface of a paraboloid (Fig. **10**) is of a specific interest with respect to the placement of slot antenna of finite width on surfaces of revolution. The flat problem in the form of a line of two coaxial cylindrical shells corresponds to it. The electrostatic field of such line is shown in Fig. **3.3***b*. The field structure in the case of two shells located on the paraboloid surface is of a similar nature, but with surfaces of equal potential *U*=const and surfaces of field strength *V*=const coinciding with parabolic surfaces rather than the cylindrical ones.

As seen from (3.11), capacitance C_2 per unit length and wave impedance W_2 of a cylindrical line are dependent only on angular width 2β of the slot and, accordingly, on angular width $2\alpha = \pi - 2\beta$ of the metal shell. For this reason, expressions (3.11) are valid also for the parabolic envelopes. Quantities C_3 and W_3 are constant along the line, *i.e.* the line of two wires is a uniform one, and the impedance of a metal radiator with angular width 2α of the arm (a slot antenna of angular width 2β):

$$Z_{AB}(\alpha) = 120\pi K(k)/ K\left(\sqrt{1-k^2}\right).$$
(4.22)

If $\alpha = \beta$, the radiator is self-complementary, and $Z_{AB}(\pi/4) = 60\pi$.

Compare the lengths of self-complementary radiators of different shapes and the same height *H*. The arm length of a vertical flat radiator is $L_1 = H$, the one of the conic radiator shell for angle θ_0 between the shell and cone axes is $L_2 = H/\sin\theta_0$, the one of the shell for the paraboloid with height *H* and projection length *S* is

$$L_3 = 2S\sqrt{1+\frac{H^2}{4S^2}} + \frac{H^2}{2S} sh^{-1}\frac{2S}{H} .$$
(4.23)

In particular, if $\theta_0 = 30^0$, then $L_2 = 2H$ and $S = 1.73H$. For the same projection length *S*

$$L_3 = 2.08H ,$$
(4.24)

i.e. the arm length of a parabolic radiator shell is larger than that of a conic radiator. It that the increase of the radiator length is expected to increase the *SWR* of the radiator.

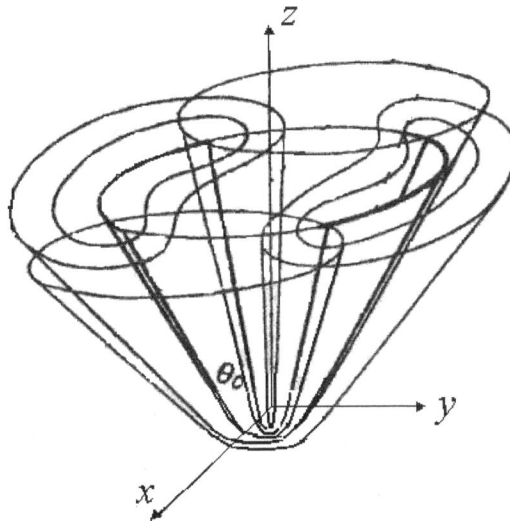

Figure 10: Two coaxial shells on the paraboloid.

4.4. MUTUAL COUPLING BETWEEN SLOT ANTENNAS ON CURVILINEAR METAL SURFACES

The calculation procedure of slot antennas situated on curvilinear surfaces with axial symmetry opens new prospects of rigorous analysis of the electrical characteristics of slots cut in screens other than planar.

The antenna theory employs the concept of ideal slot antenna. This is the slot cut in indefinitely large, perfectly conductive, infinitely thin metal screen. In accordance with the duality principle, the characteristics of the ideal slot antenna are easy to find, if those of a metal radiator with similar shape and dimensions are known. As it is shown in [40], the appearance of integral equation for voltage U across the slot edges coincides with that of the corresponding equation for the current in equivalent metal antenna, identical in shape and dimensions. But operator $G(U)$ (integrodifferential functional), involved in that equation, is linear in U and dependent on the screen shape, *i.e.,* differs, in the general case, from operator $G(I)$ of a metal antenna in free space.

Consider two identical magnetic radiators located in parallel in free space and excited in anti-phase (Fig. **11a**). By means of expressions (4.1) and (4.2), one can calculate radiation resistance R_M and input impedance Z_M of each magnetic radiator in terms of radiation resistance R_E and input impedance Z_E of an electrical radiator, similar in shape and dimensions. To perform the transition from magnetic radiators to slot antennas, we shall pass the metal surface shaped as a circular cylinder through the magnetic radiators. Since the shape of magnetic lines of force coincides with that of the surface, the field of radiators undergoes no change due to insertion of the metal surface. The metal surface divides each magnetic radiator into two radiators situated on its either side different (inside and outside of the cylinder). Since magnetomotive force e_M, exciting the radiator, and oscillating power of the radiator $P = e_M J_M$ remain unchanged at that, the share of magnetic current J_M in each newly formed radiator is equal to the share of the power radiated into each subspace.

As shown in Section 4.2, in the case of two parallel, infinitely long filaments with linear charges opposite in sign, the share of the energy flux of each filament inside the volume limited by the circular surface passing through the filaments is equal to a half of the total flux. According to the conformity principle, this means that the half of the energy flux propagating along a two-wire line of parallel wires is concentrated inside the circular cylinder passing through the wires. One can be quite certain that a similar supposition is correct for the wires of finite length (for electrical radiators) and also for magnetic radiators, though the magnitude of electrical or magnetic current varies along the wire in these cases.

If one considers the half of power to be radiated inside the cylindrical volume, the input admittance of inside magnetic radiator is

$$Y_1 = e_m/(J_m/2) = 2Y,$$

(4.25)

where $Y = 1/Z_m$ is the total admittance of the original radiator.

Let the radiator cross section inside the cylinder be shaped as a rectangle curved along a cylinder surface with sides b and a, at that $b \rangle\rangle a$ (Fig. **11b**). The greater side b is parallel to the metal surface and is an arc segment. Such radiator is equivalent to a one-sided slot of width b. The magnetic radiator on the outside of the cylinder is equivalent a similar one-sided slot. Its current and input admittance coincide with those of the radiator inside the cylinder:

$$Y_2 = Y_1 = 2Y. \tag{4.26}$$

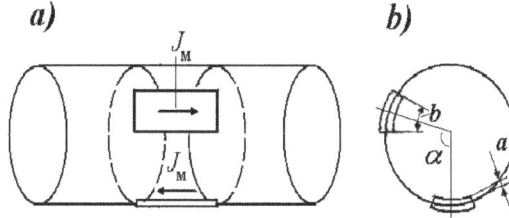

Figure 11: Two parallel magnetic radiators (*a*) and the transition to slot antennas (*b*).

If one replaces both one-sided slots with a single two-sided slot and takes into account that its admittance is equal to the sum of both slots admittances, then

$$Y_s = Y_1 + Y_2 = 4/Z_M, \tag{4.27}$$

i.e., the input admittance of the two-sided slot is according to (4.2)

$$Y_s = Z_E /(60\pi)^2. \tag{4.28}$$

Here, Z_E is the input impedance of the electrical radiator (of the metal antenna) located near another such radiator, which is parallel to the first one and is excited in anti-phase to it. So,

$$Z_E = Z_{11} - Z_{12}, \tag{4.29}$$

where Z_{11} is the self-impedance of each radiator and Z_{12} is the mutual impedance between radiators.

The expression (4.28) in view of (4.29) permits to calculate the input impedance of the slot antenna, including in the system of two anti-phased radiators situated along the generatrix of circular metal cylinder. Note that it is two arbitrary generatrices (with an arbitrary length of the arc along the cylinder circumference between them), not only located at opposite sides of the cylinder (when the arc length is equal to π). Actually, the introduction of a cylindrical metal surface dividing each magnetic antenna into two radiators requires observation of a special rule, namely, that the shape of the magnetic lines of force coincided with that of the surface. In given case, this is so, since the strength lines of the field created by two charged filaments (see Fig. **2a**) are circumferences passing through the filaments with the centers on the *y*-axis. The share of energy flux inside the volume, which is bounded by the circular metal surface passing through these filaments, is independent of the arc length between cylinder generatrices, along which the filaments are placed, *i.e.,* it is a half as before (see Section 4.2).

In accordance with the duality principle, if the current moment of a magnetic radiator is equal to that of an electrical radiator, the electric field of a magnetic radiator is equal in magnitude to the magnetic field of an electrical radiator and is opposite to it in sign. The magnetic field of a magnetic radiator is smaller than an electric field of an electrical radiator by a factor of $(120\pi)^2$. It means, in particular, that the pattern of two slot radiators system coincides with that of the system pattern of two metal radiators with the same shape and dimensions.

The pattern of an array consisting of two identical electrical radiators depends on the single radiator pattern and the array factor, *i.e.,* on the radiator spacing. Accordingly, the pattern of a system consisting of two thin slots depends on their spacing (the pattern of a single thin slot radiator has the circular shape in the equatorial plane) and is independent of the radius of the circular cylinder, in which slots are cut, *i.e.,* on the arc length (in radians) between the slot axes. As follows from (4.28), the input impedance is also independent of the cylinder radius. So, the presence of a cylinder surface has practically no effect on the electrical characteristics of two thin radiators. The cylindrical metal surface is but a structure, in which the slot is cut, and can be replaced, *e.g.,* with a metal plane passing through both radiators.

It is easily verified that the expression (4.28) permits to calculate the input impedance of the slot antenna located on infinite perfectly conducting plane not far from the other slot antenna (Fig. **12a**). If the second antenna is identical to the first one and is excited in anti-phase, magnitude Z_E can be found from (4.29). If several slots are cut in a metal plate, then, into expression (4.28), which remains valid, one must substitute as Z_E the input impedance of a metal radiator included in the radiators system with allowance for its position and phasing. The patterns of slot antennas coincide with those of metal antennas with a similar position in space.

If slot antennas have finite width, the electrical radiators to be compared with theirs must have the same width. They are metal plates with the cross section in the shape of a rectangle curved along the cylindrical surface (see Fig. **11**). Angle α between perpendiculars to the plates is equal to the arc between their axes, *i.e.,* it dependent on the cylinder radius. If the arc is π, the plates rotate in the opposite direction. When calculating Z_E, it is necessary to take into account the cross-section dimensions of metal plates and their relative rotation. The pattern changes accordingly. It means that the electrical characteristics of slot antennas in given case will depend on the radius of the cylinder, on which they are set.

One can extend all of the aforesaid to the case of slot radiators situated on the surface of a cone (Fig. **12a**), or a paraboloid (Fig. **12b**), or an arbitrary surface with the axial symmetry. In all cases the shape of metal radiators, with the patterns coinciding with those of slot radiators, must coincide with the shape of the corresponding segments of the given surface generatrix.

Metal bodies of finite length are of particular interest. As the experiments and calculations have shown, the patterns and input impedance of a whip antenna located near a circular metal cylinder of finite height, exceeding an antenna height by more than $\lambda / 4$, are practically the same if the antenna were situated near an infinitely long cylinder. Therefore, if the ends of slot antennas are distanced from the end of the metal cylinder with the axial symmetry by more than $\lambda / 4$, one may determine the characteristics of two anti-phased slots cut in the metal body in accordance with the characteristics of two radiators of the same shape and dimensions located in free space.

It is substantial that the validity of expression (4.28) for the input admittance and coincidence of slot and metal antennas patterns are independent of the radius of the metal cylinder, at which slot radiators are placed. In derivation of expression (4.28), one imposed no restrictions relating the radius of the cylinder with the antenna length or the wavelength. This means that it is enough for the transition from the magnetic radiator to two-sided slot antenna cut in a curvilinear metal surface, and also for the transition from the slot antenna to the magnetic radiator that the metal surface is smooth, and the antenna surface coincides with a part of the curvilinear surface.

The equivalence of a slot radiator to a magnetic radiator is valid, if the field structure at the radiator surface is formed by the exciting source and is independent of the electromagnetic field structure in the surrounding space.

At present, the mathematical modeling method, in accordance with which a metal surface is changed by a system of thin wires or metal strips, is used widely to calculate the electrical characteristics of radiators located not far from metal bodies [23, 41]. If stated so the problem reduces to the calculation of the currents distribution in a structure of randomly directed segments of wires or stripes. If the currents along the wires (stripes) are known, one can calculate all radiator characteristics.

The above equivalence of a slot radiator to a magnetic radiator permits to extend the mathematical modeling method to the case of calculation of the electrical characteristics for slot antennas situated on metal bodies of a complex shape. In this case, the magnetic radiator with the magnetomotive force it includes is the source of the electromagnetic field rather than the electrical radiator with the electromotive force it includes.

As said above, the coincidence of the electrical characteristics of slot radiators located on an axially symmetric metal surface with those of metal radiators permits to increase the area of slot antennas, where the rigorous methods of solving the electrodynamics problem are applicable.

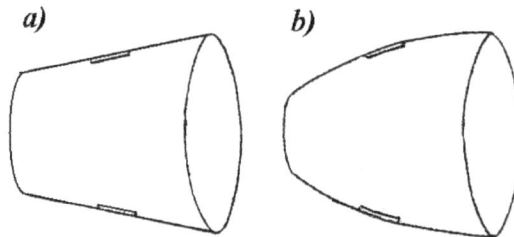

Figure 12: Slot antennas at the cone (*a*) and the paraboloid (*b*).

4.5. SHAPE AND DIMENSIONS OF THE PHANTOM MODEL

The phantom as a model of a human head was developed in order to measure and to study the parameters, including SAR of fields, established by various radiators, *i.e.,* to determine the level of human body irradiation. But the aspiration to create phantoms with characteristics closest to those of a head has led to a multitude of phantom shapes and dimensions and, accordingly, to variations in the measurement results in different systems. The scatter of output data caused necessity to develop a method for evaluating the impact of the phantom shape and dimensions on the measurement results.

Parameters of the antenna field under study and SAR depend on the capacitance between the locations of this antenna and the measuring one. An approximate calculation procedure of the said capacitance is based on the similarity between the shapes of the phantom and the paraboloid and on the reduction of the parabolic problem to the cylindrical one. The human body is simulated by a simple model filled with a liquid with dielectric permittivity and conductivity equal to their average values for the human head tissues at radio frequencies. A probe is immersed into the liquid to record the electric field strength inside the phantom. The measurements results are then processed to calculate the maximal point SAR, the local SAR (loss per unit volume), and the total SAR (loss in the whole head). Fig. **13** shows the measurement setup.

The effect of the phantom volume and shape on the accuracy of the tissue dosimetry confirms the circumstance that there is an interrelation between the phantom shapes and dimensions, on the one hand, and the measurement results, on the other hand.

Phantoms are usually constructed as vessels filled with a simulation liquid. The vessel wall is thin and made of a fiberglass material with low relative permittivity and conductivity. The vessel is open from top; the simulating liquid is homogeneous. The setup consists of a three vessels in the form of two heads (left and right) and a body.

Calculation of the phantom head field established by a radiator located outside the phantom, is a complicated three-dimensional problem. The near region of the transmitting antenna is situated in two different media, and its boundary is an intricately shaped. One needs to calculate the near field in the adjacent medium at the point of the second antenna location and take into account the mutual coupling of two radiators.

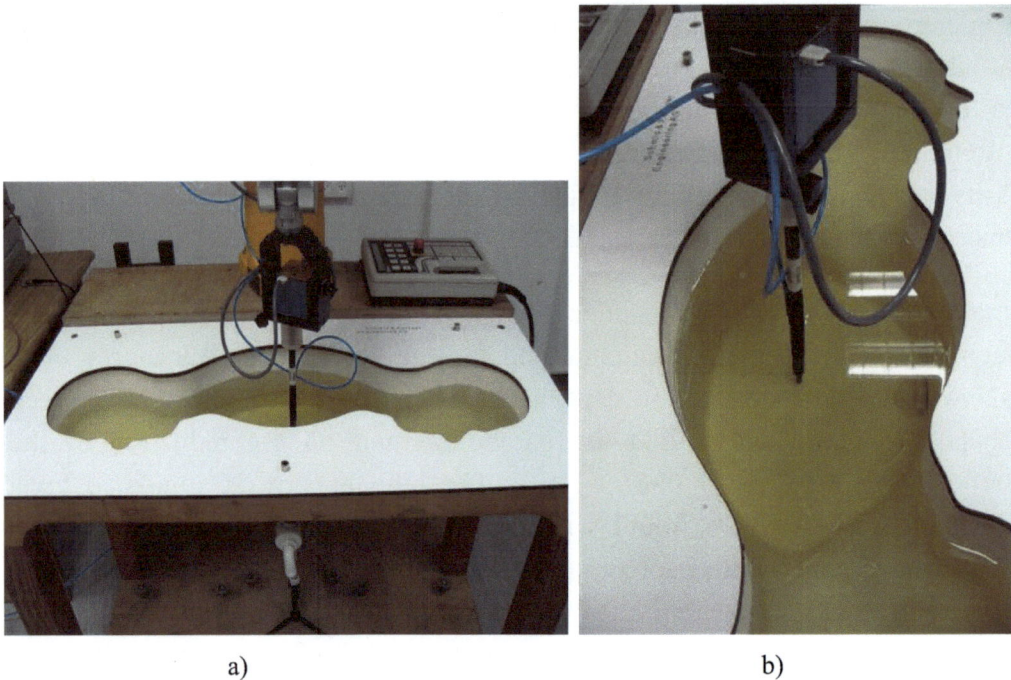

a)   b)

Figure 13: The SAR measurement setup – front (*a*) and side (*b*) view

The problem of calculating the electromagnetic field reduces, as a rule, to the electrostatic problem, *i.e.,* to the calculation of the electric fields of charged conducting bodies. There is similarity between the structure of the quasi-stationary electric field of alternating linear currents and the structure of the electrostatic field.

As known from the antenna theory, the near field of an antenna is in the first order approximation of a quasi-stationary nature. Therefore, one can reduce the calculation of the antenna near field to that of the electrostatic field of a charged conducting body.

Further, it is expedient to reduce the problem of calculating the electrostatic field to a plane (cylindrical) problem. As a result, the calculation simplifies substantially. Originally, the conical and cylindrical problems were compared with each other. But the problem studied here is not a conical one, since the vessel shape has little in common with a cone: the axis of such cone must coincide with the *z*-axis of the vessel, and vertices of all cones must lie at single point. This implies that all surfaces of equal potential in the conical problem should pass through one point, since they coincide with those of metal cones. In the considered problem, one equipotential surface coincides with the vessel shell, which is the interface of two different media, and other such surfaces pass through the points located on the z-axis at some distance from each other.

We assume that the wall is shaped as a paraboloid, *i.e.*, the generatrix of the wall surface is in the first approximation a parabola. The parabolic and cylindrical problems are compared with each other in Section 4.3: in addition, the system of parabolic coordinates (σ, τ, ψ) is described there. The parabolic problem is more similar to the real phantom structure, yet more complicated for calculations. In the general case, the paraboloid sections may be not circular, since the horizontal cross-section of the wall is not circular. If the wall is not shaped as a paraboloid at all, one can approximated it with a paraboloid closest to the actual wall in shape.

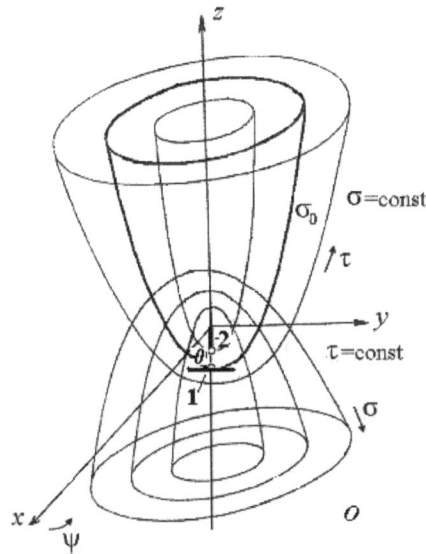

Figure 14: The system of parabolic coordinates.

Fig. **14** presents the circuit of a vessel with the parabolic coordinate system. The antenna under study, denoted with 1, operates as the transmitting antenna. As seen from Fig. **15**, that antenna, which in this case is a horizontal dipole, is situated below the vessel tangent to its wall and pass through the origin. The measuring antenna is a small probe placed inside the vessel (see Fig. **13**). It operates as the receiving antenna and is denoted in Fig. **14** with 2 it is expedient that both antennas should be axially symmetric. As is clear from Fig. **14**, the vertical axis of the structure (the z-axis) passes through radiator 1 and the vessel center 0. Probe 2 is located at the measurement point.

The wall is the interface of different media, and, for this reason, its surface $\sigma_0 = const$ is a surface of constant potential. Other surfaces $\sigma = const$ are surfaces of constant potential too. The probe is situated along the structure axis, and the radiator is located along the horizontal line.

In order to reduce the parabolic problem to the plane one, the Laplace equation should, in accordance with the theorem of uniqueness, hold true at the transition from one problem to the other, and metal wires should coincide with the equipotential lines. As shown in Section 4.3, the Laplace equation holds true at the transition, if parabolic (σ, τ, ψ) and cylindrical (ρ, φ_c, z) variables are related by

$$\rho = \sigma, \qquad \psi = \varphi_c. \tag{4.30}$$

Figure 15: Antenna under study (a horizontal dipole).

If the paraboloid is circular, the circular cylinder corresponds to it in the cylindrical coordinate system. The segment of the paraboloid in the vicinity of its vertex also passes to a segment of the circular cylinder. And, if the horizontal cross section of the paraboloid is close to circular, the error is small.

The transformation of the parabolic problem to the cylindrical one results, in particular, in the transition from parabolic surface $\tau = const$ to plane (ρ, φ). The equivalent cylindrical problem is presented in Fig. **16**. Vessel boundary $\rho_0(\varphi)$ is of rather intricate nature. Assume its cross section perimeter to be an ellipse with major axis of length $2a$ and minor axis of length $2b$, and the measuring antenna to be a thin metal filament with constant potential. The permittivity of the medium in the range $\rho_2 \langle \rho \langle \rho_0$ is equal to permittivity ε_2 of the liquid filling the vessel.

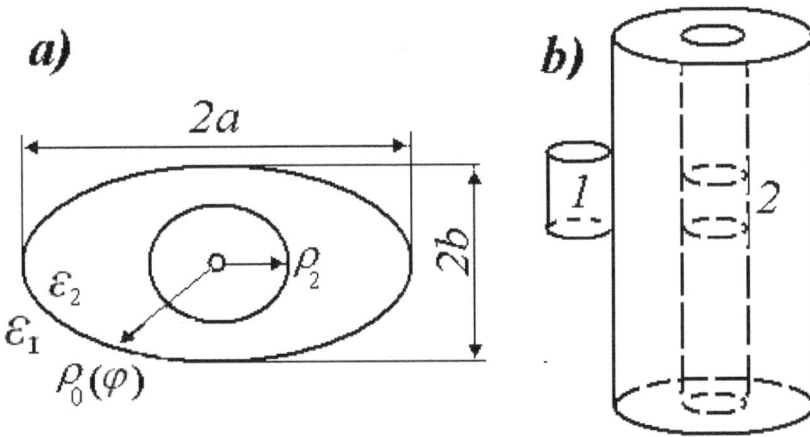

Figure 16: The equivalent cylindrical problem: a – top view, b – axonometric view.

The reduction of the three-dimensional problem to the plane (two-dimensional) one simplifies the calculation of the capacitance between equipotential surfaces. The capacitance magnitude determines the coupling level between the surfaces. The value can be used for estimating the strength of the field created by the antenna under study at the observation point. As is known, the electric component of the field at a given distance from an antenna is inversely proportional to the medium permittivity:

$$|E_z| = AJ_A F(R)/(\varepsilon_2 R),$$

$$\tag{4.31}$$

where J_A is the current in its base, $F(R)$ is a function of the distance R from the antenna axis. The value of ε_2 depends on the capacitance between the antennas.

As seen from Fig. **16**, the capacitance between the antennas is the capacitance per unit length between the inner and outer cylinders, that is (see, for example, [27])

$$C_2 = 2\pi\varepsilon_2 / \ln\left(\rho_{0av} / \rho_{2av}\right),$$

(4.32)

where ρ_{iav} is the average value of cylinders radius, since all cylindrical cross sections are similar ellipses. In order to calculate the value of C_2 for the parabolic structure, we should use (4.30) to replace the cylindrical coordinates with the parabolic ones. For this end, we have to determine the shape and dimensions of the phantom in the parabolic coordinates system in accordance with the drawing and then perform the transition to the cylindrical coordinates.

If the vessel is shaped as shown in Fig. **17**, and the x- and the y- axes are directed along the major and the minor axes of the ellipse, respectively, the coordinates of the points situated on the vessel wall are

$$x = \sigma f(\psi)\tau \cos\psi, \quad y = \sigma f(\psi)\tau \sin\psi, \quad z = 0.5\left[\tau^2 - \sigma^2 f^2(\psi)\right].$$

(4.33)

Here, function $f(\psi)$, which varies from 1 to b/a, defines the dependence of the ellipse radius on angular coordinate ψ.

Compare a paraboloid, having the elliptic cross section in the horizontal plane, with the paraboloid of revolution of the same perimeter. In the latter case, coordinate σ_0 will be constant along the entire paraboloid surface. At the lower point of the surface, where $x = y = 0$, coordinate τ is in accordance with (4.33) zero too, and therefore $z_0 = -0.5\sigma_0^2$, i.e., $\sigma_0 = \sqrt{2|z_0|}$. Quantity z_0 is the coordinate of the lowest point of the vessel. Point $z = 0$ (where $\tau_0 = \sigma_0$) is the focus point of the paraboloid of revolution. And the distance between the points is the focal length $|z_0|$, which is equal to a half of the focal parameter. The relationship, similar to $\sigma_0 = \sqrt{2|z_0|}$, is valid for the each paraboloid, including the one with the elliptic cross section.

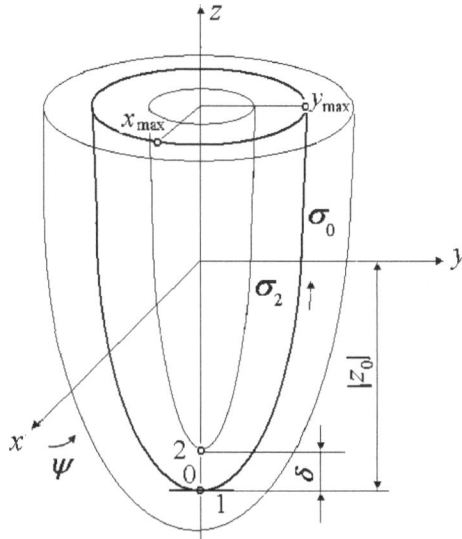

Figure 17: The vessel coordinates.

If the probe is located on the z-axis at distance δ from the wall, *i.e.,* it lies on the parabolic surface σ_2, then $\sigma_2 = \sqrt{2|z_0| - \delta}$. In the cylindrical system of coordinates, in accordance with (4.30)

$$\rho_{0av} = \sqrt{2|z_0|}, \quad \rho_{2av} = \sqrt{2(|z_0| - \delta)}.$$

$$(4.34)$$

The substitution of these values in (4.32) permits to determine capacitance C_2. Using expression (4.32), one can compare the fields measured in phantoms of different shapes and dimensions. Having measured the field value in one phantom, one can calculate the fields in other sets.

To exemplify, show the results for the side and central vessels of the phantom shown in Fig. **13**. The drawings and the dimensions of the set are given in [42]. In the calculation, the following dimensions of the left and medium vessels (the head and the body) were originally taken into account and used:

for the left vessel (head) - H30 (chin-top of head), 62 (head length), 61 (head circumference), 60 (head breadth), which are equal to 243.3, 206, 587.3 and 158.6 mm, respectively;

for the central vessel (body) – chin-top and length, which are equal to 370 and 255 mm, respectively;

for both vessels – the liquid level 140 mm.

Fig. **18** gives an example of the cross-sections of the side vessel in the lateral (1) and longitudinal (2) vertical planes. The measurement of vessel dimensions allows determining the parameters of parabolic curves. In order to calculate the parameters of the paraboloid, which is the closest to the vessel wall, we use the parabolic equation $z = ax^2$ in order to add the values of z and x^2 point by point. Factor $1/a$ is defined as the ratio of the sums,

$$ 1/a = \sum_{n=1}^{N} x_n^2 \Big/ \sum_{n=1}^{N} z_n $$

The focal parameter is equal to a half of this factor: $p = 1/(2a)$, the focal length is a half of the focal parameter. For the side vessel, we obtain: $1/a = 113$ mm, $p = |z_0|$ $= 56.5$ mm. Accordingly, $\rho_{0av} = 10.63$, and $\rho_{2av} = 9.64$ mm for $\delta = 10$ mm. If ε_r $= 41.5$, the capacitance per unit length is $C_2 = 2.45 \cdot 10^{-8}$ F.

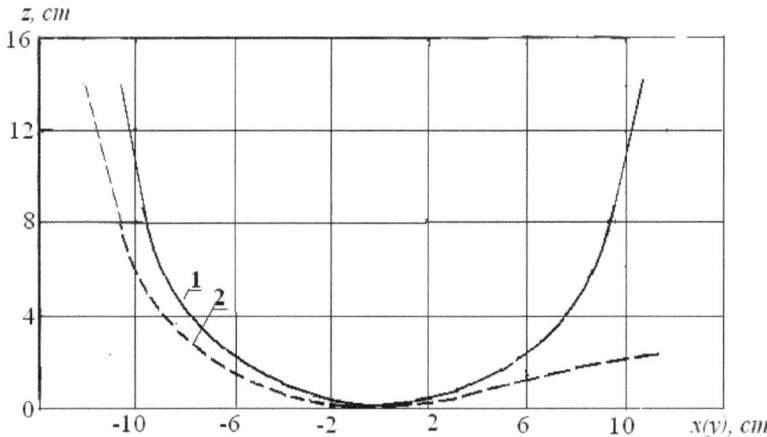

Figure 18: The cross-sections of the side vessel in the lateral (1) and longitudinal (2) vertical planes.

Evaluating the quantities for different vessels, it is expedient to take into account the following facts. If the height of the vessel will increase n times, the value of

$|z_0|$ will decrease n times. If the diameter or the perimeter of the vessel will increase n times, the value of $|z_0|$ will increase n^2 times.

For the central vessel, we obtain: $|z_0| = 113$, $\rho_{0av} = 15.03$, $\rho_{2av} = 14.35$, C_2 $= 5.03 \cdot 10^{-8}$ F.

The calculations show the capacitance per unit length of the central vessel to be at $\delta = 10$ mm twice as high as the side vessel capacitance. The results of the field measurement in the phantom at frequency 0.903 GHz agree on the whole with the calculation results.

The performed analysis clearly show that the measurements results for the fields and SAR depend substantially on the phantom shape.

Send Orders for Reprints to reprints@benthamscience.net

CHAPTER 5

Multi-Conductor Cables and Multi-Radiator Antennas

Abstract: The theory of electrically coupled lines is applied as the rigorous method to the calculation of multi-conductor cables in order to determine the cause of the emergence of the electromagnetic interference (crosstalk) in communication channels and the common mode currents in the lines. The causes of the cable asymmetry, which leads to crosstalk, are shown to be the construction of each line as a twisted pair and the different distances between wire pairs of two adjacent long lines. The cause of emergence of the common mode is the asymmetry of excitation and loads. Calculation of capacitance between two wire elements in the presence of the third one permits to determine the equivalent lengths of a long line and monopole with unequal lengths of wires.

Keywords: Calculation of capacitance between two elements in the presence of the third one, Cause of asymmetry and crosstalk, Cause of common mode currents, Common mode current, Communication channel, Complex loading in the central radiator, Crosstalk, Different distances between wire pairs, Electromagnetic interference, Loads between wires and shield, Monopole of wires with different length, Multiple-wire radiator, Multi-radiator antenna, Open-end long line with different wire lengths, Reasons of changing wave impedance, Rigorous method of calculation, Stray capacitance of a transformer winding to ground, Twisting of each line, Wave impedance of a line inside a metal cylinder.

5.1. CROSS TALKS IN CABLES

The theory of electrically coupled lines (see Section 3.2) permits to show, that the mutual coupling between lines in multi-conductor cables results in the emergence of the electromagnetic interference (crosstalk) in communication channels and that the asymmetry of excitation and loads causes the emergence of the common mode currents in the lines.

Determine the signal magnitude at the end of a multi-conductor cable located inside a metal shield. For that end, it is necessary to calculate the electrical characteristics of the lines. The values of voltage across loads placed at the ends of an adjacent line can be used as a measure of such distortions [43]. The rigorous calculation method of the mutual coupling between lines enables the development of a simple and effective procedure of preventing interference.

Electromagnetic interference in communication channels (a cable imbalance) is caused by the cable asymmetry as well as by excitation and load asymmetry, which provokes the emergence of the common mode currents in cables. The rigorous calculation method of the electrical characteristics for multi-conductor cables enables to determine the common mode currents. Compensation of the common mode currents permits to decrease the EM radiation and its susceptibility to external fields.

Employ a rigorous method to calculate first the characteristics of a two-wire line located inside a metal shield and then for mutual coupling between lines. The lines are considered as uniform ones. The electromagnetic waves are regarded as transverse (*TEM*) waves, and the cable diameter is considered small in comparison with the wavelength.

A single pair of wires (twisted pair) inside a metal cylinder can be modeled as two wires of radius a, situated at a distance b from each other inside a metal cylinder of radius R and length L (Fig. **1**). Wire radius a and distance b in multi-conductor cables are small in comparison with cylinder radius R ($a, b \langle\langle R$), so the characteristic (wave) impedance of the line is constant along its length (when the axial lines of the twisted pair and the cylinder fail to coincide, and the inequality is not valid, the wave impedance varies along the line). We assume the wires to be straight and take into account the twisting by increasing length L of the equivalent line.

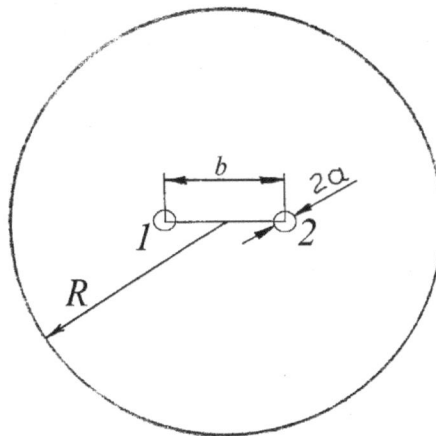

Figure 1: Two wires inside a cylinder.

Since the pitch of the helix along which each wire is located is greater than helix diameter b, inductance Λ per unit length undergoes a slight change at the replacement of a helical wire with a straight one. The wire capacitance per unit length also varies only slightly, *i.e.*, the wire twisting has no effect on the wave impedances of a structure. The line asymmetry in a real cable can cause the wave impedance and the two-wire line input impedance to change.

Another cause of the cable asymmetry is that each two-wire line is made in the form of a twisted pair (helix), a design that leads to a difference in the average distances between different wires and to the mutual coupling (crosstalk) between two two-wire lines surrounded by a single shield, even if both the exciting emf of each line and the line load are symmetric.

The equivalent circuit of a single line inside the shield is shown in Fig. **2**. The two-wire line is located above ground (inside a metal cylinder). The current and the potential along the nth wire of an asymmetrical line of N parallel wires situated above ground, in the general case, are determined by (3.18). The boundary conditions for the currents and potentials in this circuit are

$$i_1(0)+i_2(0)=0, \quad u_1(0)=u_2(0)+i_1(0)Z, \quad i_1(L)+i_2(L)=0, \quad u_1(L)=e+u_2(L). \quad \textbf{(5.1)}$$

Here, Z is the impedance of the line load.

Substituting expressions (3.18) in the first and second equalities of set (5.1), we find:

$$I_2 = -I_1, \quad U_2 = U_1 - I_1 Z.$$

Taking into account formulas (3.20), we find from the third equation of set (5.1) that

$$U_1 = I_1 Z \left(\frac{1}{W_{22}} - \frac{1}{W_{12}} \right) \Bigg/ \left(\frac{1}{W_{11}} + \frac{1}{W_{22}} - \frac{2}{W_{12}} \right) = I_1 Z \frac{\rho_{11} - \rho_{12}}{\rho_{11} + \rho_{22} - 2\rho_{12}}.$$

And, from the fourth equation, we obtain

$$I_1 = e\big/\big[Z\cos kL + j(\rho_{11} + \rho_{22} - 2\rho_{12})\sin kL\big].$$

The input impedance of a two-wire line inside a metal shield (the load impedance of generator e) is $Z_l = e/i_1(L)$. Substituting quantity $i_1(L)$ from expression (3.18) and using the relationships between e, I_1, I_2, U_1, U_2, we find that

$$Z_l = W \frac{Z + jW\tan kL}{W + jZ\tan kL}, \tag{5.2}$$

where $W = \rho_{11} + \rho_{22} - 2\rho_{12}$.

It is readily seen that expression (5.2) coincides with that for the input impedance of a lossless two-wire long line that is located in free space, is characterized by wave impedance W, and is loaded by impedance Z. The line asymmetry results in the difference of the electro- dynamic $(\rho_{11} \neq \rho_{22})$ and electrostatic $(W_{11} \neq W_{22})$ wave impedances of the wires.

The calculation of currents $i_1(z)$ and $i_2(z)$ shows that the currents in a two-wire line are identical in magnitude and opposite in sign:

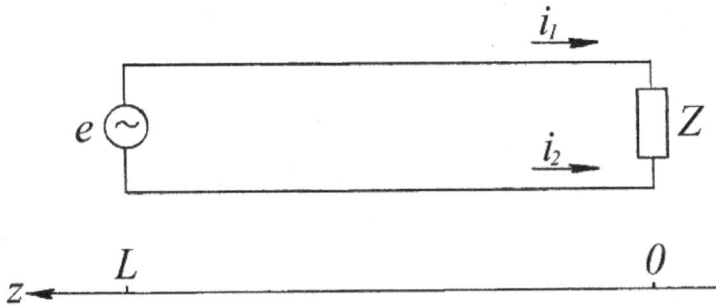

Figure 2: The equivalent circuit of a single line inside a shield.

$$i_1(z) = -i_2(z) = I_1\cos kz + jI_1(Z/W)\sin kz.$$

In a wire pair, there are only differential mode currents. There is no common mode current in the wires, because the emf and load impedance are placed between the line wires. The appearance of the common mode current can be caused by the connection of an additional emf or an additional load between one wire of a line and the shield.

To find the potential coefficients p_{ns}, one should take into account the following fact. If the system consists of two identical conductors (a wire and its image) and the structure is electrically neutral, the mutual partial capacitance coincides with the inter-conductor capacitance [27] and is equal to

$$C = 1\big/\big[2\big(p_{11} - p_{12}\big)\big],$$

where p_{11} is the self-potential coefficient, and p_{12} is the potential coefficient of the image. The conductor-to-ground capacitance is twice the capacitance between two conductors: $C_l = 2C$. For two wires of radius a at a distance b from each other, located inside the metal cylinder of a radius R symmetrically with respect to the cylinder axes (see Fig. **1**), we can write, using (4.20) from [27],

$$p_{11} = p_{22} = \frac{1}{2\pi\varepsilon} ch^{-1} \frac{R^2 + a^2 - b^2/4}{2Ra}.$$

Here ε is the permittivity of the medium inside the cable. If wire radius a and distance b are small in comparison with the cylinder radius R, then, in the air,

$$\rho_{11} = \rho_{22} \approx 60 \ln\big(R/a\big). \tag{5.3}$$

Similarly, using (4.22) from [27], we find:

$$\rho_{12} \approx 60 \ln\big(R/\sqrt{ab}\big), \tag{5.4}$$

i.e., the wave impedance of a lossless two-wire line, symmetrically situated inside a metal cylinder, is a half of the wave impedance of the same line in free space:

$$W_0 = \rho_{11} + \rho_{22} - 2\rho_{12} \approx 60 \ln\big(b/a\big). \tag{5.5}$$

If the wires inside a metal cylinder of a radius R are located asymmetrically, *e.g.*, they are displaced to the right by distance Δ (Fig. **3**), then

$$p_{11} = \frac{1}{2\pi\varepsilon} ch^{-1} \frac{R^2 + a^2 - (b/2 - \Delta)^2}{2Ra},$$

so at $a, b \langle\langle R$

$$p_{11} = \frac{1}{2\pi\varepsilon} \ln\left\{ \frac{R}{a}\left[1 + \frac{\Delta(b-\Delta)}{R^2} \right] \right\} = \frac{1}{2\pi\varepsilon}\left[\ln\frac{R}{a} + \frac{\Delta(b-\Delta)}{R^2} \right],$$

$$p_{22} = \frac{1}{2\pi\varepsilon}\left[\ln\frac{R}{a} - \frac{\Delta(b+\Delta)}{R^2} \right].$$

In this case, the wave impedance of the line is

$$W = W_0 - \frac{120\Delta^2}{R^2}. \tag{5.6}$$

That is one of the possible causes of lines wave impedance changing inside the shield.

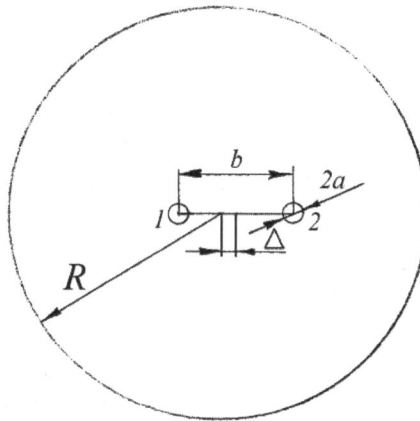

Figure 3: Offset wires inside a cylinder.

If the distance between wires is increased by value Δ, then at $\Delta\langle\langle b$

$$\rho_{12} \approx 60\ln\frac{R}{\sqrt{a(b+\Delta)}} \approx 60\left(\ln\frac{R}{\sqrt{ab}} - \frac{\Delta}{2b} \right). \tag{5.7}$$

This is the second cause. As can be seen from (5.6) and (5.7), a change in the distance between wires has a greater effect on the wave impedance of the line than the wire displacement relative to the cylinder axis.

Fig. **4** shows the equivalent circuit of two coupled two-wire lines inside a shield. One of the lines is excited by generator e and is loaded by complex impedance Z_1, the loads Z_2 and Z_3 being connected in the wires at both ends of the other line. It is necessary to emphasize that such circuit is of the most general nature. If, for example, generator e_1 is located at the end of the second line (at point $z = L$), the currents and voltages created by generator e are found considering that Z_3 is equal to the input impedance of generator e_1.

Consider inequalities $a \langle\langle b \langle\langle d, R$ (here d is the distance between the axes of twisted pairs). In many cases, the diameter of a wire bunch is small in comparison with the diameter of the cable metal shield. When there are many wires in the bunch, its diameter is close to the shield diameter. However, it is necessary to take into account that the maximum mutual coupling exists between adjacent lines. Therefore, analyzing mutual coupling between them is possible by considering in the first approximation that $d \langle\langle R$.

As was stated at the section beginning, the cable asymmetry results in mutual coupling (crosstalk) between two two-wire lines. The reason of such asymmetry is the construction of each line as a twisted pair (helix). The placement of the line

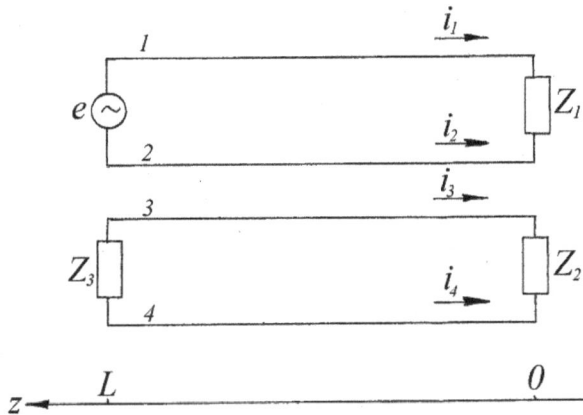

Figure 4: The equivalent circuit of two coupled lines placed inside a shield.

conductors in different variants of winding is shown in Fig. **5**. If, in the initial cross section of the cable, the starts of helices 1 and 3 are located at the same point of their section (we shall call it the initial one) and the starts of helices 2 and 4 are displaced along the cross section perimeter by π from this point, it means that the distance between wires 1 and 3 (and also between wires 2 and 4) is $D_{13} = D_{24} = d$ along all their length, whereas the distance between wires 1 and 4 (and also between wires 2 and 3) varies along wires from $d+b$ to $d-b$. For example, the distance between wires 1 and 4 (see Fig. **5a**) is

$$D_{14} = \sqrt{(d+b\cos\alpha)^2 + b^2\sin^2\alpha} \approx d + b\cos\alpha + \frac{b^2\sin^2\alpha}{2d}$$

(here, α is angular displacement of points 1 and 4 along the cross section perimeter), *i.e.*, the average distance between these wires,

$$(D_{14})_0 = \frac{1}{\pi}\int_0^\pi D d\alpha = d + \frac{b^2}{4d}, \tag{5.8}$$

differs from quantity d. The potential coefficients as well as the electrodynamic and electrostatic wave impedances vary accordingly.

If, at the initial cross section of the cable, the starts of helix 3 and 4 are displaced along the cross section perimeter by $\pi/2$ and $3\pi/2$ from the initial point, respectively, the distance between wire 1 and wire 3(or wire 4) is

$$D_{13(4)} \approx d + \frac{b}{2}(\cos\alpha \pm \sin\alpha) + \frac{b^2}{8d}(\sin\alpha \mp \cos\alpha)^2. \tag{5.9}$$

Here, the top sign applies to wire 3, and the lower sign – to wire 4. The average distance between the wires from this equation is

$$(D_{13(4)})_0 \approx d \pm \frac{b}{\pi} + \frac{b^2}{8d}, \tag{5.10}$$

i.e., the displacement of the helix starts of the cable by $\pi/2$ changes substantially the average distance between wires. Difference between $\left(D_{13}\right)_0$ and $\left(D_{14}\right)_0$ increases from value $b^2/4d$ to $2b/\pi$, with $b\langle\langle d$.

For average distance D_0 between wires 1 and 4 not to differ from d, it is necessary to wind wire 4 in the opposite sense to the other wires. In this case (see Fig. **5b**)

$$D = d + b\cos\alpha,\ D_0 = d.\tag{5.11}$$

The electrodynamic wave impedances of the structure at the same sense winding become

$$\rho_{11} = \rho_{22} = \rho_{33} = \rho_{44} = \rho_1 = 60\ln\left(R/a\right), \rho_{12} = \rho_{34} = \rho_2 = 60\ln\left(R/\sqrt{ab}\right),$$
$$\rho_{13} = \rho_{24} = \rho_3 = 60\ln\left(R/\sqrt{ad}\right), \rho_{14} = \rho_{23} = \rho_4 = 60\ln\left[R/\sqrt{a\left(d+b^2\pi/4d\right)}\right]..\tag{5.12}$$

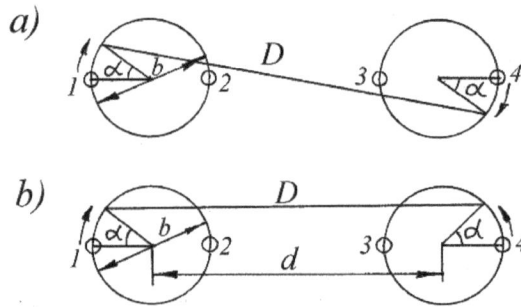

Figure 5: Distance between wires 1 and 4 at the same sense (*a*) and the opposite sense winding of wire 4 (*b*).

In the case of lines placed at finite distance H from the cable axis, we find

$$\rho_1 = 60\ln\frac{R\left(1-H^2/R^2\right)}{a}.$$

The expressions for other quantity ρ_n retain validity. This means that the wave impedance of a lossless two-wire line situated inside a metal cylinder at a distance H from its axis is, in accord with equation (5.6)

$$W = 60 \ln \left[b \left(1 - H^2/R^2 \right)^2 / a \right],$$

i.e., the line wave impedance decreases as the result of its displacement from the cable axis. When H is small and equal to Δ, we arrive at expression (5.7).

According to (3.19), the electrostatic wave impedances are

$$W_{ns} = \begin{cases} \Delta_N / \Delta_{ns} , n = s, \\ -\Delta_N / \Delta_{ns} , n \neq s, \end{cases} \tag{5.13}$$

where $\Delta_N = |\rho_{ns}|$ is the $N \times N$ determinant, and Δ_{ns} is the cofactor of the determinant Δ_N. For a structure made of four wires, in accord with (5.12) and (5.13),

$$W_{11} = W_{22} = W_{33} = W_{44} = W_1 = \Delta_4 / \Delta_{11}, \; W_{12} = W_{34} = W_2 = -\Delta_4 / \Delta_{12},$$

$$W_{13} = W_{24} = W_3 = -\Delta_4 / \Delta_{13}, \; W_{14} = W_{23} = W_4 = -\Delta_4 / \Delta_{14}. \tag{5.14}$$

The current and potential of the nth wire of an asymmetric line from N parallel wires located above ground are found from expression (3.18). The boundary conditions for the currents and voltages in the circuit shown in Fig. **4** are

$$i_1(0) + i_2(0) = 0, \; i_3(0) + i_4(0) = 0, \; u_1(0) = u_2(0) + i_1(0)Z_1,$$
$$u_3(0) = u_4(0) + i_3(0)Z_2, \; i_1(L) + i_2(L) = 0, \; i_3(L) + i_4(L) = 0, \; u_1(L) = e + u_2(L),$$
$$u_3(l) = u_4(L) + i_3(L)Z_3. \tag{5.15}$$

Substituting expressions (3.18) in the equations of system (5.15), we find

$$I_1 = \frac{e}{Z_1 \cos kL + 2j \left[\rho_1 - \rho_2 + (\rho_3 - \rho_4) A \right] \sin kL}, \; I_3 = AI_1, \tag{5.16}$$

where

$$A = \frac{4 (\rho_3 - \rho_4) + Z_1 Z_3 (1/W_3 - 1/W_4)}{-4 (\rho_1 - \rho_2) + Z_2 Z_3 (1/W_1 + 1/W_2) + j2 (Z_2 - Z_3) \cot kL}.$$

If $\rho_3 = \rho_4$ (and, accordingly, $W_3 = W_4$), then $A=0$, the current at the beginning of the second line is zero. In this case, the presence of the second two-wire line has no effect on the first line. This result obviously corroborates the fact that the cable asymmetry results in mutual coupling (crosstalk) between two two-wire lines.

Knowing all parameters in expressions (3.18), one can calculate the loading impedance of the generator e

$$Z_l = e/i_1(L) = 2\frac{Z_1 + 2j\left[\rho_1 - \rho_2 + A(\rho_3 - \rho_4)\right]\tan kL}{2 + j\left[Z_1(1/W_1 + 1/W_2) - AZ_2(1/W_3 - 1/W_4)\right]\tan kL} \qquad (5.17)$$

and the currents in the wires of the second (unexcited) line

$$i_3(z) = I_1 A\cos kz + j\frac{I_1}{2}\left[AZ_2(1/W_1 + 1/W_2) - Z_1(1/W_3 - 1/W_4)\right]\sin kz, \quad i_4(z) = -i_3(z). \quad (5.18)$$

The sum of the currents is zero, *i.e.*, as well as with one line placed into the shield, there is no common mode current since the emf and the loading impedances are connected only between wires of each line.

The voltages across passive loads are

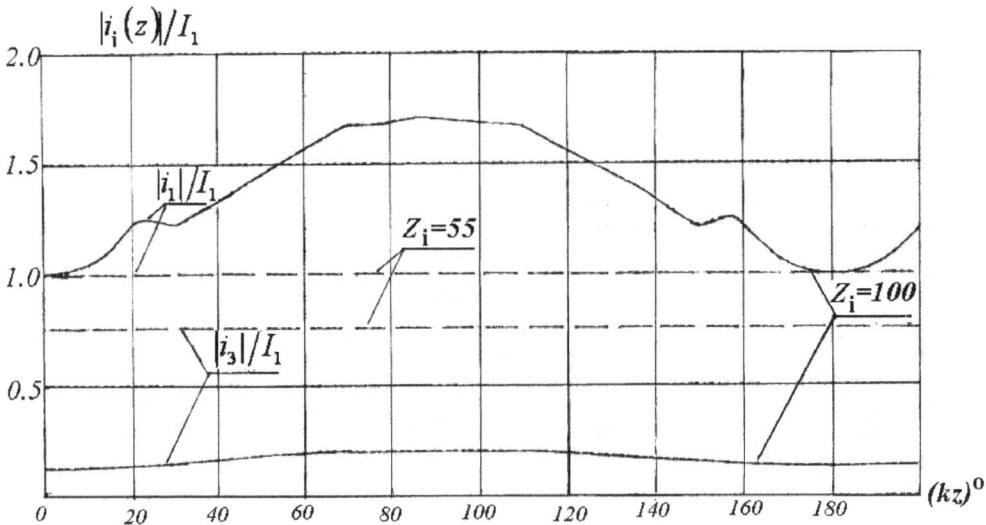

Figure 6: The absolute values of the currents in the excited and unexcited wires.

$$V_1 = i_1(0)Z_1 = I_1 Z_1, V_2 = i_3(0)Z_2 = I_1 A Z_2,$$

$$V_3 = i_3(L)Z_3 = I_1 Z_3 \left\{ A \cos kL + j\frac{1}{2}\left[AZ_2\left(1/W_1 + 1/W_2\right) - Z_1\left(1/W_3 - 1/W_4\right)\right] \sin kL \right\}. \tag{5.19}$$

As an example, consider a structure from two pairs of wires inside the shield with sizes (in millimeters): $a=0.2$, $b=0.5$, $d=2$, $R=2$. For the identical loads $Z_1 = Z_2 = Z_3 = 100$ Ohm, ratio A of the currents at the beginning of the second (unexcited) and the first line amounts to 0.13. If the values of the loads are equal to the wave impedance of the single two-wire line inside the metal shield, *i.e.*, in accordance with equation (5.6), $Z_1 = Z_2 = Z_3 = 55$ Ohm, the ratio of the currents is substantially increased ($A = -0.76$).

The absolute values of the currents as functions of kz are plotted in Fig. **6**. Here, k is the propagation constant of a wave in a medium, z is the coordinate along the line (see Fig. **4**).

5.2. IN-PHASE CURRENTS AND LOSSES

Consider the effect of loads placed between the wires and the shield, using a two-wire line as an example (Fig. **7**). It differs from the circuit shown in Fig. **2** by connection of its wires at the line end (near the generator) with a shield through complex impedances Z_1 and Z_2, whose values depend on the line excitation circuit. In a realistic circuit the secondary winding of the transformer can act as emf e, exciting a two-wire line. In this case, stray capacities of the winding to ground (to the cable shield) act as impedances Z_1 and Z_2.

The boundary conditions for the currents and potentials in the circuit shown in Fig. **7** are

$$i_1(0)+i_2(0)=0, \quad u_1(0)=u_2(0)+i_1(0)Z, \quad i_1(L)+i_2(L)+\frac{u_1(L)}{Z_1}+\frac{u_2(L)}{Z_2}=0, \quad u_1(L)=e+u_2(L). \tag{5.20}$$

Substituting expressions (3.18) in the equations of set (5.20), we find the input impedance of a two-wire line

$$Z_l = \frac{e}{i_1(L)+u_1(L)/Z_1} = \frac{Z \cos kL + j\left(\rho_{11}+\rho_{22}-2\rho_{12}\right)}{1+U_1/I_1\left[1/Z_1 + j\left(1/W_{11}-1/W_{12}\right)\tan kL\right] + j\left[Z/W_{12}+\left(\rho_{11}-\rho_{12}\right)/Z_1\right]\tan kL} \tag{5.21}$$

and the sum of the currents in the line wires

$$i_s(z) = i_1(z) + i_2(z) = jI_1 \left[Z(1/W_{12} - 1/W_{22}) + U_1/I_1(1/W_{11} + 1/W_{22} - 2/W_{12}) \right] \sin kz . \quad \textbf{(5.22)}$$

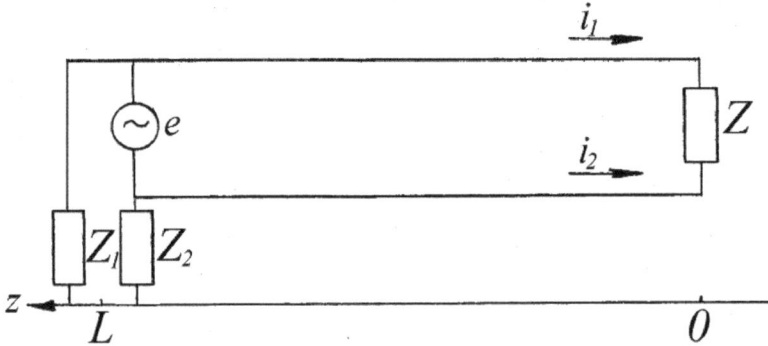

Figure 7: The equivalent circuit of a single line with loads connected between the wires and the shield.

Therefore, connection of the loads results in the emergence of the common mode current in the wires and of the current along the inner surface of the cable shield, equal in magnitude but opposite in direction.

For two wires of the same radius situated symmetrically to the cylinder axis

$$Z_l = \frac{Z + 2j(\rho_1 - \rho_2)\tan kL}{1 + j\dfrac{Z\rho_2}{\rho_1^2 - \rho_2^2}\tan kL + j\dfrac{1}{2(\rho_1 + \rho_2)}\left[Z + (\rho_1 + \rho_2)C\right]\tan kL + \dfrac{1}{2Z_1}\left[Z + (\rho_1 + \rho_2)C + 2j(\rho_1 - \rho_2)\tan kL\right]} , \quad \textbf{(5.23)}$$

and

$$i_s(z) = j\frac{eC \sin kz}{Z \cos kL + j2(\rho_1 - \rho_2)\sin kL} . \quad \textbf{(5.24)}$$

It is not difficult to verify that, if $1/Z_1 = 1/Z_2 = 0$, quantity C is zero and the expressions for U_1 and Z_l coincide with the similar expressions for the circuit without loads between wires and shield. From the presented results it is easy to obtain also the expressions for the cases when there is only one load, for example, $1/Z_1 = 0$.

The above analysis confirms that the cause of the emergence of the common mode currents in line wires is the asymmetry of its excitation due to the connection of complex impedances, *e.g.*, stray capacitances of the secondary transformer winding, to ground (to the cable shield). Asymmetry of loads at the line end farther from the generator (at $z = 0$) produces similar results. The common mode currents in the excited line induce the common mode currents in wires of the adjacent unexcited line, even if it is totally symmetric (with respect to ground and the excited line). Removal of the excitation and load asymmetry in the excited line results in the disappearance of the common mode currents in wires of both excited and unexcited lines.

In order to reduce or eliminate the common mode currents, it is necessary to violate the asymmetry, *e.g.*, to neutralize the effect of stray capacitances to ground (to the cable shield). To this end, work [44] proposes to cancel the current through stray capacitance with the current equal in magnitude and opposite in direction, which is created by an additional transformer winding.

As shown in the Section 3.2, the theory of electrically coupled lines bases on the telegraph equations and on the relationship between the wire potential coefficients and the electrostatic induction ones. The *z*-axis is selected in parallel to the wires, and the dependence of the current on coordinate z is adopted as $\exp(\gamma z)$, where γ is the complex propagation constant of the wave along the wires.

In the case of lossless the wires and the medium, where they are situated, electrostatic W_{ns} and electrodynamic ρ_{ns} wave impedances between wires n and s are real-valued quantities determined by equations (3.19), and $\gamma = jk$ is a purely imaginary quantities (k is the propagation constant of the wave in the medium).

As follows from the above, electrodynamic wave impedance ρ_{ns} is proportional to the self- or mutual inductance of wires section, *i.e.*, is proportional to reactance connected in series with wires circuits. Electrostatic wave impedances W_{ns} are proportional to the mutual capacitance between wires, *i.e.*, to the susceptance between them. Therefore, it is natural to connect, in the circuit the wire-loss resistance (*e.g.*, the skin-effect loss) in series with the inductance, and the leakage conductance - in parallel with the mutual capacitance.

Take into account the loss in the medium and in the wires considering that wave impedances W_{ns} and ρ_{ns} and the propagation constant k are complex values. If the inductance of the nth wire per unit of its length is Λ_0 and its active resistance is R_0, its impedance per unit length is $j\rho_{nn} = j\omega\Lambda_0 + R_0$, i.e., the self-electrodynamic wave impedance of a loss wire is equal to

$$\rho_{nn} = \rho_0\left(1 - j R_0/\rho_0\right), \tag{5.25}$$

where $\rho_0 = \omega\Lambda_0$ is electrodynamic wave impedance in the absence of losses, R_0 is total loss resistance in the nth wire and in the metal shield per unit length.

For the mutual electrodynamic wave impedance between wires n and s, we obtain

$$\rho_{ns} = \rho_{ns0}\left(1 - j R_{ns0}/\rho_{ns0}\right), \tag{5.26}$$

where $\rho_{ns0} = \omega M_{ns0}$, M_{ns0} is the mutual inductance between wires n and s per unit length, and R_{ns0} is the loss resistance in both wires per unit length.

Similarly we find for the admittance between wires n and s per unit length we find: $jW_{ns} = j\omega C_{ns0} + G_{ns0}$, i.e., the electrostatic wave impedance in a loss medium is

$$W_{ns} = W_{ns0}\left(1 - j G_{ns0}/W_{ns0}\right), \tag{5.27}$$

where $W_{ns0} = \omega C_{ns0}$, C_{ns0} is the mutual partial capacitance between wires n and s per unit length, and G_{ns0} is the leakage conductance per unit length.

Thus, the evaluation of the electrical performances of the coupled loss lines can use the results obtained for the lossless lines by substitution of the complex wave impedances into the earlier expressions in accordance with equations (5.25) - (5.27). Here, the loss both in wires and losses in an imperfectly conducting metallic tube (shield) is taken into account.

A rigorous method for the calculation of the characteristics of two-wire lines inside a metal shield allows us revising the mechanism of mutual coupling between lines in multi-conductors cables. It permits to determine the values of voltage (interferences) across impedances placed at the beginning and the end of the adjacent line at given power in the main line. The crosstalk is caused by the asymmetry of the wire positions (the varying average distant between wires) and, accordingly, the asymmetric wave impedances. The avoidance of the asymmetry will reduce crosstalk in multi-conductor cables, *i.e.,* will enable to increase the carrying capacity of a channel. This is valid also for multi-conductor connectors.

The cause of the emergence of the common mode currents in the lines of a multi-conductor cable is the asymmetry of excitation and loads. As noted, the compensation of the common mode currents offers to decrease the EM radiation and to reduce its susceptibility to external fields.

5.3. LONG LINE AND MONOPOLE WITH UNEQUAL LENGTHS OF WIRES

In addition to the calculation of crosstalk in a multi-conductor cable and the analysis of causes of its asymmetry, the theory of electrically coupled lines makes it possible to find electric characteristics of multiple-wire, multi-folded and multi-radiator antennas. Before proceeding to these, we shall consider an open-end long line and a linear radiator, which consist of two wires with different length. Such problem arises, in particular, when it is necessary to calculate the input impedance and electromagnetic field of two asymmetrical radiators with different height, located in near regions of each other.

The calculation of the input impedance and the electromagnetic field of a structure consisting of two parallel monopoles of equal lengths in the near region of each other can be carried out in accordance with the theory of folded radiators and the superposition principle. For this purpose, the structure is divided into two circuits: an open-end two-wire line, and a linear monopole. The procedures of calculating the currents and the input impedance of such line, and the field and input impedance of a monopole are known. In particular, the calculation of the monopole characteristics one must allow for its equivalent radius, given by $a_e = \sqrt{ab}$, if the wire radii are small and equal. Here, a is the wire radius, and b is the spacing of the wire axes.

A structure consisting of two parallel monopoles of unequal length can also be divided into a line and a monopole. But, in this case, both the line and the monopole are formed by wires of different length; hence, the calculation of their characteristics is not obvious. Such problem arises also in more complicated structures. For example, a shortwave multi-radiator antenna has different lengths of radiators. The load of a medium-frequency inverted L-antenna resulting in increasing the effective length of antenna may be implemented as a system of horizontal wires of different lengths.

As seen from Fig. **8a**, a long line formed by two parallel wires of different lengths l_1 and l_2 has two segments labeled (1) and (2). The lengths of the upper (1) and lower (2) segments are, respectively, $l = l_1 - l_2$ and $L = l_2$. The lower segment is a two-wire line with wires of equal lengths and equal circular cross sections with radius a. The capacitance per unit length between two wires situated in a homogeneous medium with permittivity ε is given by

$$C_0 = \pi\varepsilon/\ln(b/a),\qquad\qquad\qquad (5.28)$$

where b is the distance between the wire axes. Capacitance C_0 determines the wave impedance of the two-wire line (of the lower segment).

The effect of the upper segment on the line input impedance is taken into account by calculating capacitance C between the excess length of the longer wire (of length l) and the shorter wire (Fig. **8b**). The input impedance of the whole structure is equal to the input impedance of the lower line loaded with the capacitance.

The electrostatic structure in this case consists of three conducting elements designated as I, II and III in Fig. **8a**. C is the capacitance between elements II and III in the presence of element I. It is not equal to the capacitance between isolated elements II and III, in the absence of element I. Find the capacitance as the difference of two capacitances:

$$C = C_1 - C_0 L,\qquad\qquad\qquad (5.29)$$

where C_1 is the total capacitance between the longer and shorter wires, and $C_0 L$ is the capacitance between the wires of the lower segment. In fact, C_1 is the capacitance of an electrically neutral system consisting of two conductors (see, *e.g.*, [27]):

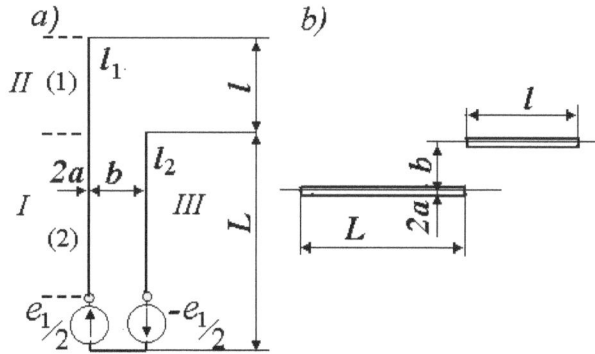

Figure 8: Two-wire line formed by the two parallel wires of different lengths (*a*) and the capacitive elements (*b*).

$$C_1 = \left(p_{11} + p_{22} - 2 p_{12} \right)^{-1},$$ (5.30)

where p_{ik} are the potential coefficients, calculated by the following formulas:

$$p_{11} = \frac{1}{2\pi\varepsilon L}\left\{ \ln\left[L/a + \sqrt{1+\left(L/a\right)^2} \right] + a/L - \sqrt{1+\left(a/L\right)^2} \right\},$$

$$p_{22} = \frac{1}{2\pi\varepsilon\left(L+l\right)}\left\{ \ln\left[(L+l)/a + \sqrt{\left(L+l\right)^2/a^2 +1} \right] + a/(L+l) - \sqrt{a^2/\left(L+l\right)^2 +1} \right\},$$

$$p_{12} = \frac{1}{4\pi\varepsilon\left(L+l\right)}\left\{ \ln\left[\left(L+\sqrt{L^2+b^2}\right)\big/b\right] + (L+l)\ln\left[\left(L+l+\sqrt{\left(L+l\right)^2+b^2}\right)\big/b\right]\right\}\bigg/L -$$
$$- \sqrt{L^2+b^2}\big/L + \sqrt{l^2+b^2}\big/L - l\ln\left[\left(l+\sqrt{l^2+b^2}\right)\big/b\right]\bigg/L + b/L - \sqrt{\left(L+l\right)^2+b^2}\big/L \ \}.$$

Since $L/a, l/a \gg 1$,

$$p_{11} = \frac{1}{2\pi\varepsilon L}\left(\ln\frac{2L}{a} -1 \right), \quad p_{22} = \frac{1}{2\pi\varepsilon\left(L+l\right)}\left[\ln\frac{2\left(L+l\right)}{a} -1 \right].$$

If $L/b, l/b \gg 1$, the expression for p_{12} simplifies:

$$p_{12} = \frac{1}{4\pi\varepsilon(L+l)}\left[\ln\frac{2L}{b} + \ln\frac{2(L+l)}{b} + \frac{l}{L}\ln\frac{L+l}{l} - 2\right].$$

Calculations show C to be small compared to $C_0 L$. In particular, if wires are located in the air, *i.e.*, $\varepsilon = 1/(36\pi\cdot10^9)$, then for L=7.5, b=1.0, $2a$=0.05 with l varying from 1 to 4 (all dimensions are in centimeters), we have $C_0 L$=7.07 pF, and C changes from 0.05 to 0.1 pF. Thus, the excess length creates a capacitive load at the two-wire line end. The load is equivalent to prolongation of the line with length l_0:

$$l_0 = (1/k)\cot^{-1}\left[1/(\omega C W_l)\right]. \tag{5.31}$$

where k is the phase constant, ω is the circular frequency, and W_l is the wave impedance of the line. The calculation results for capacitance C and equivalent lengths l_0 for the above-mentioned dimensions at 1 GHz are given in Table **1**.

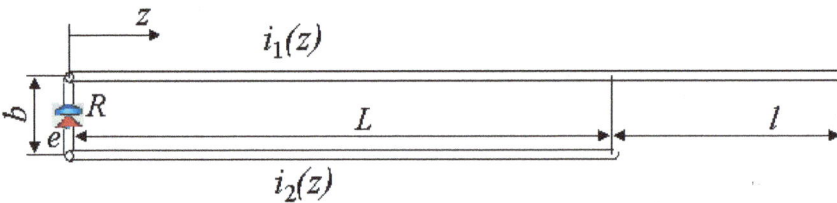

Figure 9: The simulation model for a two-wire transmission line.

The calculations are verified by simulations using the CST program. The model structure is shown in Fig. **9**, where e is a generator with output impedance R=50 Ohm. The simulation results for the equivalent lengths denoted as l_{01} are also presented in Table **1**. Since distance b between the wires is finite, the calculated values of l_0 and l_{01} for $l=0$, based on the two-wire line approximation, are other than 0. In order to clearly demonstrate how excess length l affects the equivalent lengths, l_0 and l_{01} are decreased by their values at $l=0$. As seen from Table **1**, the calculation and simulation results agree well for $l\leq0.1\lambda$. It turns out that the input impedance of a line with wires of unequal lengths differs comparatively little from that of a two-wire line, having the same lengths as the shorter wire.

Table 1: Capacitive loads and equivalent lengths l_0 and l_{01}

l, cm	2a=0.05 cm			2a=0.2 cm		
	l_0, cm	l_{01}, cm	C, pF	l_0, cm	l_{01}, cm	C, pF
0.0	0	0	0.020	0	0	0.047
0.5	0.22	0.19	0.037	0.21	0.15	0.073
1.0	0.41	0.39	0.050	0.37	0.30	0.093
1.5	0.56	0.52	0.063	0.49	0.45	0.108
2.0	0.69	0.86	0.073	0.58	0.61	0.119
2.5	0.80	1.10	0.081	0.65	0.79	0.128
3.0	0.90	1.38	0.089	0.71	1.00	0.135
3.5	0.98	1.66	0.095	0.75	1.24	0.140
4.0	1.05	1.94	0.101	0.78	1.48	0.144
4.5	1.12	2.17	0.107	0.81	1.64	0.148

Similar results for 2a=0.2 cm are also given in Table **1**.

In accordance with the obtained results, one can write the current distributions along the line wires:

$$i_1(z) = \begin{cases} I_0 \sin kl_0 \sin k(L+l-z)/\sin kl \\ I_0 \sin k(L+l_0-z) \end{cases} \quad i_2(z) = \begin{cases} 0 & L \le z \le L+l \\ -I_0 \sin k(L+l_0-z) & 0 \le z \le L. \end{cases} \tag{5.32}$$

where I_0 is the generator current. A long line with equal wire lengths located in free space can radiate only if the wire spacing is not too small compared to the wavelength. In case of wires of unequal length, excess segment l of the longer wire radiates, as it follows from expressions (5.32) for the wires currents.

Another important problem is calculating the input impedance of a linear radiator (monopole) composed of two parallel wires with different lengths (Fig. **10a**). Fig. **10b** shows an equivalent asymmetric line for the structure. The current distribution along the monopole wires is calculated in accordance with the theory of electrically coupled lines.

In this case, it is necessary to divide the equivalent line into two segments. The segments are labeled with (1) and (2), and the wires – with 1 and 2 at the wire bases. Segment 1 consists of a single wire, and segment 2 consists of two wires. The currents and potentials at segment m of wire n of the asymmetric line are

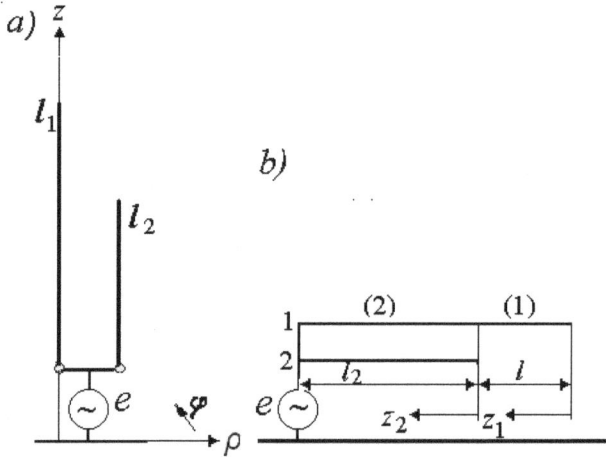

Figure 10: Monopole formed by the wires of different lengths (a) and equivalent long line (b).

$$i_n^{(m)} = I_n^{(m)} \cos kz_m + j \left[\frac{2U_n^{(m)}}{W_{nn}^{(m)}} - \sum_{s=1}^{M} \frac{U_s^{(m)}}{W_{ns}^{(m)}} \right] \sin kz_m,$$

$$u_n^{(m)} = U_n^{(m)} \cos kz_m + j \sum_{s=1}^{M} \rho_{ns}^{(m)} I_s^{(m)} \sin kz_m,$$

(5.33)

where $I_n^{(m)}$ and $U_n^{(m)}$ are the current and potential of wire n at the beginning of segment m (at point $z_m = 0$) respectively; $n=1, 2$, $m=1, 2$, M is the total number of wires of segment m, and $W_{ns}^{(m)}$ and $\rho_{ns}^{(m)}$ are the electrostatic and electrodynamic wave impedances between wires n and s at segment m. If the wire spacing is small compared to the wire lengths, one can assume that $\rho_{nn}^{(m)} = const(n) = \rho_1^{(m)}$, $\rho_{ns}^{(m)}\big|_{n \neq s} = const(n) = \rho_2^{(m)}$, $W_{nn}^{(m)} = const(n) = W_1^{(m)}$, $W_{ns}^{(m)}\big|_{n \neq s} = W_2^{(m)}$.

The zero currents at the wire ends and the continuity of the current and potential along each wire determine the boundary conditions:

$$i_1^{(1)}\big|_{z_1=0} = i_2^{(2)}\big|_{z_2=0} = 0, \ i_1^{(1)}\big|_{z_1=l} = i_1^{(2)}\big|_{z_2=0}, \ u_1^{(1)}\big|_{z_1=l} = u_1^{(2)}\big|_{z_2=0}, u_1^{(2)}\big|_{z_2=l_2} = u_2^{(2)}\big|_{z_2=l_2} = e,$$

where from we derive:

$$I_1^{(1)} = I_2^{(2)} = 0, \ I_1^{(2)} = j\left[U_1^{(1)}/W_1^{(1)}\right]\sin kl, \ U_1^{(2)} = U_1^{(1)}\cos kl,$$

$$U_2^{(2)} = U_1^{(1)}\left[\cos kl - \frac{\rho_1^{(2)} - \rho_2^{(2)}}{W_1^{(1)}}\sin kl \cdot \tan kl_2\right],$$

$$U_1^{(1)} = \frac{e}{\cos kl \cos kl_2}\left[1 - \frac{\rho_1^{(2)}}{W_1^{(1)}}\tan kl \tan kl_2\right]^{-1}.$$

The current distribution along the first section of the longer wire is given by

$$i_1^{(1)} = j\left[U_1^{(1)}/W_1^{(1)}\right]\sin k(l_1 - z), \tag{5.34}$$

The current along the second section is

$$i_1^{(2)} = jU_1^{(1)}\left\{\frac{\sin kl \cos kz_2}{W_1^{(1)}} + \cos kl\left[\frac{1}{W_1^{(2)}} - \frac{1}{W_2^{(2)}}\left\langle 1 - \frac{\rho_1^{(2)} - \rho_2^{(2)}}{W_1^{(1)}}\right\rangle\tan kl \tan kl_2\right]\sin kz_2\right\}. \tag{5.35}$$

And the current along the shorter wire is

$$i_2^{(2)} = jU_1^{(1)}\cos kl\left\{1/W_1^{(2)} - 1/W_2^{(2)} \cdot \left[1 - \frac{\rho_1^{(2)} - \rho_2^{(2)}}{W_1^{(1)}}\right]\tan kl \tan kl_2\right\}\sin kz_2. \tag{5.36}$$

Thus, the total current along the second section is

$$i_1^{(2)} + i_2^{(2)} = jU_1^{(1)}\left\{\frac{\sin kl \cos k(l_2 - z)}{W_1^{(1)}} + \cos kl\left[\frac{1}{W_1^{(2)}} - \frac{1}{W_2^{(2)}}\left\langle 1 - \frac{\rho_1^{(2)} - \rho_2^{(2)}}{W_1^{(1)}}\right\rangle\tan kl \tan kl_2\right]\sin k(l_2 - z)\right\} \tag{5.37}$$

The expressions show that the current distribution along both sections of the monopole is sinusoidal, *i.e.,* similar to that of a monopole consisting of two segments with different wave impedances (*e.g.,* with different wire diameters).

Write the expression for the total current along the monopole in the form

$$J_{Am}(z) = \sum_{n=1}^{M} i_n^{(m)}(z), \ \ l_{m+1} \leq z \leq l_m, \tag{5.38}$$

where $i_n^{(m)}(z) = A_{nm} \cos k(l_m - z) + jB_{nm} \sin k(l_m - z)$. In accordance with (5.34)-5.36),

$$A_{11} = A_{21} = A_{22} = B_{21} = 0, \quad B_{11} = U_1^{(1)}/W_1^{(1)}, \quad A_{12} = jU_1^{(1)} \sin kL/W_1^{(1)},$$

$$B_{12} = B_{22} = U_1^{(1)} \cos kl \left[\frac{1}{W_1^{(2)}} - \frac{1}{W_2^{(2)}} \left\langle 1 - \frac{\rho_1^{(2)} - \rho_2^{(2)}}{W_1^{(1)}} \right\rangle \tan kl \tan kl_2 \right]. \tag{5.39}$$

The input reactance of the monopole is equal to the input impedance of the equivalent line:

$$jX_A = Z_l = \frac{e}{J_A(0)} = -j\frac{\left[W_1^{(1)} - \rho_1^{(2)} \tan kl \tan kl_2 \right] \cos^2 kl_2}{\tan kl \cos^2 kl_2 + DW_1^{(1)} \sin 2kl_2}, \tag{5.40}$$

where

$$D = \frac{1}{W_1^{(2)}} - \left[1 - \frac{\rho_1^{(2)} - \rho_2^{(2)}}{W_1^{(1)}} \right] \frac{\tan kl \tan kl_2}{W_2^{(2)}}.$$

The radiation resistance of the monopole is

$$R_\Sigma = 40k^2 h_e^2 \tag{5.41}$$

where h_e is the effective monopole height given by

$$h_e = \frac{1}{kJ(0)} \sum_{m=1}^{2} \left\{ (A_{1m} + A_{2m}) \sin k(l_m - l_{m+1}) + j(B_{1m} + B_{2m}) \left[1 - \cos k(l_m - l_{m+1}) \right] \right\} =$$

$$= \frac{jU_1^{(1)}}{kJ(0)W_1^{(1)}} \left\{ 1 + \sin kl \sin kl_2 + \cos kl \left[2DW_1^{(1)} (1 - \cos kl_2) - 1 \right] \right\}. \tag{5.42}$$

The wave impedances $W_{ns}^{(m)}$ and $\rho_{ns}^{(m)}$ used in the equations presented above are determined by potential coefficients:

$$\rho_{nn}^{(m)} = \rho_1^{(m)} = p_{nn}^{(m)}/(2\pi\varepsilon c) = 60p_{nn}^{(m)}, \quad \rho_{ns}^{(m)}\big|_{s\neq n} = \rho_2^{(m)} = p_{ns}^{(m)}/(2\pi\varepsilon c) = 60p_{ns}^{(m)}.$$

The wave impedances for the one-wire segment are

$$W_1^{(1)} = \rho_1^{(1)} = p_1^{(1)} \big/ \left(2\pi\varepsilon c\right) = 60 p_1^{(1)},$$

while we have for the two-wire segment

$$\frac{1}{W_1^{(2)}} = \frac{\rho_1}{\rho_1^2 - \rho_2^2}, \quad \frac{1}{W_2^{(2)}} = \frac{\rho_2}{\rho_1^2 - \rho_2^2}.$$

Potential coefficients $\rho_n^{(m)}$ are calculated by the method of mean potentials in accordance with the actual position of antenna wires. The simplest variant of the method is the Howe's method. It is easy to show that the mutual potential coefficient of two parallel equal-length wires with the dimensions and position shown in Fig. **7** is given by (3.23). The expression implies that the self-potential coefficient of wire n at segment m is taking into account the mirror image, equal to (3.24), and the mutual potential coefficient between wires n and s of segment m is equal to (3.25).

Expressions (5.38) and (5.39) determine the currents along each wire of the monopole and permit to calculate its input impedance more accurately. We get, in accordance with the second formulation of the emf method, for the input impedance of a linear monopole, its current being the total current of both wires

$$Z_A = -\frac{1}{J_A^2(0)} \int_0^{l_1} E_\varsigma J_A(\varsigma) d\varsigma . \tag{5.43}$$

Here, E_ς is the tangential component of the electric field, established by current $J(\varsigma)$ along the radiator axis. In calculating E_ς it is necessary to take into accounts that the derivative $dJ/d\varsigma$ has a jump at the segment boundaries. For this reason the tangential component of the electric field is (see, for example, Chapter 1)

$$E_\varsigma = j\frac{15}{k} \cdot \left\{ \left[\frac{\exp(-jkR_{11})}{R_{11}} + \frac{\exp(-jkR_{12})}{R_{12}} \right] \frac{dJ_A(l_1)}{d\varsigma} + \left[\frac{\exp(-jkR_{21})}{R_{21}} + \frac{\exp(-jkR_{22})}{R_{22}} \right] \cdot \right.$$
$$\left. \cdot \left[\frac{dJ_A(l_2+0)}{d\varsigma} - \frac{dJ_A(l_2-0)}{d\varsigma} \right] - \frac{2\exp(-jkR_0)}{R_0} \frac{dJ_A(0)}{d\varsigma} \right\}. \tag{5.44}$$

Here $R_{p1} = \sqrt{a_e^2 + (l_p - \varsigma)^2}$, $R_{p2} = \sqrt{a_e^2 + (l_p + \varsigma)^2}$, $R_0 = \sqrt{a_e^2 + \varsigma^2}$, $\dfrac{dJ_A(l_2 + 0)}{d\varsigma}$ and

$$\frac{dJ_A(l_2 - 0)}{d\varsigma}$$

are the values of the right- and the left-hand derivatives at point $z = l_2$, and a_e is the equivalent radius of the antenna:

$$a_e = \begin{cases} a, & l_2 \le z \le l_1, \\ \sqrt{ab}, & 0 \le z \le l_2. \end{cases}$$

However, the electrical field of a two-wire monopole (as considered here) differs from that of a single-wire monopole of equivalent radius and the same total current, due to the fact that the former consists of fields from two radiators. Assuming the base of the first radiator at the origin of the cylindrical coordinate system (see Fig. **10**), the electrical field strength in the far region at distance ρ is given by

$$E_{z1} = -j\frac{15\exp(-jk\rho)}{k\rho}\int_0^{l_1} i_1(\varsigma)d\varsigma. \tag{5.45}$$

The field strength of the second radiator along the line passing through the radiators' bases is

$$E_{z2} = -j\frac{15\exp\left[-jk(\rho - b\cos\phi)\right]}{k(\rho - b\cos\phi)}\int_0^{l_2} i_2(\varsigma)d\varsigma \tag{5.46}$$

The structure that was used in the compensation method and consists of two radiators with unequal length is analyzed as in the case of identical radiators. As a result, we obtain, rather than (3.45), the following expressions for the self- and the mutual impedances of radiators:

$$Z_{11} = Z_{22} = Z_m(l_1, a_e) - j(W_l/4)\cot k(l_2 + l_0) + R, \quad Z_{12} = Z_{21} = Z_m(l_1, a_e) + j(W_l/4)\cot k(l_2 + l_0). \tag{5.47}$$

The results of calculating the impedances are presented in Fig. **11**. They are determined in accordance with the described method for the variant with L=7.5, b=1.0, $2a$=0.05 (dimensions in centimeters) as a function of l.

5.4. MULTIPLE-WIRE AND MULTI-RADIATOR ANTENNA

In order to increase matching level in a wide frequency range, radiators must have low input reactance. For this purpose, antennas of hectometer, decameter and metric waves are constructed typically of several wires situated along the cylinder generatrices. The wires converge to the common point at the radiator ends forming cones (Fig. **12a**). Such radiator is called a multiple-wire one, or a cage radiator.

The basis of a multiple-wire radiator analysis is usually the assumption that it may be replaced with a solid metal tube of equivalent radius

$$a_e = \sqrt[N]{Na\rho^{N-1}}, \tag{5.48}$$

where N is the number of wires, a is wire radius, ρ is the radius of a cylinder, along generatrices of which the wires pass. In this case, the fields established in the far region by a multiple-wire radiator and by the equivalent solid radiator are identical, if their lengths are identical. As seen from (5.48), radius a_e of the equivalent radiator is substantially exceeds radius a of the singular wire. Accordingly, wave impedance W_r of the radiator decreases, and, consequently, reactive component X_A of the input impedance decreases also.

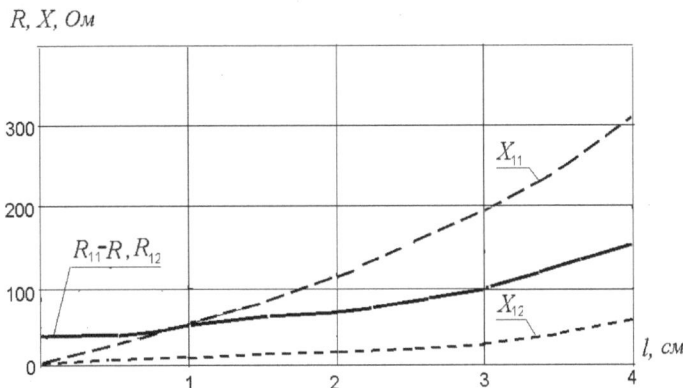

Figure 11: The self and mutual impedances of radiators with different lengths.

The equality of fields in the far region does not imply, as a matter of fact, the equality of the rest of electrical characteristics (*e.g.*, input impedances). As calculations show, the curves of the input impedances of the multiple-wire radiator and its equivalent radiator are of the same nature, but those for the former are displaced towards to greater values of L/λ. When N increases, the displacement decreases. The current distribution depends on N and differs from that along an ordinary linear radiator ($N=1$): the current magnitude in the singular wire decreases, the propagation constant of the wave along the radiator grows, the electrical length of the radiator increases [45]. As an example, the impedance of the multiple-wire cylindrical radiator with conic end sections is shown in the Fig. **12b** with solid curves, and that of the equivalent cylindrical radiator with the radius determined in accordance with (5.48) – by the dotted curves.

The multiple-wire radiators are constructed of wires with the equal lengths. If the antenna wires have different lengths, one can consider such antenna as a system of several radiators excited by a single generator (Fig. **7b**) of Chapter 1. The number and lengths of radiators are selected so as to ensure the mutual compensation of the reactances produced by the antenna elements and the radiation at small angles to horizon in the operation frequency band. Such antenna is called a multi-radiator one.

The calculation of characteristics of multi-radiator antenna uses, as a rule, the programs based on the solution of integral equations by the numerical methods. Another method of such antenna calculation gives greater physical clarity and required less computer time. The currents distributions along antenna wires are found by means of the theory of electrically coupled lines, and the pattern and the input impedance are calculated by common methods, which are valid for a line radiator with the current equal to the total current of a multi-radiators antenna.

The multi-radiators antenna (see Fig. **7**) consists of central radiator 1 with complex loading impedance Z_0 and side radiators 2 placed around the central radiator along the cylinder generatrices and connected to the base of radiator 1. To simplify, assume identical geometric dimensions of the side radiators to be, though the problem can be solved in the general case. If the side radiators are identical, one can reduce the asymmetrical line to a two-wire one and obtain the solution for the current in an explicit form. Here, the first wire of the equivalent asymmetrical line is the central radiator, and the second wire is a system of *N-1* side radiators (*N* is the total number of radiators).

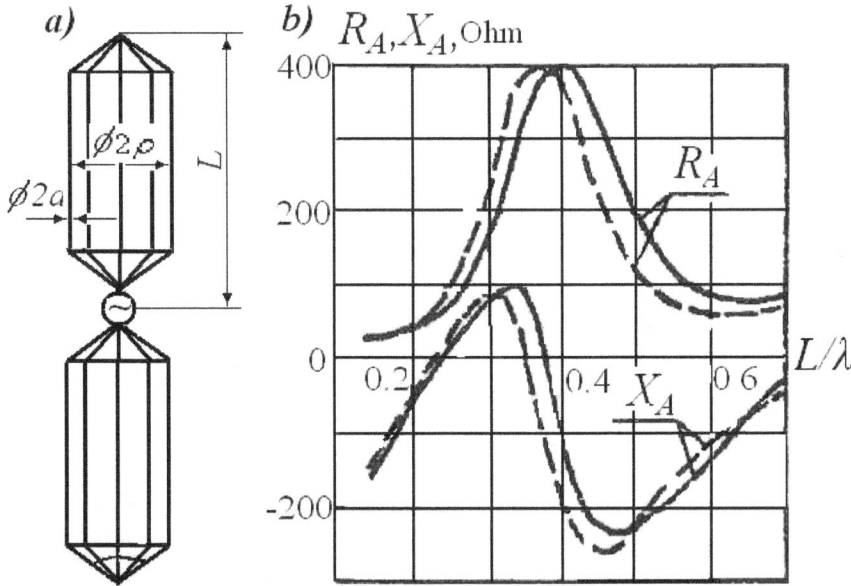

Figure 12: Multiple-wire radiator (*a*) and its input impedance (*b*).

If the wires of the line have different lengths, and the loading impedance is connected in the first wire, it is necessary to divide the line into three segments. The expressions for the current and potential of the *n*th wire at *m*th segment are given by (3.21). Here, *m, n* =1, 2, 3, the maximum number of wires at segment *m* is *M*=2. The direction of z_i-increase is given in Fig. **7**.

The boundary conditions for the two-wire asymmetrical line shown in this figure appear as follows:

$$i_1^{(1)}\Big|_{z_1=0} = i_2^{(3)}\Big|_{z_3=0} = 0, \; i_1^{(1)}\Big|_{z_1=l_1-l_2} = i_1^{(2)}\Big|_{z_2=0}, \; i_1^{(2)}\Big|_{z_2=l-l_3} = i_1^{(3)}\Big|_{z_3=0},$$

$$u_1^{(1)}\Big|_{z_1=l_1-l_2} = u_1^{(2)} - i_1^{(2)}Z_0\Big|_{z_2=0}, \; u_1^{(2)}\Big|_{z_2=l_2-l_3} = u_1^{(3)}\Big|_{z_3=0}, \; u_1^{(3)}\Big|_{z_3=l_3} = u_2^{(3)}\Big|_{z_3=l_3} = e. \quad \textbf{(5.49)}$$

The conditions mean the absence of the currents at free ends of the wires and continuity of the current and the potential along each wire with the exception of the point where load Z_0 is placed and the potential step occurs.

Substituting (3.21) into (5.49) and solving the equation set, we find

$$I_1^{(1)} = I_2^{(3)} = 0, \; I_1^{(2)} = j\frac{U_1^{(1)}}{\rho_{11}^{(1)}}\sin k\left(l_1 - l_2\right), \; I_1^{(3)} = j\frac{U_1^{(1)}}{\rho_{11}^{(1)}}\frac{\sin\left(l_1 - l_2\right)\sin k\left(l_{1e} - l_3\right)}{\sin k\left(l_{1e} - l_2\right)},$$

$$U_1^{(2)} = \frac{U_1^{(1)}\rho_{11}^{(2)}}{\rho_{11}^{(1)}}\sin k\left(l_1 - l_2\right)\cot k\left(l_{1e} - l_3\right),$$

$$U_1^{(3)} = \frac{U_1^{(1)}\rho_{11}^{(2)}}{\rho_{11}^{(1)}}\frac{\sin k\left(l_1 - l_2\right)\cos k\left(l_{1e} - l_3\right)}{\sin k\left(l_{1e} - l_2\right)},$$

$$U_2^{(3)} = \frac{U_1^{(1)}\rho_{11}^{(2)}}{\rho_{11}^{(1)}}\frac{\sin k\left(l_1 - l_2\right)\cos k\left(l_{1e} - l_3\right)}{\sin k\left(l_{1e} - l_2\right)}\left[1 - \frac{\rho_{11}^{(3)} - \rho_{12}^{(3)}}{\rho_{11}^{(2)}}\tan k\left(l_{1e} - l_3\right)\tan kl_3\right],$$

$$U_1^{(1)} = -e\rho_{11}^{(1)}\sin k\left(l_1 - l_2\right)\left\{\rho_{11}^{(3)}\sin k\left(l_1 - l_2\right)\sin k\left(l_{1e} - l_3\right)\sin kl_3\left[1 - \frac{\rho_{11}^{(2)}}{\rho_{11}^{(3)}}\cot k\left(l_{1e} - l_3\right)\cot kl_3\right]\right\}^{-1}$$

Here, one has taken in account that $W_{11}^{(1)} = \rho_{11}^{(1)}$, $W_{11}^{(2)} = \rho_{11}^{(2)}$, l_{1e} is a complex quantity, obtained from the expression

$$Z_0 - j\rho_{11}^{(1)}\cot k\left(l_1 - l_2\right) = -j\rho_{11}^{(2)}\cot k\left(l_{1e} - l_2\right). \tag{5.50}$$

The total current along the antenna as function of coordinate $\varsigma = l_m - z_m$ is

$$J_A\left(\varsigma\right) = \sum_{n=1}^{M}i_n^{(m)} = \begin{cases} j\dfrac{U_1^{(1)}}{\rho_{11}^{(1)}}\sin k\left(l_1 - |\varsigma|\right), l_2 \leq |\varsigma| \leq l_1, \\[3mm] j\dfrac{U_1^{(1)}}{\rho_{11}^{(1)}}D_1\dfrac{\sin k\left(l_{1e} - |\varsigma|\right)}{\sin k\left(l_{1e} - l_3\right)}, l_3 \leq |\varsigma| \leq l_2, \\[3mm] j\dfrac{U_1^{(1)}}{\rho_{11}^{(1)}}\left\langle D_1\cos k\left(l_3 - |\varsigma|\right) + D_2\sin k\left(l_3 - |\varsigma|\right)\right\rangle, 0 \leq |\varsigma| \leq l_3, \end{cases} \tag{5.51}$$

where

$$D_1 = \frac{\sin k\left(l_1 - l_2\right)\sin k\left(l_{1e} - l_3\right)}{\sin k\left(l_{1e} - l_2\right)},$$

$$D_2 = \rho_{11}^{(2)} \frac{\sin k\left(l_1 - l_2\right)\cos k\left(l_{1e} - l_3\right)}{\sin k\left(l_1 - l_2\right)}\left\{\left[\frac{1}{W_{11}^{(3)}} - \frac{1}{W_{12}^{(3)}} + \left[\frac{1}{W_{22}^{(3)}} - \frac{1}{W_{12}^{(3)}}\right]\right]\left[1 - \frac{\rho_{11}^{(3)} - \rho_{12}^{(3)}}{\rho_{11}^{(3)}}\tan k\left(l_{1e} - l_3\right)\tan kl_3\right]\right\}.$$

The input impedance of the asymmetrical line is

$$Z_l = e/J_A(0). \tag{5.52}$$

The expression permits in view of (5.51) to calculate approximately the reactance of a multi-radiators antenna, in the same way as the expression for the impedance of the equivalent long line permits to calculate approximately the reactance of the linear radiator. One can find the antenna impedance with greater accuracy, if the antenna is considered as a linear radiator with the current along it being the total current of the multi-radiators antenna.

Field E_ς is calculated in accordance with expression (3.41). Function $J_A(\varsigma)$ is continuous in the entire interval $0 \le \varsigma \le l_1$ and behaves sinusoidally in each antenna segment. However, the function $\dfrac{dJ_A}{d\varsigma}$ has a jump at the segment boundaries. Therefore, one must use (1.61). Substitution (5.52) into (1.61) gives

$$E_\varsigma = \frac{15U_1^{(1)}}{\rho_{11}^{(1)}}\left[\sum_{m=1}^{3} D_{1m}\left(\frac{\exp\left(-jkR_{m1}\right)}{R_{m1}} + \frac{\exp\left(-jkR_{m2}\right)}{R_{m2}}\right) + D_{14}\frac{\exp\left(-jkR_0\right)}{R_0}\right], \tag{5.53}$$

where

$$D_{11} = 1, D_{12} = D_1\frac{\cos k\left(l_{1e} - l_2\right)}{\sin k\left(l_{1e} - l_3\right)} - \cos k\left(l_1 - l_2\right), D_{13} = D_2 - D_1\cot k\left(l_{1e} - l_3\right), D_{14} = 2\left(D_1\sin kl_3 - D_2\cos kl_3\right)$$

The electric field in the far region at the distance r is

$$E_\theta = j15kJ_A(0)\exp\left(-jkr\right)H(\theta)/r, \tag{5.54}$$

where $H(\theta)$ is the generalized effective height, namely

$$H(\theta) = \frac{\sin\theta}{J_A(0)} = \int_0^{l_1} J_A(\varsigma)e^{jk\varsigma\cos\theta}d\varsigma.$$

If

$$J_A(\varsigma) = A_m \cos k\left(l_m - |\varsigma|\right) + jB_m \sin k\left(l_m - |\varsigma|\right), l_{m+1} \le |\varsigma| \le l_m, \tag{5.55}$$

then $H(\theta) = \dfrac{1}{kJ_A(0)\sin\theta}$.

$$\cdot\sum_m A_m\left[-\cos\theta\sin\left(kl_m\cos\theta\right)+\sin k\left(l_m-l_{m+1}\right)\cos\left(kl_{m+1}\cos\theta\right)+\cos\theta\cos k\left(l_m-l_{m+1}\right)\sin\left(kl_{m+1}\cos\theta\right)\right]+ \tag{5.56}$$
$$+jB_m\left[\cos\left(kl_m\cos\theta\right)-\cos k\left(l_m-l_{m+1}\right)\cos\left(kl_{m+1}\cos\theta\right)-\cos\theta\sin k\left(l_m-l_{m+1}\right)\sin\left(kl_{m+1}\cos\theta\right)\right]$$

Hence, the effective height of an asymmetric multi-radiator antenna is

$$h_e = H(\pi/2) = \frac{1}{kJ(0)}\sum_m A_m \sin k\left(l_m-l_{m+1}\right)+jB_m\left[1-\cos k\left(l_m-l_{m+1}\right)\right]. \tag{5.57}$$

Comparing (5.51) and (5.55), we find

$$A_1 = 0, \; B_1 = \frac{U_1^{(1)}}{\rho_{11}^{(1)}}, \; A_2 = j\frac{U_1^{(1)}D_1}{\rho_{11}^{(1)}}\frac{\sin k\left(l_{1e}-l_2\right)}{\sin k\left(l_{1e}-l_3\right)},$$

$$B_2 = \frac{U_1^{(1)}D_1}{\rho_{11}^{(1)}}\frac{\cos k\left(l_{1e}-l_2\right)}{\sin k\left(l_{1e}-l_3\right)}, \; A_3 = \frac{jU_1^{(1)}D_1}{\rho_{11}^{(1)}}, \; B_3 = \frac{U_1^{(1)}D_2}{\rho_{11}^{(1)}},$$

that is,

$$h_e = \frac{\left\{1-\cos k\left(l_1-l_2\right)+D_1\sin kl_3 + D_1\dfrac{\cos k\left(l_{1e}-l_2\right)-\cos k\left(l_{1e}-l_3\right)}{\sin k\left(l_{1e}-l_3\right)}+D_2\left(1-\cos kl_3\right)\right\}}{k\left(D_1\cos kl_3 + D_2\sin kl_3\right)}. \tag{5.58}$$

The antenna radiation resistance is

$$R_\Sigma = R_A - R_{ls}, \tag{5.59}$$

where R_A is the active component of the input impedance, and R_{ls} is the loss resistance in load Z_0 referred to an antenna input:

$$R_{ls} = \mathrm{Re}\left[J_A^2\left(l_2\right) Z_0 \big/ J_A^2\left(0\right)\right].$$ (5.60)

Quantities $W_{ns}^{(m)}$ and $\rho_{ns}^{(m)}$ are found, as a rule, with the help of the potential coefficients.

DISCLOSURE

Some of the material of the chapter are published by the author in the article "Calculation of electrical parameters of two-wire lines in multiple-conductor cables," - IEEE Trans. Electromagn. Compat., Vol. 50, No. 3, 2008, pp. 697-703.

Send Orders for Reprints to reprints@benthamscience.net

CHAPTER 6

Antenna with Loads

Abstract: Application of the impedance line method to radiators with concentrated loads permits to create antennas with required characteristics, in particular, the wideband monopole, an antenna with the given current distribution, *etc.* Wideband radiators must have an exponential or linear in-phase current distribution, created by capacitive loads that vary, in particular, along the radiator length in accordance with linear law. In order to retain in-phase current distribution in a wider frequency range, the capacitances of these loads should vary in inverse proportion to the squared frequency. Using of capacitive loads in V-dipoles yields similar results.

Keywords: Antenna with required current distribution, Capacitance of a bottom capacitor, Decrease of antenna reactance, Exponential current distribution, Freedom of antenna length choice, Frequency-dependent capacitance, Frequency ratio, High matching level, In-phase current distribution, Inverse problem of the radiators theory, Load as parallel connection of a resistor and a capacitor, Maximal matching level, Negative inductance, Optimization of 12-meter antennas, V-dipole, Weakening the effect of metal structures upon the pattern, Wide-band monopole, Wide frequency range.

6.1. INVERSE PROBLEMS OF THIN ANTENNA THEORY

As said in Section 3.3, loads can be used to solve the inverse problem of the thin antennas theory - to create an antenna with required electrical characteristics. Of a great practical importance is a particular case of the problem: creation of a radiator that ensures, in a wide frequency range, a high matching level and the radiation maximum in the plane, perpendicular to the radiator axis.

A typical linear radiator (thin, without loads) fails to meet the requirements. The reactive component of its input impedance is great everywhere, excepting the vicinity of the series resonance. This results in the antenna mismatch with a cable. If the radiator arm is longer than 0.7λ, the radiation in the plane, perpendicular to its axis, decreases, since the current distribution along a thin linear monopole without loads (Fig. **1a**) is close to the sinusoidal one and anti-phased segments on the current curve are formed at high frequencies (Fig. **1b**, curve 1).

By connecting concentrated loads across the radiator length, one can, depending on their magnitudes and points of connection, obtain the current distribution other than the sinusoidal one. The experimental results show that a radiator with linear in-phase current distribution exhibits good performance (high matching level, the directivity pattern of required shape) in a wide frequency range. In particular, such distribution is created by capacitive loads. The results confirm the known fact that the radiation maximum in the direction, perpendicular to the dipole axis, is attained, if the current is in-phase along the entire length of the antenna. Moreover, with the in-phase current, a long radiator has high radiation resistance, which allows increasing the matching level.

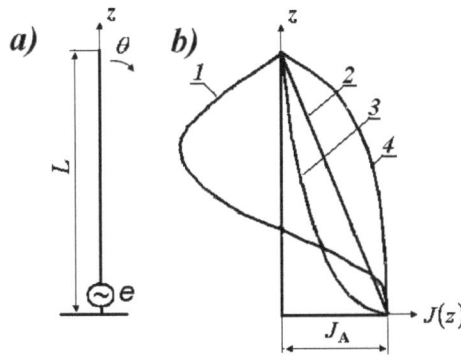

Figure 1: A linear monopole (*a*) and laws of the current variation along it (*b*).

Consider a monopole of height L with N loads Z_n (Fig. **6b**). Let loads be spaced uniformly at distance b along the antenna. If the distance is small ($kb \ll 1$), the current distribution along the antenna undergoes practically no change at replacing the concentrated loads with surface impedance $Z(z)$ distributed across each segment length. Assume the surface impedance of each antenna segment to be constant and equal to $Z^{(n)}$ then

$$Z_n = bZ^{(n)} / (2\pi a_n),$$

(6.1)

where a_n is the antenna radius at the *n*th segment.

The current distribution along the antenna with a piecewise constant surface impedance is, in the first approximation, identical to that along an open-end piecewise uniform long line, *i.e.,* a line with stepwise variation of propagation constant (see Fig. **6b**). In this case, wave propagation constant γ_n at the *n*th segment of the line is related to surface impedance $Z^{(n)}$ by relationship

$$-\gamma_n^2 = k^2 - j2k\chi_n Z^{(n)}\big/\left(a_n Z_0\right).$$
(6.2)

Here, $\chi_n = \dfrac{1}{2\left[\ln\left(2L/a_n\right)-1\right]}$ is a small parameter value at the *n*th-segment. In the general case, γ_n is a complex quantity. In a particular case, when the quantity is purely imaginary ($\gamma_n = jk_n$), the current distribution at the *n*th segment of the line is of sinusoidal nature.

If the law of variation of the propagation constant, which allows ensuring the required current distribution, is found, expressions (6.2) and (6.1) can be used to calculate, first, the surface impedance and, second, concentrated loads Z_n, respectively.

The current at the *n*th segment of a stepped line at arbitrary γ_n is

$$J\left(z_n\right) = I_n \sinh\left(\gamma_n z_n + \phi_n\right), \qquad 0 \leq z_n \leq b,$$
(6.3)

where I_n and ϕ_n are the current amplitude and phase at the *n*th segment, respectively, and z_n is the coordinate counted off from the segment end, *i.e.,* $z_n = \left(N-n+1\right)b - z$.

Suppose we want to obtain the given current distribution along a line,

$$J\left(z\right) = J_A f\left(z\right), \qquad 0 \leq z \leq L,$$
(6.4)

where J_A is the current at the line input (generator current), $f\left(z\right)$ is a real and positive distribution function, which corresponds to the in-phase current. Equate currents $J\left(z\right)$ and $J\left(z_n\right)$ at the beginning and the end of each line segment. In this case, if the segment lengths are small, the current distribution along a line is close to the required one.

In accordance with (6.3) and (6.4), at $z_n = b$ and $z_n = 0$,

$$I_n \sinh\left(\gamma_n b + \phi_n\right) = J_A f\left[(N-n)b\right], \qquad I_n \sinh\phi_n = J_A f\left[(N-n+1)b\right]$$

Divide the left- and right-hand sides of the first equation into the respective sides of the second one. Assuming b a small magnitude and retaining the first terms of expansion of hyperbolic functions of small arguments into a series, we get

$$\tanh\phi_n = \gamma_n b \bigg/ \left\{\frac{f[(N-n)b]}{f[(N-n+1)b]} - 1\right\}. \tag{6.5}$$

For the $(n+1)th$ segment, similarly to (6.5),

$$\tanh\phi_{n+1} = \gamma_{n+1} b \bigg/ \left\{\frac{f[(N-n-1)b]}{f[(N-n)b]} - 1\right\}. \tag{6.6}$$

The voltage and current are continuous along a stepped line, hence

$$\tanh\phi_{n+1} = \left(\gamma_{n+1}/\gamma_n\right)\tanh\left(\gamma_n b + \phi_n\right). \tag{6.7}$$

Equations (6.5)-(6.7) represent a set of equations that allow relating γ_n and γ_{n+1} to each other.

The set solution shows that quantity γ_n is independent of γ_{n+1}:

$$\gamma_n = \frac{1}{b}\sqrt{1 - \frac{2f[(N-n)b] - f[(N-n-1)b]}{f[(N-n+1)b]}}. \tag{6.8}$$

Function $f(z)$ characterizes the law of the current amplitude variation along the radiator. At linear distribution of the current amplitude (see Fig. **1b**, curve 2),

$$J_2(z) = J_A\left(1 - z/L\right),$$

i.e.,

$$f_2(z) = (L-z)/L = \left[(n-1)b + z_n\right]/(Nb). \tag{6.9}$$

The linear distribution is a particular case of the exponential one (see Fig. **1b**, curves 3 and 4):

$$J_{3,4}(z) = J_A \frac{\exp(-\alpha z) - \exp(-\alpha L)}{1 - \exp(-\alpha L)},$$

i.e.,

$$f_{3,4}(z) = \frac{\exp(-\alpha z) - \exp(-\alpha L)}{1 - \exp(-\alpha L)} = \frac{\sinh\{(\alpha/2)[(n-1)b + z_n]\}}{\sinh(\alpha Nb/2)}, \tag{6.10}$$

where α is the logarithmic decrement. At positive α, the current curve is concave, *i.e.,* the current quickly decreases from the maximum value near the generator to zero near the free end of the antenna. At negative α, the current curve is convex, *i.e.,* the current is more uniformly distributed along the dipole. The curve steepness depends on the value of α. It is easy to show that at $\alpha \to 0$, the expression for $J_{3,4}(z)$ passes into $J_2(z)$.

Note two particular cases of the exponential distribution. At $\alpha \to -\infty$, the current distribution along the antenna becomes rectangular. At $\alpha \to \infty$, the current quickly decreases with growing z and is zero everywhere, except for the section near the generator. The use of the exponential distribution allows covering a wide class of distributions.

Substituting (6.10) into (6.8) yields

$$\gamma_n = \frac{\alpha}{\sqrt{2}}\sqrt{1 + \coth[\alpha(n-1)b/2]}. \tag{6.11}$$

In particular, if $0 < \alpha << 1/L$,

$$\gamma_n = \sqrt{\alpha/[(n-1)b]}. \tag{6.12}$$

Thus, in order to obtain the concave exponential current distribution along a stepped line, the propagation constant should be real and vary along the line according to (6.11). In particular, in order to obtain a distribution close to linear one, the value of γ_n should be small and inversely proportional to the square root of distance from a given point to the free end of the line.

At $\alpha < 0$,

$$\gamma_n = j\frac{|\alpha|}{\sqrt{2}}\sqrt{\coth\left[|\alpha|(n-1)b/2\right]-1}\,, \tag{6.13}$$

i.e., in order to obtain the convex exponential distribution, the propagation constant should be purely imaginary along the entire line.

At large positive α, quantity γ_n is real, constant along the line length and equals to the decrement; at large negative α, the propagation constant tends to zero. The family of curves for γ depending on coordinate z is given in Fig. **2**.

In the general case of the in-phase current distribution along the antenna, its pattern in the vertical plane has the form

$$F(\theta) = \sin\theta \int_{-L}^{L} f(z)\exp(jkz\cos\theta)dz.$$

The meaning of function $f(z)$ is clear from (6.4).

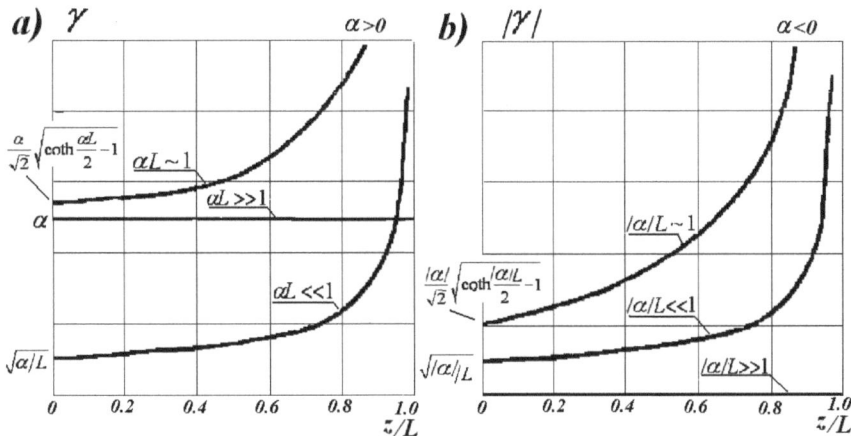

Figure 2: Propagation constant at positive (*a*) and negative (*b*) logarithmic decrements.

For the exponential distribution, the integral calculation results in the expression

$$F_{3,4}(\vartheta) = \frac{\tan\vartheta}{k^2\cos^2\vartheta+\alpha^2}\left\{k\cos\vartheta - e^{-\alpha L}\left[k\cos\vartheta\cos(kL\cos\vartheta)+\alpha\sin(kL\cos\vartheta)\right]\right\}. \tag{6.14}$$

In particular, for the linear distribution:

$$F_2(\theta) = \left(\sin\theta / \cos^2\theta\right)\left[1 - \cos\left(kL\cos\theta\right)\right].$$

Fig. **3** demonstrates the patterns of antennas with linear and exponential distribution of the current amplitude. The antenna arm length varies from $3\lambda/4$ to 4λ. For the sake of clarity, the curves are plotted in the rectangular coordinate system. Here, for comparison, patterns $F_1(\theta)$ of a radiator with the sinusoidal current distribution are presented as well.

As seen from the Figure, for linear and exponential distribution in contrast to sinusoidal one, the radiation maximum does not deviate with growing frequency from the perpendicular to the radiator axis. The increase of L/λ makes the main lobe narrower and increases the directivity. When α decreases (including taking on negative values), the main lobe beam width decreases.

The antenna input impedance in the first approximation is equal to the input impedance of a stepped long line:

$$Z_l = -jW_N \coth\left(\gamma_n b + \phi_n\right),$$

where $W_N = \left(\gamma_N / k\right)W$, and W is the wave impedance of a metal monopole without loads (of the same dimensions).

Using equations (6.6) and (6.7), we find

$$Z_l = -j\left(W/kb\right)\left[f(-b) - 1\right]. \tag{6.15}$$

As seen from the expression, reducing the reactive component of the input impedance requires a slow variation of function $f(z)$ near the antenna base, so that the difference in square brackets should be a small quantity – on the order of kb. Otherwise, the reactive component of input impedance will be great.

For the exponential distribution, replacing $f(-b)$ with $f_3(-b)$ yields

$$Z_{A3} = -j\left(W/kL\right)f_x\left(\alpha L/2\right) \tag{6.16}$$

where $f_x(x) = x(1 + \coth x)$. The plot of function $f_x(x)$ is given in Fig. **4a**. In particular, $f_x(x) = 1$ for linear distribution, *i.e.*,

$$Z_{A2} = -jW/(kL).$$

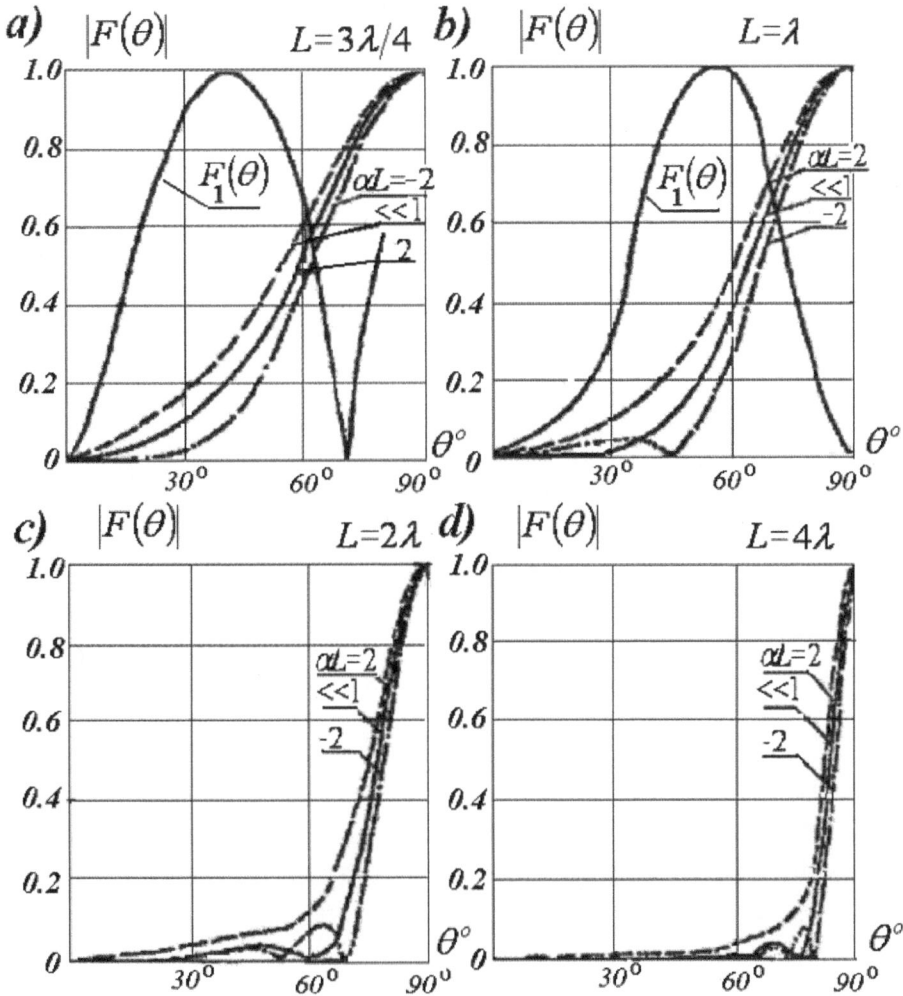

Figure 3: Patterns of antennas with linear, exponential and sinusoidal distribution of current amplitude for different arm lengths: (*a*) $L = 3\lambda/4$; (*b*) $L = \lambda$; (*c*) $L = 2\lambda$; (*d*) $L = 4\lambda$.

Fig. **4b** compares input impedances X_{A1} and X_{A3} of a uniform line with sinusoidal current distribution and a non-uniform line with exponential current distribution as functions of frequency. Here, magnitude α is assumed constant. In the first case the input impedance is a cotangent curve, in the second case, curve smoothly tends to the axis with growing frequency, and that allows ensuring good matching in a wide range.

As seen from Fig. **4a** and expression (6.16) for Z_{A3}, if α decreases, the input impedance of a long line at a given frequency diminishes. It means that a decrease of α results in a decrease of the reactive component of the antenna input impedance. The effective height grows concurrently, since the area bound by the curve of the current increases; hence, the radiation resistance grows, too. Thus, at exponential distribution, it is expedient to decrease α, in particular, into the region of negative values.

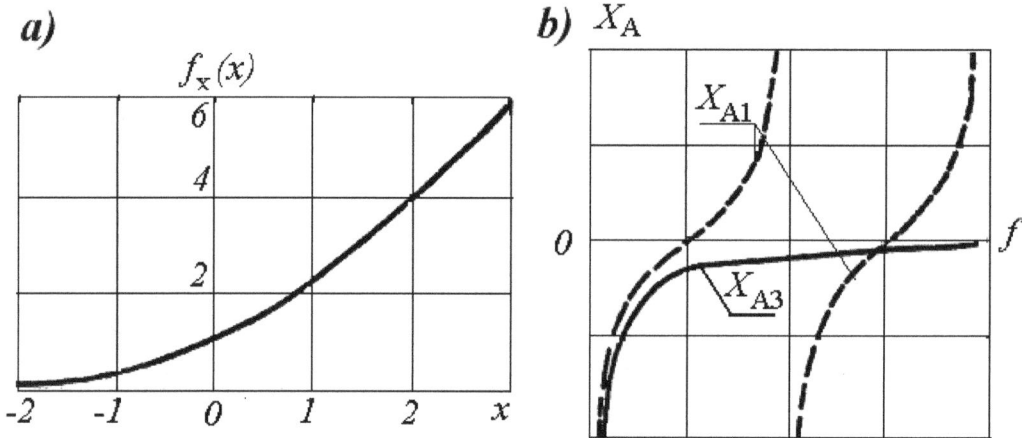

Figure 4: Plot of function $f_x(x)$ (a) and input impedances of uniform and non-uniform lines (b).

The positive values of α correspond to real propagation constant γ_n, negative values – to purely imaginary one. The possibility of realizing this or that propagation constant (and, consequently, this or that in-phase distribution) depends on the realizability of the surface impedance (or concentrated loads, equivalent to it), which corresponds to the propagation constant.

According to (6.1) and (6.2),

$$-\gamma_n^2 = k^2 - jk\chi_n Z_n/(30b),$$ (6.17)

In order to create the in-phase current distribution, quantity γ_n should be purely real or purely imaginary along the entire antenna, and quantity γ_n^2, correspondingly, only positive or only negative:

$$\text{sign } \gamma_n^2 = \text{const } (n).$$

For the in-phase distribution to take place in a wide frequency range, quantity γ_n should be real:

$$\gamma_n^2 > 0. \tag{6.18}$$

Indeed, were the values of γ_n (and also ϕ_n) purely imaginary, the hyperbolic sine in the formula (6.3) would become the trigonometric one. With frequency growth, the increase of argument will result in exceeding π, and the sine will change sign. If γ_n is real, then, as seen from (6.5), if function $f(z)$ decreases monotonically, the sign of ϕ_n coincides with that of γ_n. It follows from (6.8) that condition

$$f\left[(N-n)b\right] \le \frac{1}{2}\left\{f\left[(N-n+1)b\right] + f\left[(N-n-1)b\right]\right\} \tag{6.19}$$

should hold at real γ_n, i.e., function $f(z)$ cannot be convex.

Two options for condition (6.18) to be valid in a wide frequency range follow from (6.17).

The first one takes place at

$$k^2 \langle\langle jk\chi_n Z_n/(30b), \tag{6.20}$$

i.e.,

$$\gamma_n^2 = jk\chi_n Z_n/(30b) \tag{6.21}$$

If one takes into account that parameter χ_n is, strictly speaking, a complex magnitude ($\chi_n = \chi_{n1} - j\chi_{n2}$), then the load admittance is according to (6.21),

$$Y_n = 1/Z_n = j\omega C_n + 1/R_n \,, \tag{6.22}$$

where

$$C_n = 4\pi\varepsilon\chi_{n1}/(b\gamma_n^2), \qquad R_n = b\gamma_n^2/(4\pi\varepsilon\omega\chi_{n2}).$$

As follows from (6.22), for quantity γ_n^2 to be positive, each load should be executed as a parallel connection of a resistor and a capacitor (Fig. **5a**). The resistance of the resistor should then vary in inverse proportion to frequency, and the capacitance of the capacitor should remain constant. When creating an actual antenna, it is expedient to choose the value of R_n for the middle frequency of band.

To achieve required current distribution $f(z)$ along the antenna, quantity γ_n should correspond to (6.8). Its substitution into (6.22) gives

$$C_n = 4\pi\varepsilon\chi_{n1}b\left\{1 - \frac{2f[(N-n)b] - f[(N-n-1)b]}{f[(N-n+1)b]}\right\}^{-1}, \quad R_n = \frac{\chi_{n1}}{\chi_{n2}\omega C_n}. \tag{6.23}$$

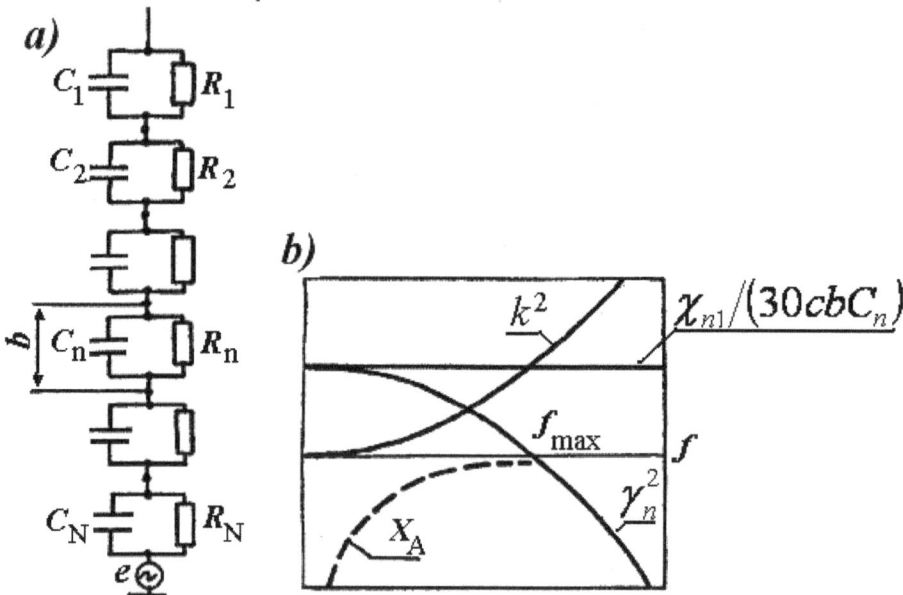

Figure 5: An antenna circuit with capacitive and resistive elements (*a*) and frequency dependence of propagation constant (*b*).

By comparing (6.19) and (6.23), one can easily verify that, when inequality (6.19) holds, the values of C_n are nonnegative. For exponential and linear distribution,

$$C_{n3} = \frac{8\pi\varepsilon\chi_{n1}}{\alpha^2 b\{1+\coth[\alpha(n-1)b/2]\}}, \quad C_{n2} = \frac{4\pi\varepsilon}{\alpha}\chi_{n1}(n-1), \quad R_n = \frac{\chi_{n1}}{\chi_{n2}\omega C_n}. \quad (6.24)$$

One can see from (6.24) that in the particular case, if it is necessary to obtain a law of current distribution, close to the linear one, capacitances of loads should decrease towards the free end of the antenna in proportion to the distance from it:

$$C_{n2} = C_{N2}(n-1)/(N-1), \quad (6.25)$$

where C_{N2} is the capacitance of the capacitor near the antenna base. The resistances of resistors should grow towards the free end of the antenna:

$$R_{n2} = R_{N2}(N-1)/(n-1) \quad (6.26)$$

Thus, to create the in-phase current distribution ensuring high electrical performance of an antenna in a wide frequency range, each load should represent parallel connection of a resistor and a capacitor. The expediency of using a complex load to create a linear current distribution was for the first time indicated in [46]. But later on, most attention was given to antennas with capacitive loads. The calculation results show that, if resistors are connected in parallel with capacitors, the linear law of current distribution along the radiator is followed more closely and the operating frequency range increases. However, connection of resistors results in decreasing the antenna efficiency, so the issue of their application should be decided in each particular case.

Fig. **5b** shows a plot of γ_n^2 against frequency for an antenna with capacitive loads:

$$\gamma_n^2 = -k^2 + \chi_{n1}4\pi\varepsilon/(bC_n). \quad (6.27)$$

For propagation constant to be real at given frequency f, capacitances of capacitors should not exceed

$$C_n \leq \frac{\chi_{n1}}{30k^2 bc} = \frac{2.54 \cdot 10^5 \chi_{n1}}{f^2 b}. \tag{6.28}$$

Here, c is the speed of light, capacitance C is measured in farads, if frequency f is in Hertz's, and in picofarads, if f is in megahertz's. Under the linear distribution, capacitance C_{N2} of the capacitor near the antenna base is greater than others and should be chosen in accordance with (6.28). Similarly, under other distributions, this expression determines the maximum capacity.

As follows from (6.27), propagation constant at low frequencies is real along the entire antenna. As the frequency increases, the values of γ_n become purely imaginary (first of all, on segments adjoining the generator), *i.e.*, the current distribution along these segments of the radiator becomes sinusoidal, and the main lobe of the pattern deviates from the perpendicular to the dipole axis. This effect limits the antenna frequency range from above. From below, the range is limited by frequencies, where the reactive component of input impedance is still great. Lest magnitude γ_n should become purely imaginary with growing frequency, the capacitances of capacitors should decrease with its increase (*e.g.*, vary in inverse proportion to its square).

Considering the second option for realization of condition (6.18) in a wide frequency range, one can make a similar conclusion. It takes place, if the second term of the right-hand side of (6.17) is proportional to k^2:

$$-\gamma_n^2 = k^2 \left[1 - j \chi_n Z_n / (30kb) \right], \tag{6.29}$$

i.e., load Z_n represents negative inductance Λ_n:

$$Z_n = -j\omega |\Lambda_n| \tag{6.30}$$

with $|\Lambda_n| = 30b\left(1 + \gamma_n^2 / k^2\right) / (\chi_n c)$. At small γ_n / k the inductance is independent of frequency f.

In this case, the value of $\gamma_n^2 = k^2 \left[\chi_n |\Lambda_n| c / (30b) - 1 \right]$ is positive, if

$$|\Lambda_n| > 30b / (\chi_n c). \tag{6.31}$$

The negative inductance is a circuit element, whose impedance is purely reactive, negative and proportional to f. It is equivalent to frequency-dependent capacitance:

$$-j\omega|\Lambda_n| = 1/(j\omega C_n),\tag{6.32}$$

where

$$C_n = 1/(\omega^2|\Lambda_n|) = C_{n0}\, f_0^2/f^2,\tag{6.33}$$

C_{n0} is the magnitude of capacitance C_{n0} at frequency $f = f_0$ and is independent of f.

Thus, to retain the in-phase current distribution in a wide frequency range, the capacitances of concentrated loads connected into antenna should vary in inverse proportion to the squared frequency. As one can easily verify, at connection of negative inductances $\Lambda_n = -30b/(\chi_n c)$ in series with loads determined by expression (6.22), inequality (6.18) is valid at all frequencies and the restriction (6.20) is lifted.

The proposed method allows making a number of practical conclusions. For the loads to efficiently affect the law of current distribution, the distance between them should be small in comparison with the wavelength. Only capacitors are to be used as reactive elements when creating a wide-range dipole, since connection of reactive two-terminal networks of a more complex type, their structure including inductance coils, results in narrowing of the working range.

Capacitors enable one to create, in a wide frequency range, an electromagnetic wave with real propagation constant along an antenna, which corresponds to the exponential distribution of the current amplitude with positive decrement (concave curve of the current). Obtaining a convex curve of the current with the help of simple concentrated elements (resistors, capacitors, inductance coils) is impossible. Among distributions with positive α, the antenna with distribution close to linear, which is created by capacitances decreasing to the free end of the antenna in proportion to the distance from it, has a higher matching level and narrower main lobe of the directivity pattern.

6.2. CALCULATIONS AND MEASUREMENTS

The described method of the analysis is based on the theory of a radiator with surface impedance varying with the length. On the one hand, the method is a sufficiently general one and allows deriving analytical expressions for magnitudes of load impedances, which ensure different laws of the current change along the radiator. On the other hand, it has, to a certain extent, of approximate nature, *i.e.,* requires verification by calculations and experiments.

Consider, as an example, a monopole of height 6 m and radius $6.67 \cdot 10^{-3}$ m with 10 capacitors, which are connected along it at distance 0.6 m from each other. The capacitance of the capacitor near the base is adopted to be 17.7 pF (then, propagation constant γ_n is real along the entire antenna up to frequency 40 MHz), and the capacitances of other capacitors are calculated by (6.25).

The calculation of the radiator electrical characteristics uses the general-purpose programs, which are, as a rule, based on the principles described in Section 2.5, *i.e.,* on the generalized method of induced emf. For an antenna with loads, they use the set of equations (2.71).

Fig. **6a** shows the frequency dependence of active R_A and reactive X_A components of antenna input impedance as well as TWR in cable with wave impedance 75 Ohm. Here, for the sake of comparison, the experimental data obtained on a full-scale mock-up are given. Fig. **6b** also presents the calculated vertical directivity patterns.

Calculations and experiment confirm that characteristics of radiators with loads are much better than those without loads. Under identical requirements to electrical characteristics, the frequency ratio is from 1.3 to 1.5 for a thin whip antenna, from 1.5 to 3 – for radiators of the same dimensions with capacitive loads, and from 3 to 4 – for radiators with capacitive-resistive loads. Here, the upper limit of the frequency range for antennas with loads is defined by the deviation of the main lobe of the directivity pattern from the perpendicular to the dipole axis. As to the matching level, it remains high in a wider range (with frequency ratio about 10).

The antennas with frequency-dependent capacitances have a wider frequency range than those with constant ones. Fig. **7a** gives TWR of three antennas variants in cable with wave impedance 75 Ohm. The calculations are performed for an antenna of height 12 m and radius 0.03 m, with 10 capacitors connected at distance 1.2 m from each other along it (the top and bottom ones are away from the antenna ends for 0.6 m).

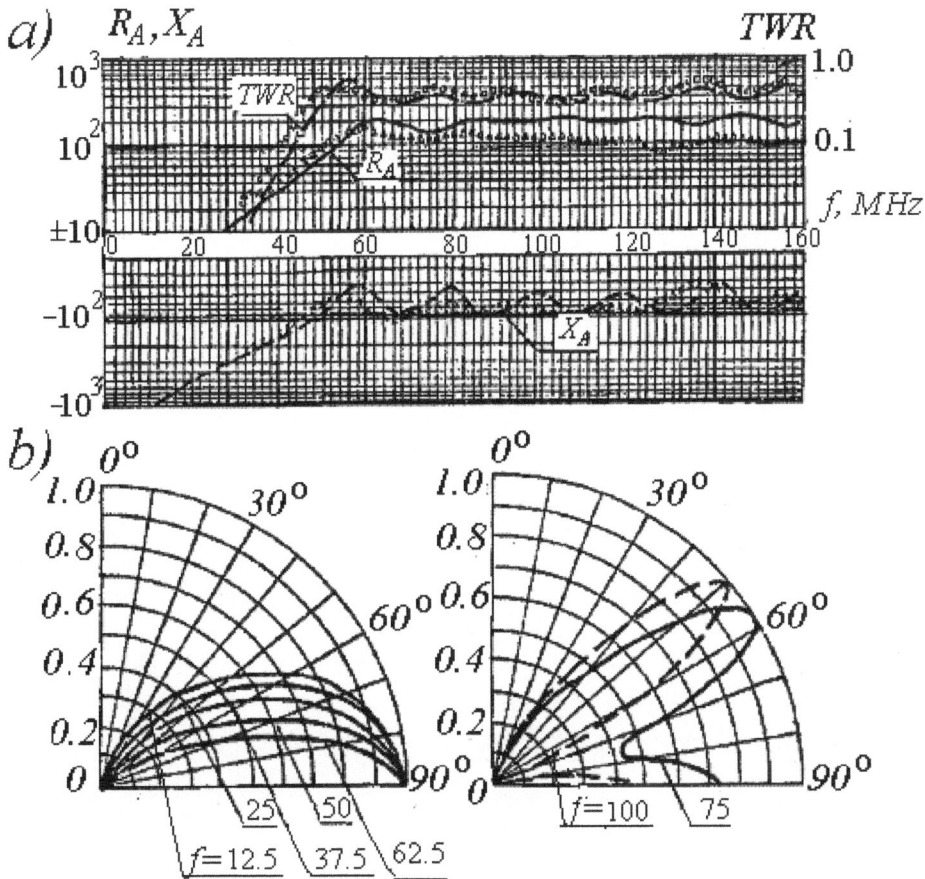

Figure 6: Input characteristics (*a*) and patterns (*b*) of a radiator with constant capacitive loads.

Curve 1 in Fig. **7a** corresponds to a radiator without loads (a whip antenna), curve 2 – to a radiator with loads, whose capacitances are frequency independent. Here, value C_{N0} of the bottom capacitor is chosen to be 177 pF (from the condition of γ_n being real at frequencies up to 10 MHz), and capacitances C_{n0} of others capacitors decrease to the free end of the antenna in proportion to the distance from it, which allows obtaining a law of current distribution along the radiator close to the linear one. Curve 3 is plotted for a radiator with frequency-dependent capacitive loads corresponding to (6.33), with $f_0 = 20$.

Table **1** gives lower f_1 and upper f_2 frequencies of the working range of each antenna, with f_1 being the frequency, where TWR becomes greater than 0.2, and f_2 – the one, where the field strength along the perpendicular to the axis becomes less than 0.7 of the maximum (as a rule, it corresponds to the second maximum on the traveling-wave ratio curve). The TWR of a whip antenna quickly decreases with growing frequency below the level of 0.2, with the value of frequency corresponding to this point taken as f_2. Besides, range width $\Delta f = f_2 - f_1$ and frequency ratio $k_f = f_2 / f_1$ are presented.

As seen from Fig. **7a** and Table **1**, variant 3 at low frequencies approaches an antenna without loads as to the matching level, and upper limit f_2 displaces to the right, since the main lobe of directivity pattern deviates from the perpendicular to the dipole axis at a higher frequency. In addition, the minimum TWR at the operating range center increases.

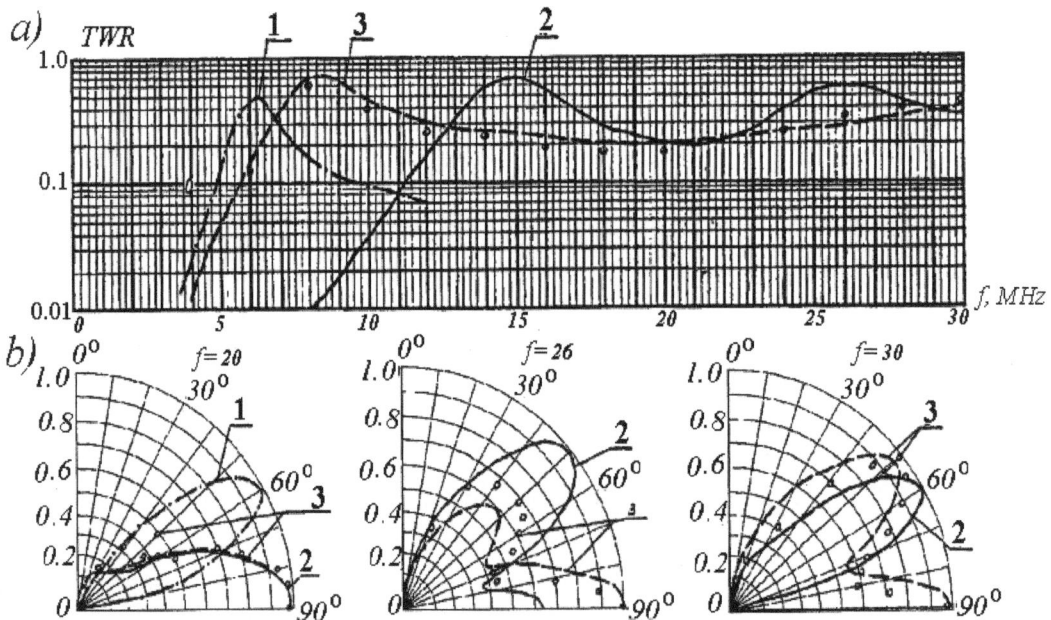

Figure 7: Input characteristics (*a*) and patterns (*b*) of a radiator with frequency-dependent capacitive loads.

Fig. **7a** gives also the results of experimental verification of variant 3 performance. The measurements are performed on an antenna mock-up at scale 1:10. The frequency range was split into short intervals, with capacitor capacitances in each of them selected equal to those of variable capacitances at the section center. The agreement of calculated data with experimental ones is good.

Fig. **7b** gives calculated curves for patterns of the same antennas variants (in the vertical plane) as well as experimental values for variant 3.

Table 1: Bandwidth ratio of radiators

Variant of antenna	Frequency, MHz		Range width Δf, MHz	Frequency ratio k_f
	f_1	f_2		
1	5.2	7.7	2.4	1.5
2	12.3	26.0	13.7	2.1
3	6.3	34.0	27.7	5.4

The task of implementation of frequency-dependent capacitances is not easy, but has prospects. In principle, it can be accomplished by placing some material, whose permeability varies with frequency under given law (*e.g.*, manufactured dielectric) between capacitor plates. Yet, the use of tunable capacitors seems to be more realistic. In this case, (6.33) determines the optimum frequency dependence of the capacitances. The smooth tuning of elements can be replaced with stepwise one, which is easier to implement.

The method of analysis, based on the theory of an impedance dipole, allows finding potential capabilities of antennas with loads. Furthermore, the results obtained with its help can be used to solve the optimization problem of an antenna with loads by the mathematical programming method, described in Section 3.3.

6.3. SYNTHESIS OF A WIDE-BAND MONOPOLE

As shown in Section 3.3, the software for antenna synthesis by the mathematical programming method can be used to choose the optimum capacitive loads that allow to ensuring the maximum level of TWR and PF in given frequency range f_1 - f_2. The number of loads and places of their connection are fixed.

Fig. **8** gives the basic dimensions of an antenna with loads. It is a monopole of height 12 m with nine capacitors spaced equidistantly from each other, with insulator at the base.

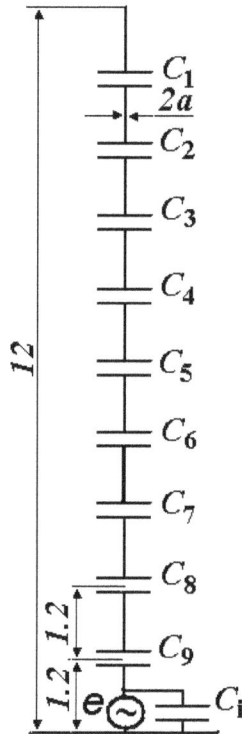

Figure 8: Basic dimensions of optimized antennas with nine capacitors.

Capacitor C_i connected in the antenna base in parallel to its input is equivalent of a typical ceramic insulator, and its capacitance is 15 pF. The basic parameters of the antenna at hand are summarized in Table **2**. Optimal capacitive loads are given in Table **3**.

Frequency ratio of the variants 1-6 is two. As seen from Table **2**, at frequencies up to 30 MHz, the increase of antenna radius from 0.03 m to 0.15 m results in the minimum TWR growing approximately 1.5 times. The variation of radius has a slighter effect upon the minimal PF magnitude.

Fig. **9** shows the electrical characteristics of variants 1-3 as well as of a whip antenna with the same geometrical dimensions and the same capacitance C_i. The characteristics of the antenna with radius 0.15 m have a similar nature. As seen from the Figure, the TWR curve can have two maximums at high frequencies. The PF curve does not decrease monotonically with frequency, but has extremums, too.

Table 2: Main parameters antenna

Variant	Height L, m	Radius a, m	Numbers of capacitors N	Range width Δf, MHz	N_f	M	TWR min	PF min
1	12	0.03	9	7.5-15	16	4	0.123	0.819
2	12	0.03	9	15-30	16	4	0.273	0.610
3	12	0.03	9	30-60	16	5	0.360	0.562
4	12	0.15	9	7.5-15	16	4	0.205	0.813
5	12	0.15	9	15-30	16	5	0.414	0.680
6	12	0.15	9	30-60	16	4	0.380	0.605
7	12	0.03	9	8.5-13	10	3	0.217	0.870
8	12	0.03	9	13-22	10	5	0.216	0.790
9	12	0.03	9	22-60	20	8	0.204	0.437
10	12	0.15	9	8.5-13	10	5	0.314	0.892
11	12	0.15	9	13-22	10	4	0.278	0.859
12	12	0.15	9	22-60	20	5	0.322	0.565

Table 3: Optimal capacitive loads

Variant	Optimal capacitive load, pF								
	C_1	C_2	C_3	C_4	C_5	C_6	C_7	C_8	C_9
1	37.3	81.1	127	181	258	369	516	691	883
2	8.7	20.6	36.5	51.1	58.7	53.3	50.6	88.1	156
3	2.0	3.9	5.4	6.1	9.5	12.2	15.7	21.1	18.3
4	51.2	134	219	340	477	633	804	981	1150
5	10.7	28.4	57.0	76.7	86.4	86.8	53.5	216	409
6	4.5	19.0	11.4	15.8	26.4	29.8	32.4	23.0	35.7
7	33.9	72.8	115	164	223	296	385	492	608
8	8.4	18.4	30.3	40.4	47.1	53.3	82.8	151	248
9	1.7	5.6	12.2	11.7	17.6	41.0	30.5	56.4	240
10	21.1	0.2	39.2	231	519	909	1380	1900	2450
11	20.3	76.3	122	78.5	0.1	107	351	761	1340
12	2.9	26.9	0.3	42.1	22.6	59.1	35.8	55.0	73.1

In addition to calculated curves, the figures give the results of experimental verification, carried out on mock-ups at scale 1:5 (by dots and other symbols). The calculation and experiment agree well.

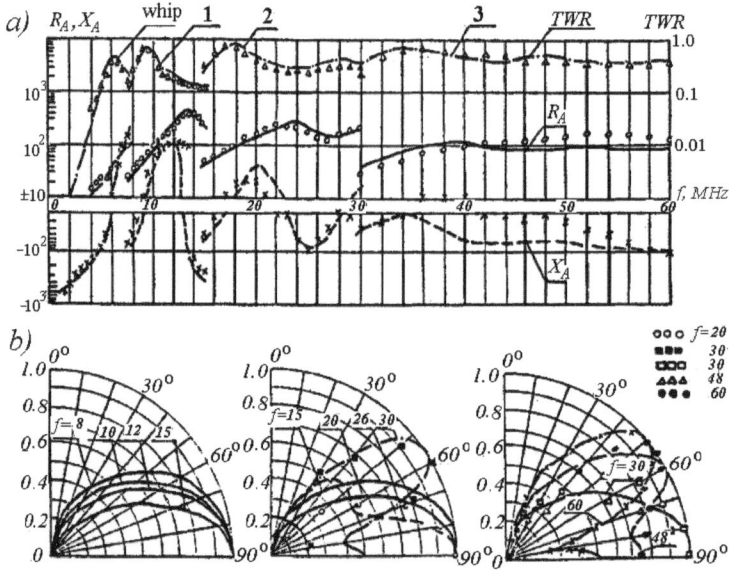

Figure 9: Input characteristics of 12-meter antennas of radius 0.03 m (*a*) and their patterns (*b*).

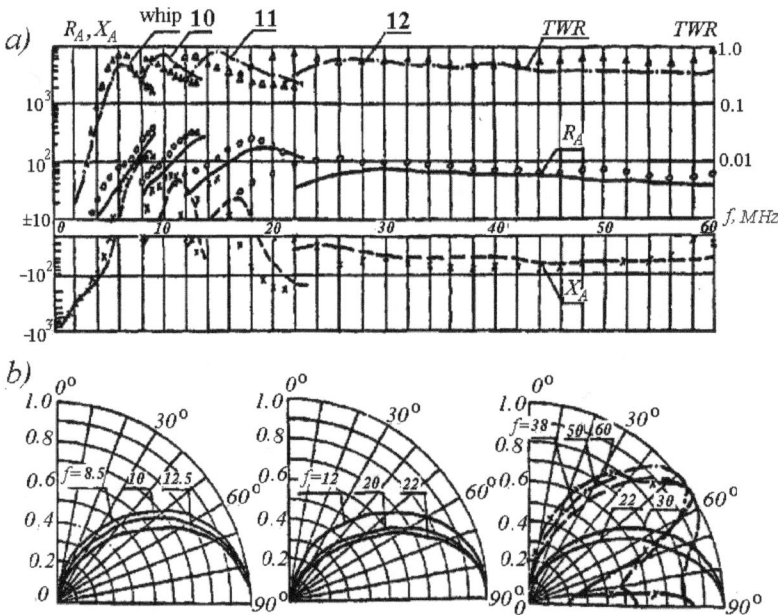

Figure 10: Input characteristics of 12-meter antennas of radius 0.15 m (*a*) and their patterns (*b*).

As follows from Table **2** (variants 1-6), at identical frequency ratios, the level of antenna matching with a cable in various ranges differs substantially and increases as the frequency grows. To obtain more uniform and better, on the whole, characteristics over the entire frequency range (at constant number of subranges), it is expedient to split it into such sections that the frequency ratio of a subrange should increase with growing frequency. The results of solving the problem are presented in Table **3** as variants 7-12.

The electrical characteristics of variants 10-12 as well as of a monopole of radius 0.15 m without loads (with capacitance C_i in the base) are given in Fig. **10**. Data of Table **3** confirm a general increase of the TWR level in comparison with variants 4-6. In all subranges, the larger antenna radius increase causes the minimum TWR level to become higher (approximately 1.5 times), together with PF level, especially at high frequencies.

The results of optimization of 12-meter antennas with capacitances $C_i = 15$ pF in the base are used to plot the curves for the minimum TWR values as a function of relative antenna length L/λ_{max} (λ_{max} is the maximum wavelength of the range) at various frequency ratios k_f and antenna radii a in Fig. **11**. These curves determine the maximum obtainable characteristics, which can be attained with the help of antennas with constant capacitive loads.

The calculation results show that, if necessary, the antenna range can be expanded in the direction of high frequencies at a sufficiently high TWR level, but the patterns in the additional (high-frequency) range deteriorate substantially. In this connection, the frequency ratio of an antenna with capacitive loads does not exceed 3 (at PF ≥ 0.5 and TWR≥ 0.2).

Thus, the synthesis software is an efficient method of optimization of antennas with capacitive loads. The softwear can be used to optimize antennas with loads of other kinds. It can also be applied to find loads ensuring the required current distribution along the radiator.

6.4.   CREATION   OF   A   DESIRED   CURRENT   DISTRIBUTION. WEAKENING OF THE SUPERSTRUCTURE EFFECT

When synthesizing a wideband radiator with loads by the mathematical programming method, the values calculated by the method of an impedance long line are taken as the initial ones for the loads. One can use the method also to choose the initial values of the loads in the synthesis of a radiator with given current distribution. In addition to it, another approximate method of calculating the value of loads is known. With the help of the method, given current distribution is used to find the law of increase of the equivalent length of a line along the latter, which allows calculating the loading impedance of each section. The use of the method of a long line with loads and the method of an impedance long line yields similar results.

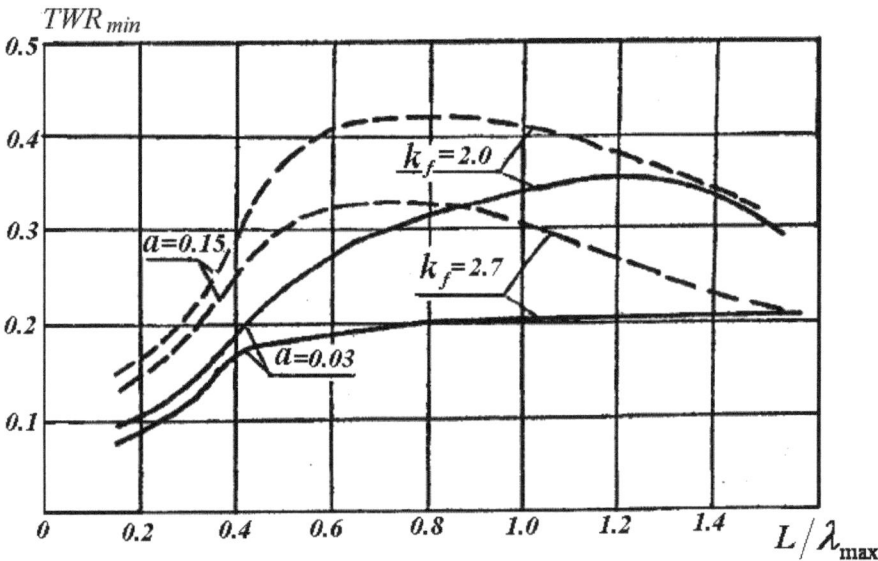

Figure 11: The maximal obtainable matching level of an antenna loaded by constant capacitances.

If loads are uniformly distributed along an antenna at small distance b from each other ($kb \ll 1$), the method of a long line with loads allows arriving at the following expression for loading impedance (see Fig. **6b**):

$$Z_n = -j\frac{W}{kb}\left\{ \frac{f[(N-n-1)b]}{f[(N-n)b]} - 2 + (1+k^2b^2)\frac{f[(N-n+1)b]}{f[(N-n)b]} \right\}, \tag{6.34}$$

where W is the wave impedance of an equivalent long line, and $f(z)$ is the function of the current distribution along it, which is assumed to be real-valued - see (6.4).

Expression (6.34) is derived by equating current $J(z_n)$ at the beginning and the end of each line segment to given current $J_A f(z)$. Since the segment lengths are small, one can limit oneself to the first terms of series expansion for trigonometric functions. At exponential distribution described by (6.10):

$$Z_n = 1/(j\omega C_n),$$ **(6.35)**

where

$$C_n = b\left(1 - e^{-nab}\right)/\left[2Wc\left(\cosh ab - 1\right)\right].$$

Here, as seen from the last formula, the sign of C_n coincides with that of α. Thus, from the method of an equivalent line with loads it follows also that, if the exponential current distribution (with decrement α that is not small) is to be obtained, loads should be of the capacitive nature. At that the capacitors allow creating only a concave current distribution ($\alpha > 0$). In order to obtain a convex curve for the current amplitude, the capacitances should be negative.

The capacitance values found by the method of an equivalent long line with loads were used as a zero approximation for the synthesis of an antenna with given current distribution by the mathematical programming method. The calculation was carried out for the antenna of height 6 m with 10 capacitive loads described in Section 6.2. The synthesis results in frequency range 40-80 MHz are given in Fig. **12**.

The calculation assumed $N_f = 9$, $N_l = 11$; the zero approximation is found at frequency $f = 60$. As seen from the Figure, the obtained distribution is, on the whole, close to the given one, but is not identical to it.

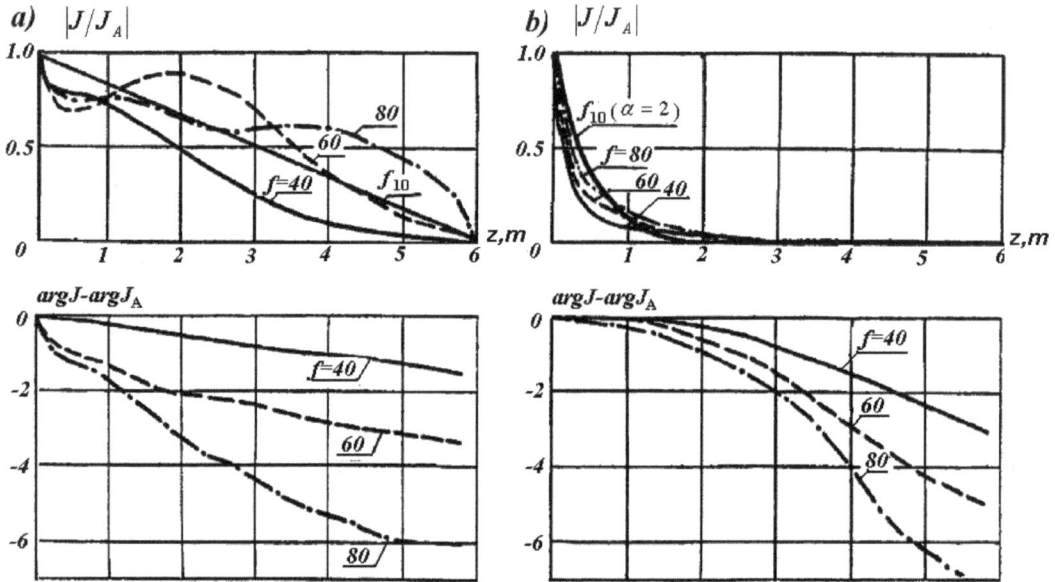

Figure 12: Amplitude and phase of the current along 6-meter antenna: distribution close to linear one (*a*) and to exponential one with $\alpha = 2$ (*b*).

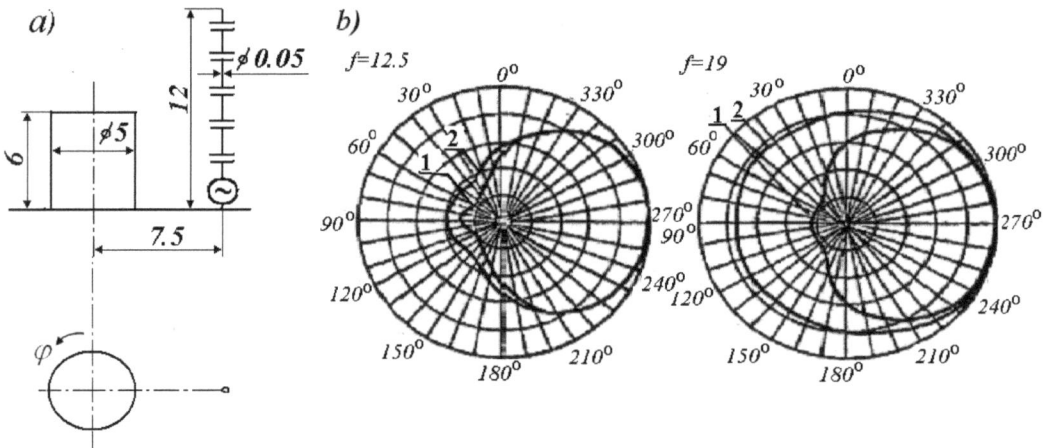

Figure 13: An antenna near a superstructure (*a*) and its pattern in the horizontal plane (*b*).

The use of loads gives also freedom in choosing the antenna length (with allowance for design options), since it permits securing the required characteristics in the necessary frequency range at given antenna length. The freedom in choosing the radiator length enables weakening the effect of adjacent metal structures, *e.g.*, of superstructures, upon the pattern of an antenna or an antenna array. Fig. **13** shows the calculation results for the pattern of a monopole situated near a metal superstructure shaped as a circular cylinder of finite length. The pattern in the horizontal plane is calculated at two frequencies of HF range.

Two variants of monopoles considered are: (1) a monopole of height 6 m and diameter 0.016 m without loads, (2) a monopole of height 12 m and diameter 0.06 m with 9 capacitive loads selected with the view to ensure the optimal electrical characteristics at frequencies from 8 to 22 MHz. The relative position of the superstructure and the monopole as well as the superstructure dimensions are shown in Fig. **13a**. The calculation had the circular cylinder replaced with a wire structure of equidistant conductors located along cylinder generatrices and the radii of its end surface. As is seen from Fig. **13b**, the radiation of an ordinary monopole in the direction of superstructure falls off sharply, and the use of a monopole with loads allows lessening this effect.

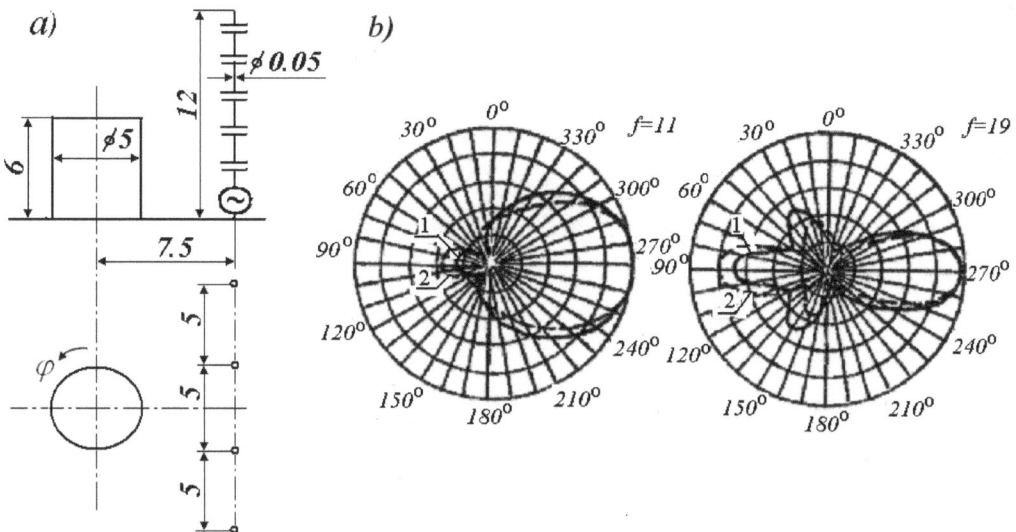

Figure 14: A linear array near a superstructure (*a*) and its pattern in the horizontal plane (*b*).

Fig. **14** shows similar results for a uniform linear array situated near a superstructure. The same two variants of monopoles are adopted as array radiators. The relative position of the superstructure and radiators as well as the superstructure dimensions are shown in Figure, the phase shift between radiators assumed zero. The calculation results show that in the upper part of the frequency range, the superstructure effect upon the directivity pattern of antenna array of monopoles without loads is slighter than upon that of an isolated monopole. This is, apparently, related to the fact that the superstructure does not hinder the propagation of electromagnetic waves from side radiators. Nevertheless, the use of monopoles with loads allows in this case, too, to weaken the superstructure effect and to increase the signal in its direction.

6.5. CAPACITIVE LOADS IN V-DIPOLES

Capacitive loads can also be used to improve the electrical characteristics of a directed V-antenna. At arm length L greater than 0.7λ, an ordinary V-antenna becomes a directed one with preferential radiation along the bisector of the opening angle. However, with growing frequency, the main lobe of the pattern in the antenna plane splits, and the radiation along the bisector decreases. Connection of capacitive loads allows expanding the frequency range, where the directed radiation along the bisector takes place, and increasing the antenna directivity in this direction.

Consider a symmetric V-dipole with arm length L and arbitrary angular aperture $\alpha = \pi - 2\theta$ (Fig. **15**). The far field along the bisector of the angular aperture, which is created by an elementary segment $d\zeta$ of the upper antenna arm with current $J(\varsigma)$, is

$$E_\theta(\varsigma)d\varsigma = E_\theta(0)\left[J(\varsigma)/J(0)\right]\exp\left(jk\varsigma\sin\theta_0\right)d\varsigma, \tag{6.36}$$

where $E_\theta(0)d\varsigma$ is the field created by a segment of the upper arm located near point O with current $J(0)$, $k\varsigma\sin\theta_0$ is the path-length difference, ζ is the coordinate counted off along the radiator axis.

For the far fields of different segments of the dipole upper arm to coincide with each other in phase, the current distribution along this arm should conform to expression

$$J(\varsigma) = J(0)f(\varsigma)\exp\left(-jk\varsigma\sin\theta_0\right), \tag{6.37}$$

where $f(\varsigma)$ is a real and positive function.

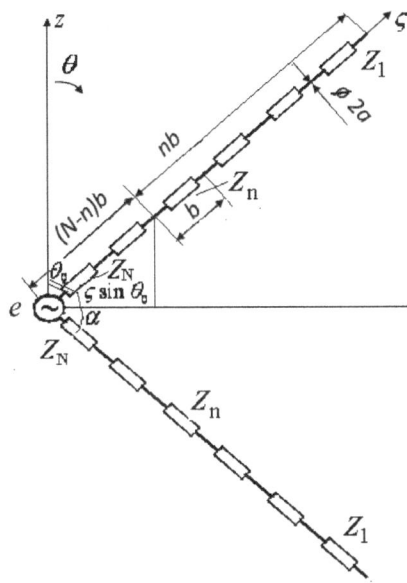

Figure 15: V-dipole with loads.

Let N loads Z_n be connected in each dipole arm, and they connected uniformly along the antenna wire at distance b from each other. If the load spacing is small ($kb \ll 1$), then, as well as for the linear dipole, the current distribution along the antenna undergoes practically no change at replacement of concentrated loads with distributed surface impedance $Z(\varsigma)$. Assume the surface impedance of each antenna segment with load Z_n constant and equal to $Z^{(n)}$.

As is said in Section 1.4, the current distribution along the antenna with piecewise constant surface impedance coincides in the first approximation with that along an open-ended piecewise uniform long line, *i.e.,* the line with stepwise variation of propagation constant (see Fig. **6b**). Here, wave propagation constant γ_n of the line nth segment is related to surface impedance $Z^{(n)}$ by (6.2). If the variation law of propagation constant, which allows ensuring the required current distribution (6.37), is known, one can use quantity γ_n to calculate, with the help of (6.17), concentrated loads Z_n needed for that.

The current along the nth segment of a stepped line is

$$J(\varsigma_n) = I_n \sinh(\gamma_n \varsigma_n + \phi_n), 0 \le \varsigma_n \le b, \tag{6.38}$$

where I_n and ϕ_n are the current at the antinodal point and its spatial phase at the nth segment, $\varsigma_n = (N-n+1)b-\varsigma$ is the coordinate counted off from the segment end. Equate current $J(\varsigma_n)$ at the beginning and the end of each line segment to current $J(\varsigma)$ ensuring the phase coincidence of far fields from all dipole segments. The current inside each line segment does not coincide with current $J(\varsigma)$. However, if the segment lengths are small, the current distribution along the line is close to $J(\varsigma)$.

According to (6.37) and (6.38), at $\varsigma_n = b$ and $\varsigma_n = 0$:

$$I_n \sinh(\gamma_n b+\phi_n) = J(0) f\left[(N-n)b\right]\exp\left[-jk(N-n)b\sin\theta_0\right],$$

$$I_n \sinh\phi_n = J(0) f\left[(N-n+1)b\right]\exp\left[-jk(N-n+1)b\sin\theta_0\right].$$

Divide the left- and right-hand sides of the first equation into the corresponding sides of the second one. Considering b a small quantity and retaining only the first terms of series expansions for trigonometric functions of small arguments, we get

$$\tanh\phi_n = \gamma_n b \bigg/ \left\{ \frac{f[(N-b)]}{f[(N-n+1)b]}(1+jkb\sin\theta_0)-1\right\}. \tag{6.39}$$

In similarity to (6.39), for the $(n+1)th$ segment:

$$\tanh\phi_{n+1} = \gamma_{n+1} b \bigg/ \left\{ \frac{f[(N-n-1)b]}{f[(N-n)b]}(1+jkb\sin\theta_0)-1\right\}. \tag{6.40}$$

Voltage and current are continuous along a stepped line. Therefore, (6.17) is valid. Together with (6.39) and (6.40), it forms a set of equations that allows relating γ_n and γ_{n+1}. It follows from its solution that quantity γ_n is independent of γ_{n+1}:

$$\gamma_n = \frac{1}{b}\sqrt{1-\frac{2f\left[(N-n)b-f\left[(N-n-1)b\right]\right]}{f\left[(N-n+1)b\right]}-2jkb\sin\theta_0\frac{f\left[(N-n)b\right]-f\left[(N-n-1)b\right]}{f\left[(N-n+1)b\right]}}. \tag{6.41}$$

This expression generalizes (6.8) for a linear dipole and transforms into it at $\theta_0 = 0$.

The possibility of implementation of propagation constant γ_n is determined by that of concentrated loads. According to (6.17), at low frequencies, when inequality (6.20) and equality (6.21) that follows from it are valid, the load value is

$$Z_n = -j30\left(\gamma_n b\right)^2 \big/ \left(kb\chi\right).\tag{6.42}$$

By substituting (6.41) into (6.42), we get

$$Z_n = R_n + 1\big/\left(j\omega C_n\right),\tag{6.43}$$

where

$$R_n = \frac{60}{\chi}\sin\theta_0\,\frac{f[(N-n-1)b]-f[(N-n)b]}{f[(N-n+1)b]},$$

$$C_n = 4\pi\varepsilon_0 b\chi \Big/ \left\{1 - \frac{2f[(N-n)b]-f[(N-n-1)b]}{f[(N-n+1)b]}\right\}.$$

As seen from (6.43), each load should be a series connection of a resistor and a capacitor, with the resistor resistance positive, if function $f(\varsigma)$ decreases monotonically with growing ς, and the capacitor capacitance positive, if function $f(\varsigma)$ is concave. Here, the resistance depends on the angular aperture of the antenna and the form of function $f(\varsigma)$, whereas the capacitance – on the latter only.

For a linear radiator with loads ensuring the maximum radiation in the plane, perpendicular to its axis, each load should, when condition (6.20) holds, represent a capacitor. Capacitors ensure real wave propagation constant γ_n and an in-phase current distribution along an antenna. For a V-dipole, a resistor is to be connected in series with a capacitor, which will result in a phase delay of the current wave along an antenna wire. Such phase delay is necessary for a V-dipole, as it compensates the path-length difference from individual dipole segments to an observation point and ensures phase coincidence of the far fields of the segments along the bisector of the angular aperture.

The use of resistors in a transmitting antenna is inexpedient. This means that the loads of a V-dipole should not differ from those of a linear radiator, which ensure an in-phase current distribution along a wire.

At high frequencies, when condition (6.20) is invalid, the in-phase current distribution along a linear radiator takes place, if the load represents a negative inductance (a capacitor with the capacitance varying in inverse proportion to frequency squared). Similarly, the load for a V-dipole should be a series connection of a capacitor with a frequency-dependent capacitance and a resistor. For propagation constant to be real and the current distribution along an antenna to be the in-phase one, the capacitances should not exceed the value determined by inequality (6.28).

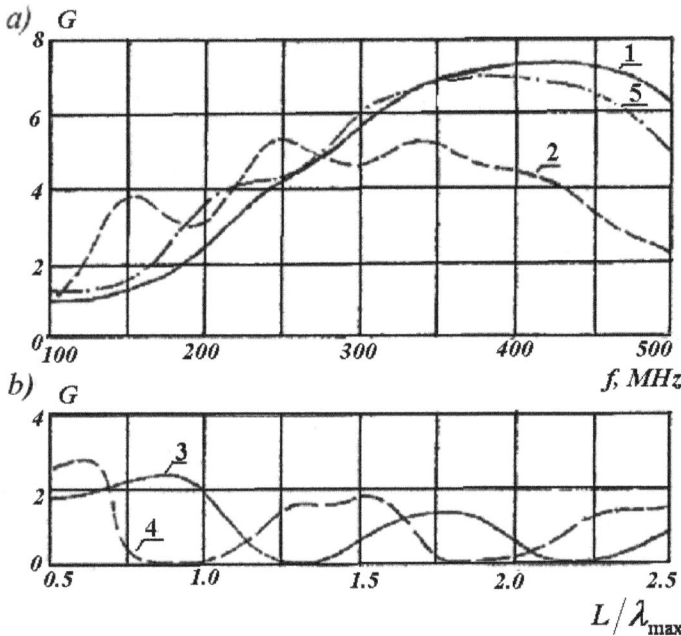

Figure 16: Directivity of a V-dipole (*a*) and of a linear dipole (*b*).

As an example, we shall consider a V-dipole with arm length $L = 1.5$ m and radius 0.025 m. Fifteen capacitors are connected in each arm with spacing 0.1 m from each other (the first and last one are 0.05 m distant from the end and center of the antenna). The capacitances of the capacitors nearest to the generator are chosen to be 33.5 pF. Then, the wave propagation constant remains real at frequencies up to

100 MHz. The capacitances of other capacitors decrease to the antenna end according to the linear law. As shown in section 6.1, a current amplitude distribution along an antenna arm, which is close to linear one, and high matching level are ensured in this case.

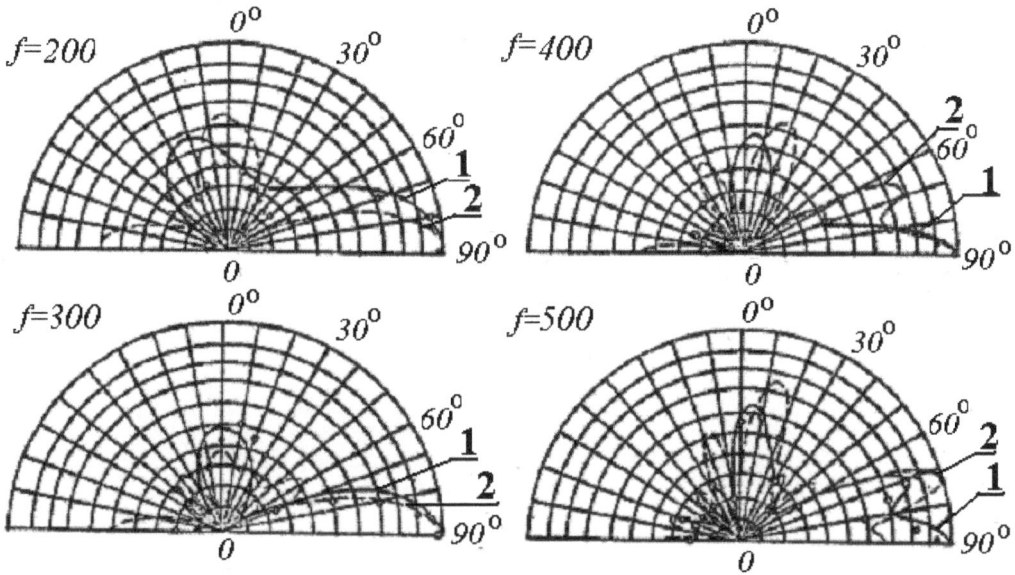

Figure 17: Patterns of V-dipole in the antenna plane.

Fig. **16a** shows the directivity of V-dipole with capacitive loads (curve 1) and without loads (curve 2) along the bisector of the angular aperture (the angle between the dipole arms is 90°). For the sake of comparison, Fig. **16b** shows similar curves for a linear dipole (curve 3 with loads, curve 4 without loads). The values of loads and the antenna arm length are the same. The calculations are performed in a frequency range from 100 to 500 MHz.

As seen from the Figure, the directivity of a linear dipole without loads in the direction, perpendicular to its axis, quickly decreases at $L \approx (0.6–0.7)\lambda$. For a linear dipole with loads, this threshold value is found as $L \approx (1–1.2)\lambda$. V-dipoles, especially with loads, allow achieving a high directivity along the bisector of the angular aperture in a substantially wider frequency range: at frequencies from 350 ($L = 1.75\lambda$) to 500 MHz ($L = 2.5\lambda$), loads raise the V-dipole directivity by a factor between 1.4 and 2.8.

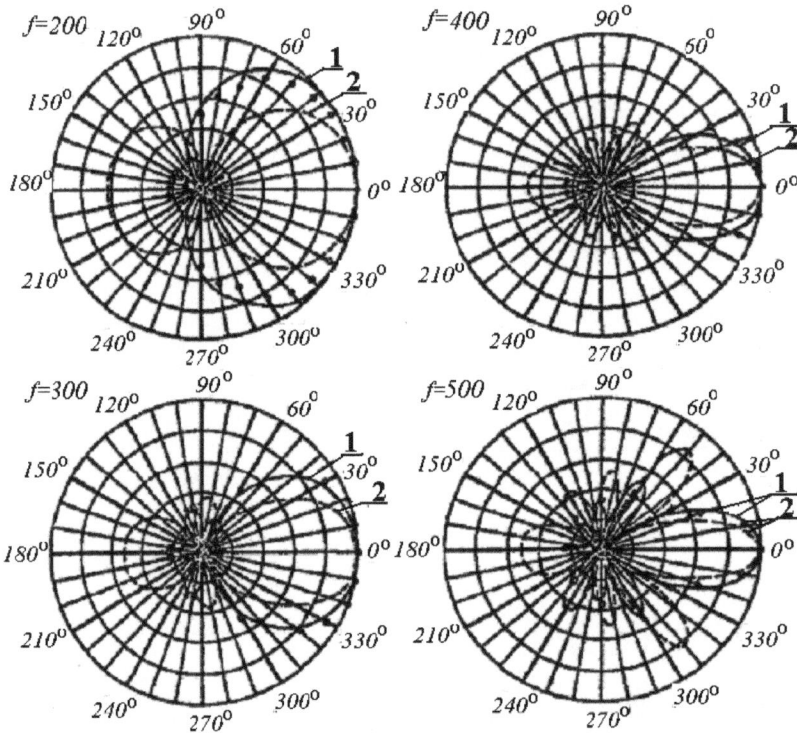

Figure 18: Patterns of V-dipole in the plane, perpendicular to the antenna plane.

Figs. **17** and **18** give the patterns of the described V-dipole with loads (curve 1) and without loads (curve 2) in the antenna plane and the plane perpendicular to it. The calculated curves are compared with the results of experiment (circles).

As shown in Section 3.3, electrical characteristics of an antenna with loads can be improved by optimal selection of the latter with the help of the mathematical programming method. Here, to calculate the zero approximation, it is expedient to use the method of a stepped (piecewise uniform) impedance long line. In the considered example, the mathematical programming method allows raising the directivity level of V-dipole with loads in the lower part of a frequency range (Fig. **16a**, curve 5).

A V-dipole with capacitive loads can be used as a directional antenna of VHF range.

The Theory of Thin Antennas and Its Use in Antenna Engineering, 2013, 197-229   197

CHAPTER 7

Compensation Problem

Abstract: Issues of applying the compensation method application are studied, including the heterogeneous nature of the media, optimal placement of main and auxiliary radiators, the dark spot dimensions, the reduction factor and variation of the total and local SAR, retention of patterns and field strength. Results of calculations, performed by different methods, are compared with each other and with experiment. A technique of protecting system against external action is described; the automatic adjustment circuit is developed. Recommendations for using the compensation method in a wide frequency range are given.

Keywords: Additional radiator, Automatic adjustment circuit, Compensation method, Dark spot, Dark spot boundary, Employment of reflectors, Equivalent relative permittivity, External actions, Heterogeneous media, Higher efficiency of fields retention, Homogeneous medium, Low efficiency of the variant of currents, Main radiator, Proximity of metallic bodies, Radiation pattern, Radiators placement at equal distances from the compensation point, Reduction factor, Retention of driving currents, Retention of radiators fields, Total and maximum local SAR, Use of two additional radiators, Wideband field compensation.

7.1. DIMENSIONS OF DARK SPOT AND THE FACTOR OF LOSS REDUCTION

Application of the compensation method requires analyzing its efficiency. To this end, one needs to calculate the field in the space surrounding a radiator and to estimate the irradiation reduction factor. The problem is complicated due to the heterogeneous nature of the medium. For example, cellular handset antennas are in close proximity to the user's head, hand and body, which consist of multiple tissues with different permittivity, much higher than that of free space. Since the field strength and hence the dissipated power are at their maximum near the antenna, the capability to calculate correctly the near field of the antenna with due account of the user's body's presence is crucial for accurate results.

Consider a linear antenna tangent to the user's head that is modeled with a vertical prolate ellipsoid (see Fig. **1a**). One may treat the structure as a linear radiator situated along the (plane) boundary between two half-spaces: air with $\varepsilon_r=1$ and the head with $\varepsilon_r \neq 1$, as shown in Fig. **1b**. It is a crude approximation, because the head dimensions are on the order of the wavelength.

The problem of finding the electromagnetic field of a linear radiator located along the boundary between two media reduces to calculation of the electrostatic field in a heterogeneous (or more exactly, piecewise-homogenous) medium [47]. Such problem arises, in particular, if an isolated wire is located at the interface of two media, *e.g.*, air and a dielectric medium with $\varepsilon_r \neq 1$, as shown in Fig. **1c**.

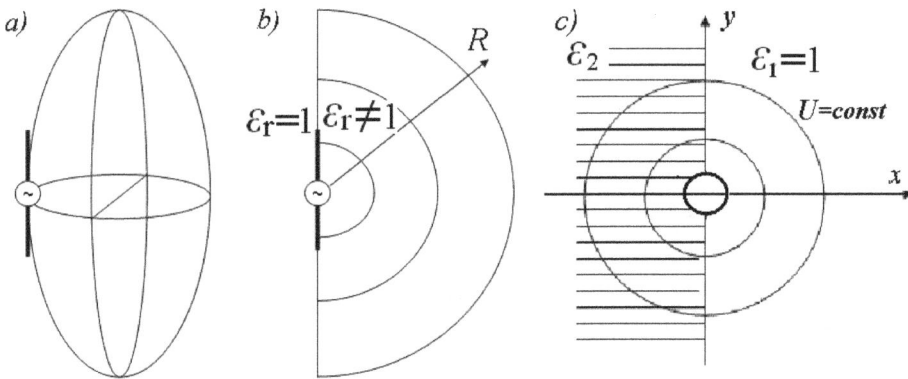

Figure 1: Model for the head and an adjacent linear antenna (*a*), the latter at the interface between two half-spaces (*b*), electrostatic field of a charge in a heterogeneous medium (*c*).

As said in Section 4.5, the known solutions relating to electrostatic field of charged conducting bodies can be used to solving similar, yet more complicated problems. Employing the correspondence principle, one can conveniently use the electric field of a line charge to find the magnetic field produced by a constant linear current, provided that the current and the line charge have the same spatial distributions. Furthermore, the structure of the quasi-stationary electrical field of alternating linear currents is known to be similar to that of the magnetic field created by constant linear currents. As known from the antenna theory, in the first order approximation, the near field of an antenna is a quasi-stationary nature.

So, if the electrostatic field of charged conducting bodies is known, one can determine the structure of the antenna near field. This analogy with the electrostatic problem allows reducing the calculation of the antenna near field in a piecewise-homogenous medium to an approximate field calculation in a homogenous medium. But, the near field of an antenna is of quasi-stationary nature in the first order approximation only, and so the results must be verified by a simulation tool such as the CST program.

The solution for the electrostatic field in a heterogeneous medium shown in Fig. **1c**, where an isolated charged straight wire is situated at the interface of two homogeneous media, is based on the fact that the potential on the wire surface is constant and the potential at a great distance from a charged wire is the same in all directions. It is natural to assume the equipotential lines to be circumferences around the wire, and the equipotential surfaces are cylinders. As shown in [48], the field in a piecewise-homogenous medium (from two homogeneous media) coincides with that in a homogenous medium with equivalent relative permittivity

$$\varepsilon_e = \left(1 + \varepsilon_r\right)/2 .$$ (7.1)

If the dielectric medium is homogeneous (the permittivity is constant) the equivalent medium is also homogeneous, and the field in the piecewise-homogenous medium coincides with that in a homogenous medium, which has equivalent relative permittivity ε_e. Thus, starting from the correspondence principle, one can show the expressions for the field components of a linear radiator located at the boundary of two media to differ from those relating to a homogenous medium only with the relative permittivity quantity.

It is known from the literature that the relative permittivity of the brain, muscles and skin has similar values and on the average is about 40-50, with only the permittivity of bones differing substantially from the value. Accordingly, equivalent permittivity ε_e is close to 20-25, and the field values in the vicinity of an antenna located near the head can be calculated in a homogenous medium with ε_e. Therefore, taking into consideration the quasi-stationary structure of the near field of the antenna, the irradiation reduction factor can be obtained in the first order approximation by calculating the field strength under the assumption of a homogenous medium.

If the distance from the antenna to the head is not zero, yet small in comparison with the wavelength, one can consider the field in the equivalent homogeneous medium as the first approximation to the field inside the head. Here, the field close to the antenna differs from that field in the free space. At the interface, the field changes its strength, since it depends on ε_r, and the field component magnitudes inside the head are weaker then outside it.

The analysis of a heterogeneous medium using an equivalent homogeneous medium simplifies substantially the problem of calculating the irradiation power and the reduction factor. The obtained results are verified with the program that bases on the Moment Method and allows taking in account detailed characteristics, including heterogeneity of a medium.

Section 3.4 dealt with the impact of the mutual coupling between radiators situated in the near region of each other on the input impedance, current distribution and fields of radiators. The mutual coupling between the radiators was shown to affect the current distribution along each radiator in such a way that the currents include anti-phased components in addition to the in-phase components. For equal radius wires, the in-phase and the anti-phased currents of the first radiator are

$$J_{A1}^{(i)}(z) = (e_1 + e_2)Y_2 \frac{\sin k(L - |z|)}{\sin kL}, \quad J_{A1}^{(a)}(z) = (e_1 - e_2)Y_1 \frac{\sin k(L - |z|)}{\sin kL}. \tag{7.2}$$

In accordance with (1.31), the total field of the first radiator in plane $\varphi = 0$ is given by

$$E_{z1}(z) = -j\,30 F_1 / \varepsilon_r \left[(e_1 + e_2)Y_2 + (e_1 - e_2)Y_1 \right]. \tag{7.3}$$

The total field of the second radiator in plane $\varphi = 0$ is

$$E_{z2}(z) = -j\,30 F_2 / \varepsilon_r \left[(e_1 + e_2)Y_2 + (e_2 - e_1)Y_1 \right]. \tag{7.4}$$

Here,

$$F_m = \frac{1}{\sin kL} \left[\exp(-jkR_{m1})/R_{m1} + \exp(-jkR_{m2})/R_{m2} - 2\cos kL \exp(-jkR_{m0}/R_{m0}) \right],$$

where m is the radiator number, $R_{m1} = \sqrt{(z-L)^2 + \rho_m^2}$, $\quad R_{m2} = \sqrt{(z+L)^2 + \rho_m^2}$, $R_{m0} = \sqrt{z^2 + \rho_m^2}$.

In order to bring the total field to zero at compensation point ($\rho_0 + b, 0, 0$), i.e., that $E_z(0) = E_{z1}(0) + E_{z2}(0) = 0$, the ratio of emf's feeding the radiators must be

$$\frac{e_2}{e_1} = -\frac{(Y_2 + Y_1)F_{10} + (Y_2 - Y_1)F_{20}}{(Y_2 - Y_1)F_{10} + (Y_1 + Y_2)F_{20}}. \tag{7.5}$$

Here, F_{m0} is the value of function F_m at the compensation point. Equation (7.5) allows finding the voltage amplitude and phase at the input of the second radiator, if those amplitude and phase of the first radiator are known. Using (3.42), we find:

$$J_{A1}/e_1 = Y_1 + Y_2 + e_2(Y_2 - Y_1)/e_1, \quad J_{A2}/e_1 = Y_2 - Y_1 + e_2(Y_1 + Y_2)/e_1.$$

Substituting (7.5) in these expressions, we obtain the expression for the driving currents ratio

$$J_{A2}/J_{A1} = -F_{10}/F_{20}. \tag{7.6}$$

The expression coincides with the expression in the absence of mutual coupling between the radiators. This occurs when the radiators are of the same length. For $L = \lambda/4$, we obtain, in the first approximation

$$F_{10} = \exp(-jkR_{10})/R_{10}, \quad F_{20} = \exp(-jkR_{20})/R_{20},$$

i.e.,

$$J_{A2}/J_{A1} = -R_{20}\exp\left[-jk(R_{10} - R_{20})\right]/R_{10}, \tag{7.7}$$

where $R_{20} = \sqrt{L^2 + \rho_0^2}, R_{10} = \sqrt{L^2 + (\rho_0 + b)^2}$.

As it follows from the above expressions, the dipole moments of linear radiators must be in inverse proportion to the distances between the radiators and the compensation point. The dark spot dimensions for dipoles of finite length can be calculated using equations similar to (7.3)-(7.4), written for the field of such radiators in the cylindrical coordinates. The dark spot boundary is determined by radius ρ_n, at which $n=|E_z/E_{z10}|$ is equal to a given value, e.g., $n=0.1$, where E_z is the total field and E_{z10} is the field of the main radiator.

To simplify, assume the electric field strength to increase linearly from compensation point $t=0$ to dark spot boundary $t=t_0$ so that $|E_z/E_{z10}|=ns$, where $s=t/t_0$, n=const(s), and hence the loss power exhibits a quadratic growth with the distance. The total losses power inside the spot is given by $P=P_0\int (ns)^2 ds =P_0 n^2/3$, where P_0 is the loss power due to the main radiator in the absence of the auxiliary radiator, within the limits of spot. For $n_0=0.2$, the loss power is smaller by a factor of $3/(0.2)^2 =75$ (or 18.8 dB). For $n_0=0.1$, the reduction factor is 300 (24.8 dB), and for $n_0=0.04$, the factor is 1875 (32.7 dB). In practice, the factor is roughly a half of the calculated value, since the field within the dark spot has a more complicated structure, i.e., the reduction factor is approximately equal to $1.5/n_0^2$.

To find the boundary surface of the dark spot, one must determine quantity ρ_n as a function of z and φ. In Fig. **2**, field ratio $n=|E_z/E_{z10}|$ is plotted as a function of P, assuming $\lambda=30$, $L=7.5$, $\rho_0 =1$ for different values of b (all dimensions are in centimeters). The field magnitudes E_{z10} and E_z are calculated at $z=0,\varphi=0$ by the above formulas.

Table 1: The dark spot length for different values of b and n_0

b	$n_0=0.01$	0.02	0.04	0.07
0.5	3.53	6.80		
1	1.89	3.89	7.67	
2	1.00	2.00	4.17	8.12
3	0.71	1.42	2.91	5.43
b	$n_0=0.07$	0.10	0.14	0.25
4	4.28	6.58	11.38	
8	2.89	4.29	6.43	49.60
16	2.44	3.56	5.24	23.63
32	2.29	3.35	4.88	11.73

a) $|E_z/E_{z10}|$

b) $|E_z/E_{z10}|$

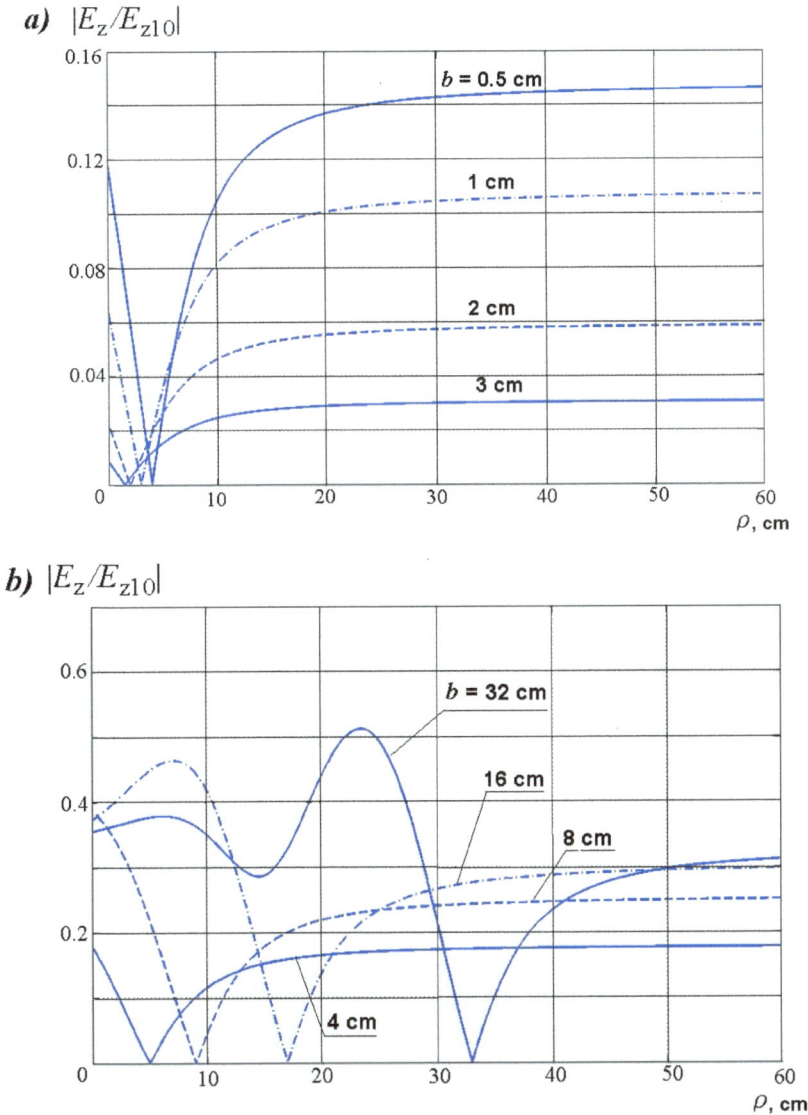

Figure 2: Variation of $n = |E_z/E_{z10}|$ along the ρ-axis.

The boundaries of the dark spot in the ρ-direction, denoted as points ρ_1 and ρ_2, are found as the intersection points of curve n with given level n_0. Along the ρ_1 ρ_2 segment between the points ρ_1 and ρ_2, n is smaller than the required value of n_0. Length $\Delta\rho=\rho_2-\rho_1$ of the segment, *i.e.*, the dark spot length, is presented in Table **1** for different values of b and n_0 (all dimensions are in centimeters), *i.e.*, for different levels of field reduction at the dark spot boundary. One can see from

Table **1** that the dark spot length for given level n_0 decreases as distance b between radiators increases. But at large distances, when n_0 at the dark spot boundary increases, the dark spot length increases also. Therefore, for small values of b, the dip of curve n becomes narrower and deeper; for great values of b, the dip of curve n is wider.

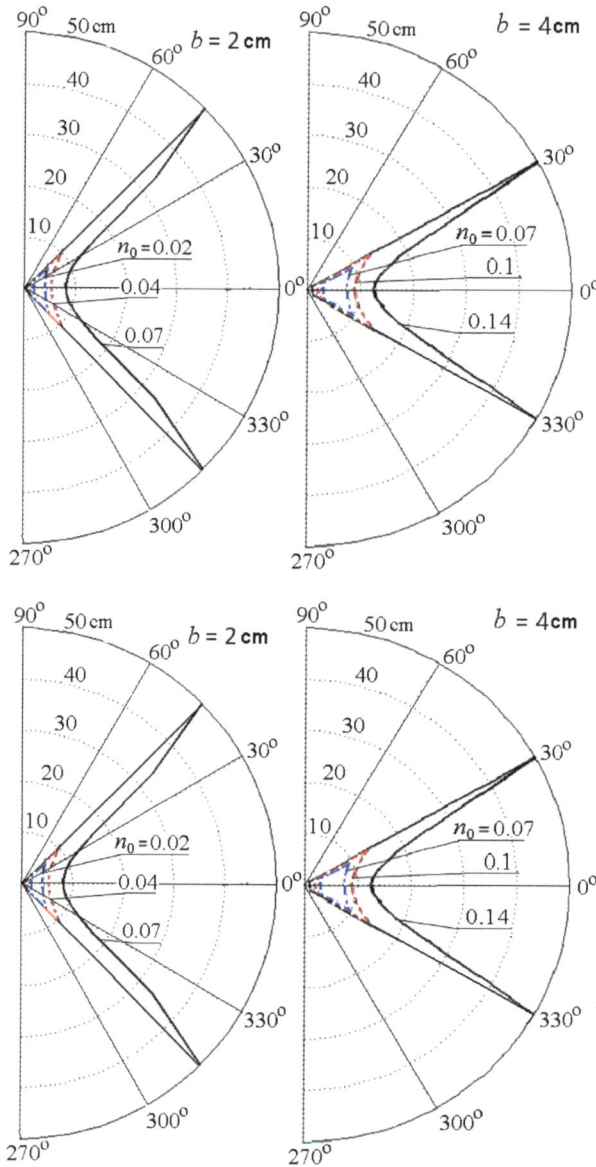

Figure 3: Dark spot boundaries in the horizontal plane.

Fig. **3** shows a few examples of the dark spot boundaries in the horizontal plane ($z = 0$) for $f{=}1$ GHz, $\rho_0 = 1$cm and different values of b and n_0. In the figure, fields E_{z10} and E_z are calculated using equations (1.31) at $z = 0$ and given angles φ, in the manner similar to the preceding case. The difference between coordinates ρ_1 and ρ_2 denoting the start and end of the corresponding straight line segment with $n \leq n_0$ determines the dark spot length in the given direction.

Fig. **3** and Table **1** show that it is sufficient on ultrahigh frequencies to separate the radiators by a small distance of the order of several centimeters. Since the auxiliary radiator power is significantly smaller than that of the main antenna, the field of the main radiator predominates in the far region, *i.e.,* the total field is only slightly different from that of the main radiator. It is seen from Fig. **3** that the dark spot width is, as a rule, greater than its length. The spot height is close to its length. This result is particularly important as it shows the head movements, in which direction are more critical in terms of the growth of the absorbed power.

One can use the method in order to plot the dark spot boundaries in the vertical plane too.

The calculations also show that at the optimal design of the structure in terms of the relative positions of the radiators and the compensation point, the volume of the dark spot increases, and the field inside this spot decreases substantially. One can choose the structure parameters so that the spot dimensions would coincide with the human head dimensions, resulting in a drastic reduction of loss power in the head, and in a high tolerance with respect to the user's head movements. Employing several auxiliary radiators allows increasing the dark spot volume further.

7.2. RADIATION PATTERN

The use of an additional radiator in the compensation method involves a contradictory relation between the requirements for field reduction in the near region and for retention of the circular pattern in the horizontal plane. Retention of the far field strength and pattern is a crucial issue for any method of SAR reduction. In order to retain the pattern shape and, in particular, to avoid a deep dip in the radiation pattern along the structure axis, the phases of the far fields created by the two radiators should not differ by more than a few degrees. The small distance between the antennas and the close values of their driving current phases ensure the preservation of the circular shape of the radiation pattern. Yet, in order to bring the total field at the compensation point to zero, the fields of radiators at the point must have opposite phases.

If the driving current of the main radiator is J_{A1}, the driving current of the auxiliary radiator can be written as $J_{A2} = J_{A1} D e^{-j\psi}$, where D is the ratio of the radiators' dipole moments, and ψ is the phase difference between the driving currents. In this case, the field of the auxiliary radiator is $E_{z2} = E_{z1} D e^{ju}$, where E_{z1} is the field of the main radiator, and $u = kl - \psi$. Here l is the path difference between rays from the radiators to a point located at angle φ (see Fig. **9**).

The total pattern in the horizontal plane can be written as

$$|F| = |1 + D \exp(ju)| \Big/ \sqrt{(1 + D \cos u_m)^2 + D^2 \sin^2 u_m} \ . \tag{7.8}$$

Here u_m is the value of u at the maximal of the denominator. Since D and u_m are constant, the denominator of (7.8) is independent of φ. If the main radiator pattern is a circular one, the ratio of the total pattern maximum to the minimum, a measure of the pattern distortion level, is

$$\frac{|F_{max}|}{|F_{min}|} = \frac{\left|1 + D e^{ju}\right|_{max}}{\left|1 + D e^{ju}\right|_{min}} = \frac{1 \pm D}{1 \mp D}, \tag{7.9}$$

where the upper sign applies when D is positive, and the lower sign applies if D is negative. It is easy to show that when the pattern maximum and minimum differ by 6 dB (deviation of 3 dB from the average level), we obtain $|D| = 0.33$, and when the difference between the maximum and the minimum is 3 dB, we have $|D| = 0.17$.

It is necessary to explain here that cellular communication is not the only area of application where one must create a weak field region near the transmitting antenna. The task is vital, if, for example, a mobile transmitter is located in a vehicle close to users and other passengers. Creating a weak field region in such cases is an efficient technique of protecting against irradiation. A problem is often complicated by a transmitter operation in a broad frequency band. The problem is considered in Section 7.4. It is essential for the analysis of radiation pattern that habitually there is some degree of freedom to choose the antenna location. This helps to keep the pattern undistorted.

From (7.7), taking into consideration that $R_1 = \rho_0 + b$, $R_2 = \rho_0$ (see Fig. **9**), we obtain $D = R_2/R_1 = \rho_0/(\rho_0 + b)$, which reduces to $\rho_0 = b/\beta$, where $\beta = 1/D - 1$. Consider the case when the main radiator is situated on the structure axis, at a distance of 3 cm from the head, *i.e.,* $\rho_0 + b = 3$, and the compensation point is placed on the head surface. In this case, if $D = 0.33$, then $\beta = 2$, $\rho_0 = b/2 = 1$, $b = 2$; for $D = 0.17$ we have $\beta = 4.88$, $\rho_0 = 0.5$, $b = 2.5$. The results point out the necessity for a tradeoff between the irradiation reduction level and the radiation pattern of the two-antenna structure, especially in cases when the antenna platform is subject to spatial restrictions. However, the pattern changes are predictable and small.

Fig. **4** shows three variants of the dark spot position relative to the head where the spot boundaries are given as a dotted line and the solid curve shows the head boundary. Feed points A_1 and A_2 of the main and the auxiliary radiators and compensation point A in Fig. **4a** and **4b** lie on the structure axis, and so the dark spot is symmetric relative to the axis. Fig. **4a** has the compensation point placed outside the head, and Fig. **4b** has it inside the head, close to its surface. In the latter case, the whole dark spot lies inside the head, resulting in a significant reduction of loss inside the head. Fig. **4c** corresponds to the case, where the auxiliary radiator is located off axis, so that the axis of dark spot would not coincide with the structure axis A_1O. This may result in such displacement of the dark spot from the head center that the loss power in the head increases. However, this variant offers more flexibility, allowing managing possible construction constraints.

Table **2** shows the total and maximum local SAR (in W/kg) with and without the auxiliary radiator, where the compensation points are located in accordance with Fig. **4a** and Fig. **4b**. The compensation method is seen to permit reducing the power absorbed in the head by a factor of 3-4 as well as reducing the maximum local power by a factor of 5-10. Here, the maximum absorbed power and the maximal local SAR exist near the main radiator. And the maximum reduction of the local power occurs here, too.

The calculations of the dark spot dimensions by the Matlab program were verified both experimentally and by means of CST simulations. The CST simulations were carried out with and without the user's head.

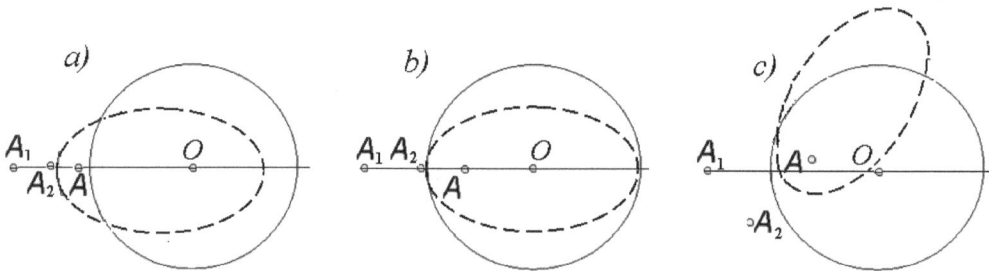

Figure 4: The dark spot: compensation point is located outside the head (*a*), inside the head (*b*) and off the symmetry axis (*c*).

Table 2: SAR levels

Number of radiators	Figure	Total SAR	Maximum local SAR (in 1 g)
1		0,00793	0,166
2	7.4*a*	0,00257	0,0235
2	7.4*b*	0,00225	0,0162

The results of the calculations, simulations and measurements with the user's head present are given in Fig. **5**. The main and the auxiliary radiators are located at points $\rho=0$ and $\rho=2$ cm, respectively. The head is situated at a distance 4 cm from the main radiator. The compensation point is located at $\rho=5$ cm.

The calculations are performed with the Matlab program. The presence of the head was taken into account by replacing $\varepsilon_r=1$ with ε_e. The calculation results are presented as curves plotted for absolute values of fields E_{z1}, E_{z2} and $E_z = E_{z1} + E_{z2}$. It is easily verified that, with ρ from 4 to 10 cm, the field E_z is substantially less than field E_{z1}.

The simulation was performed with the CST program. The user's head is included into the CST model. The rectangles, squares and empty circles mark the CST simulation results. The field measurements were carried out in laboratory conditions, and the presence of the user's head was taken into account. With allowance for this replacement, the measured values of total field $E_z = E_{z1} + E_{z2}$ are presented with full circles.

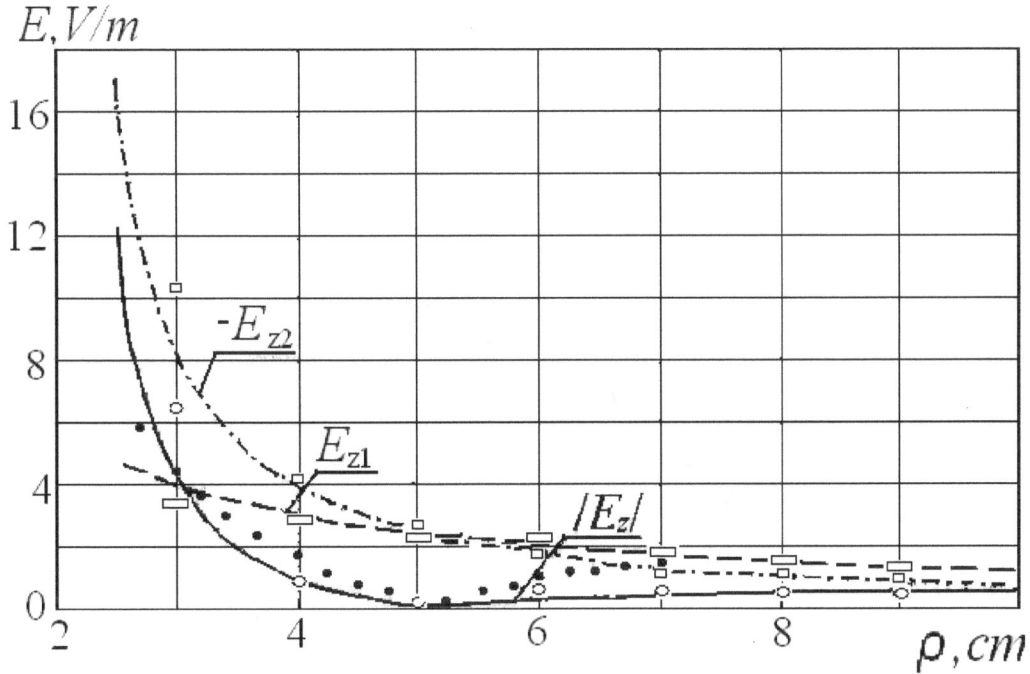

Figure 5: Field strength in the dark spot: comparison of calculation, simulation and measurements. The dashed, dash-dotted and solid curves show, respectively, the calculation results for E_{z1}, E_{z2}, and $E_z = E_{z1} + E_{z2}$ obtained with the Matlab program (the presence of the head was taken into account by replacing $\varepsilon_r = 1$ with ε_e). The rectangles, squares and empty circles are the simulation results obtained by means of CST program (the head is included in the CST model). The full circles are the measured values of the total field.

With the user's head absent, the results of calculations, simulations and measurements are in good agreement. However, as one might expect from the approximate estimation of ε_e (see Section 7.1), the coincidence between the results in the presence of head is not too good, but a qualitative agreement does exist.

The obtained results show that under realistic conditions the compensation method allows reducing the irradiation power by a factor of 3-4 and more. The pattern distortion due to the auxiliary antenna is relatively small provided that the latter is properly located.

7.3. EXTERNAL ACTIONS

Applying the compensation method under realistic conditions become difficulties due to changeable environment, since the operation of a complicated radiators structure often is disrupted by various external actions. The system disturbances are to be counteracted. In particular, proximity of metallic objects to the antenna system or relative displacement of the antennas as a consequence of user's movement can result in a tuning disturbance. In such cases, the compensation point may be displaced and the field inside the dark spot can grow significantly. Consider the impact of approaching metallic objects to an antenna as an example of the external action.

To eliminate of the consequences of the actions, one can seek to retain the amplitudes and phases of the radiator driving currents (the first method) or prevent the change of the radiator fields (the second method). It should be noted, first, that the appearance of a metal body near a radiator changes the current at all points along the radiator wire, whereas the feedback circuits are capable of adjusting the current only at a single point in each radiator, *e.g.*, at its input. And, second, one must say that a metal body causes different changes of the radiator fields in the entire space, whereas the feedback circuits are capable of adjusting the field only at one or two points. Therefore, the efficiency of both methods, especially of the first one, is inherently limited. Here, the comparative analysis of the method efficiencies is performed. It should be emphasized that cases of severe disturbance in antenna systems are considered.

The calculation model of the antenna system is given in Fig. **6**. The model consists of two monopoles (*A* is the main radiator, *B* is the additional radiator) mounted on a metal plate close to the human head. The compensation point is located inside the head, near to its front boundary. Dimensions in the figure are indicated in centimeters. Fig. **7** shows the same circuit for the case when a vertical square metal sheet is placed not far from the antenna system (at distance 0.5 m from the radiators). Presence of a metal sheet in the proximity to a cellular phone may occur, *e.g.*, when a cellular phone user enters an elevator or a car. The metal sheet affects the antenna system and causes the fields in the dark spot to grow.

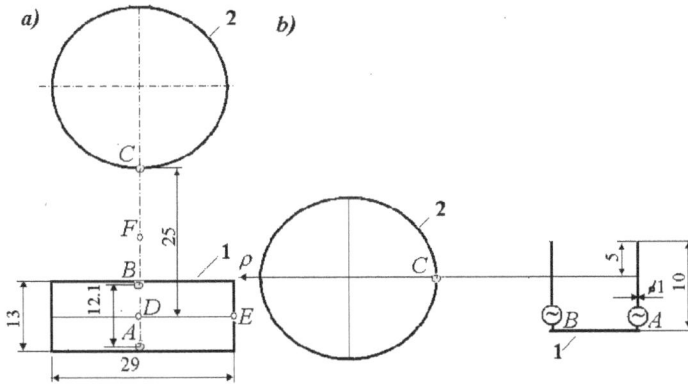

Figure 6: Model of antenna system close to human body.

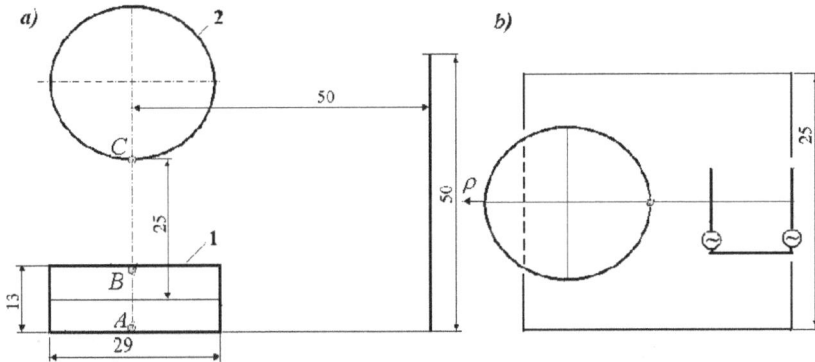

Figure 7: A metal sheet placed not far from the antenna structure.

Let us start with the relatively simple case of a single radiator. The current at the base of a single radiator in the absence of any metal sheet is $J_{A1} = e_A/Z_A$, where e_A is electromotive force at the input of the radiator, and Z_A is its input impedance. The metal sheet changes the radiator input impedance to Z'_A. To avoid the current change at the radiator base, one must change the emf at its input to

$$e'_A = J_A Z'_A = e_A Z'_A/Z_A . \tag{7.10}$$

For a single radiator of height 10 cm and diameter 1 cm we have: $e_A = 1$ V, $Z_A = 38.6 + j12.8$, $Z'_A = 41.7 + j14.3$, $e'_A = 1.09 \exp(j0.0145)$.

Fig. **8** shows the field of single radiator A along the horizontal line passing through the radiator and the head center in the absence (1) and presence (2) of a metal sheet and after adjustment of the emf (3). Fig. **8** and several others are divided into two parts to allow using different scales.

In the case of two radiators one must first calculate the amplitude and phase of the second radiator emf, which ensure the field compensation at a given point. For this purpose one can excite both radiators, in turn, by emf e_1, calculate the fields E_1 and E_2 of each radiator at the compensation point and take emf $e_2 = -e_1 E_1 / E_2$ as that for the second radiator. It should be noted that the field and the input impedance of each radiator are calculated in the presence of the other radiator, when its input is grounded.

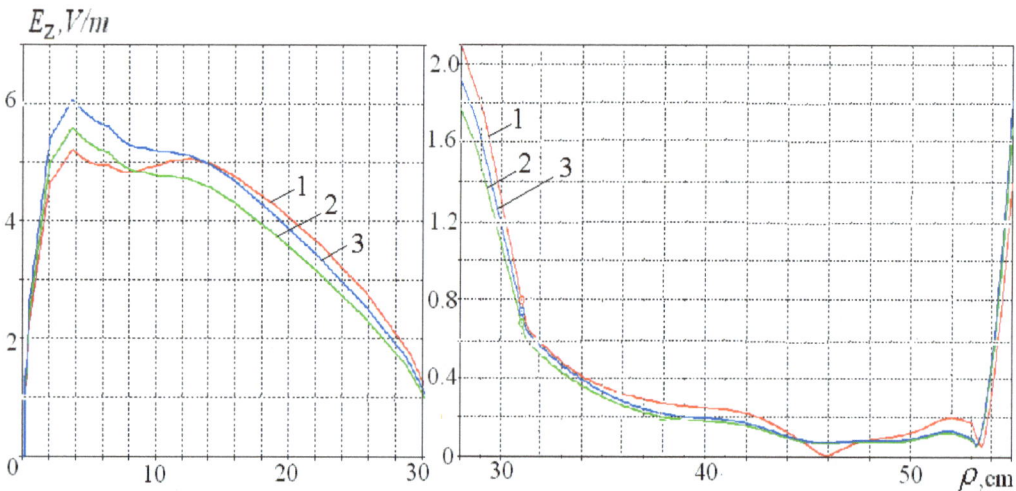

Figure 8: The field of the singular radiator A in the absence (1-red) and presence (2-green) of the metal sheet and after adjustment of the emf (3-blue).

Calculations were performed with the CST program, which permits to simulate the total circuit with both generators and find the self-impedances Z_{11} and Z_{22} of each radiator and their mutual impedance Z_{12}. The solution of the equations set $e_1 = J_1 Z_{11} + J_2 Z_{12}$, $e_2 = J_2 Z_{22} + J_1 Z_{12}$ allows to finding currents J_1 and J_2.

The metal sheet changes the self- and mutual impedances of the radiators. To avoid the change of the currents at radiators bases, the emf's must be modified to

$$e_1' = J_1 Z_{11}' + J_2 Z_{12}', e_2' = J_2 Z_{22}' + J_1 Z_{12}', \tag{7.11}$$

where the new impedances are marked by primes. Calculating the new emf's, we may find the new fields and ascertain the adjustment results.

For two radiators of the same dimensions we obtain: $e_1 = 1\,V$, $E_1 = 0.48 \exp(j1.11)$, $E_2 = 1.09 \exp(j2.54)$, $e_2 = 0.44 \exp(j1.71)$, $Z_{11} = 37.3 + j16.2$, $Z_{22} = 39.5 + j16.8$, $Z_{12} = 0.67 - j28.9$, $J_1 = 0.013 - j0.0008$, $J_2 = 0.0064 + j0.018$, $Z_{11}' = 40.6 + j17.1$, $Z_{22}' = 42.0 + j18.9$, $Z_{12}' = 2.04 - j28.02$, $e_1' = 1.04 \exp(j0.04)$, $e_2' = 0.51 \exp(j1.70)$. The curves for the field of the two-radiator structure are presented in Fig. **9**.

The verification of adjustment results involves computation of fields along the horizontal line passing through the radiators and the head center as well as of the total SAR level and the maximum local SAR (in 1 g). The results obtained for the cases of a single radiator and a two- radiator structure in the absence and presence of a metal sheet and after adjustment of the radiators emf's in accordance with expressions (7.10) and (7.11) are compared with each other. The total SAR and the maximum local SAR for the corresponding cases are given in Table **3**. The calculated amplitudes of field at the compensation point are also presented in Table **3**. The SAR level and the fields are given in units of *W/kg* and in *V/m*, respectively.

Table 3: SAR and field at adjustment of the emf in accordance with currents

Characteristic	Total SAR	Local SAR	Field	Total SAR	Local SAR	Field
	one radiator			two radiators		
Sheet is absent	$4.5 \cdot 10^{-5}$	$1.54 \cdot 10^{-3}$	0.79	$1.4 \cdot 10^{-5}$	$0.3 \cdot 10^{-3}$	0.0029
Sheet is present	$2.8 \cdot 10^{-5}$	$0.29 \cdot 10^{-3}$	0.68	$1.7 \cdot 10^{-5}$	$0.11 \cdot 10^{-3}$	0.0044
After adjustment	$3.4 \cdot 10^{-5}$	$0.34 \cdot 10^{-3}$	0.74	$2.1 \cdot 10^{-5}$	$0.18 \cdot 10^{-3}$	0.1214

As one can see from the table and Figures, the results indicate the rather low efficiency of the correction method based on retaining the radiator driving currents.

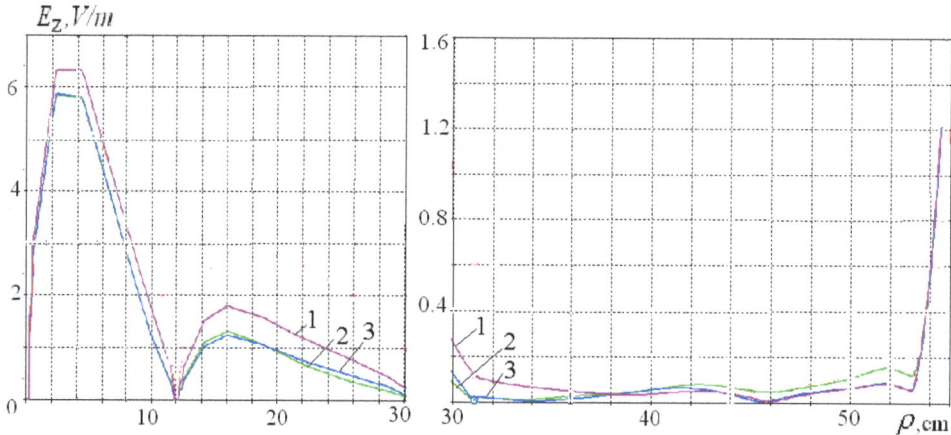

Figure 9: The field of two radiators in the absence (1-red) and presence (2-green) of the metal sheet and after adjustment of the emf (3-blue).

The implementation of the second method calls for using one of the two radiators as a measuring antenna whereas the second radiator is employed as a local transmitting antenna (*i.e.,* a field source), or using both radiators, in turn, as a measuring and a local transmitting antenna, or using the third antenna as a measuring one. The third antenna may be mounted at any suitable and convenient place. The adjustment is performed in the following way. The field at the receiving point is measured in the presence of a metal sheet and is compared with its magnitude in the absence of the metal sheet (*i.e.,* in the compensation mode), and afterwards the transmitting antenna emf is varied until the initial value of the field is obtained.

Fig. **10a** and Table **4** give the results of such adjustment for the following variants: 1)when the main antenna is used as the transmitting one and the auxiliary antenna is the receiving one, 2) the opposite case: the main antenna is used as the receiving antenna, whereas the auxiliary antenna is the transmitting one, 3) both emf's are changed. Fig. **10b** and Table **4** present the results of field adjustment using the third antenna located in the center of a metal plate, *i.e.,* at point D (see Fig. **6**) for the following variants: 4) when the emf and the field of the main antenna is changed so that its field at point D becomes equal to its original one (before disturbance), 5) when the field of the auxiliary antenna is changed for the purpose, 6) when the fields of both antennas are changed.

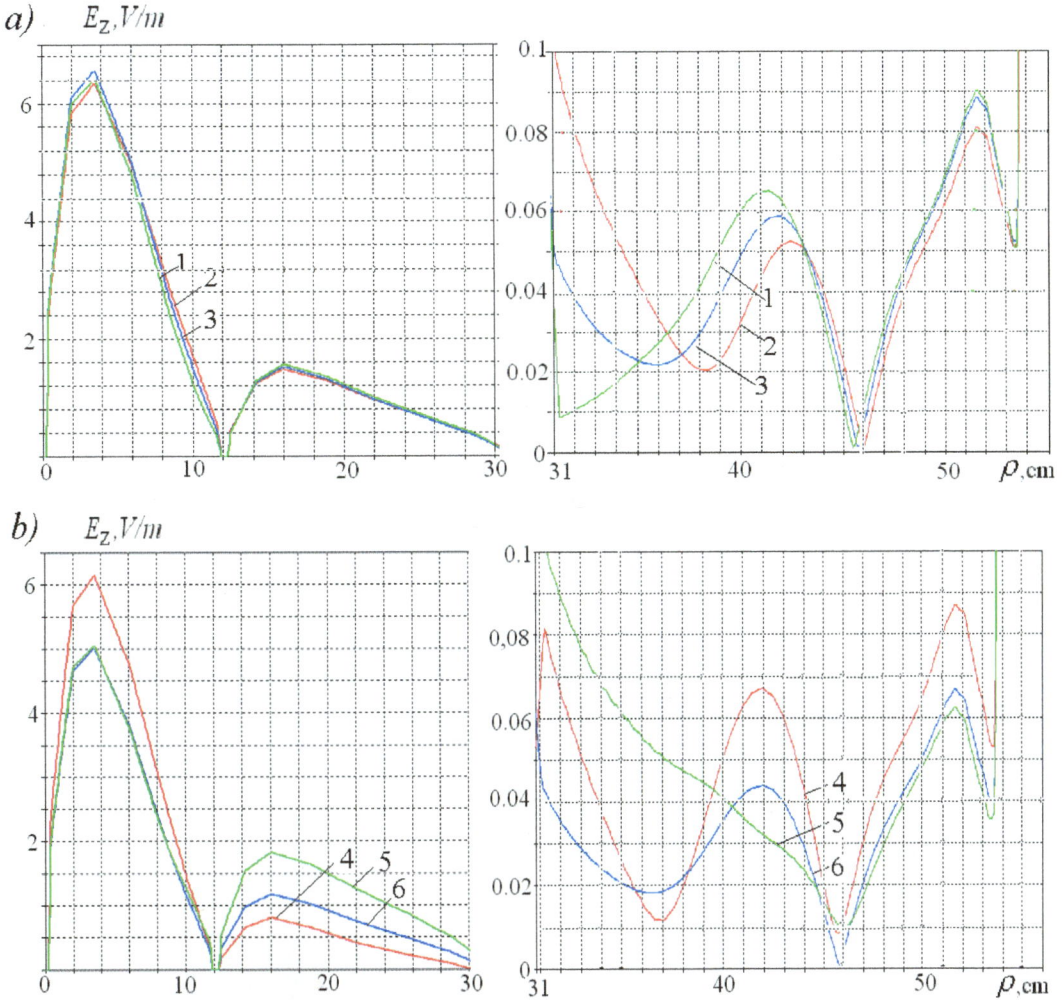

Figure 10: The fields of a two-radiator structure after emf adjustment based on the fields received by these radiators (*a*) and a third radiator (*b*).

Table 4: SAR and field at the emf adjustment in accordance with fields (same units as in Table **3**)

Variant	Total SAR	Max. Local SAR	Field
1	$1.99 \cdot 10^{-5}$	$0.137 \cdot 10^{-3}$	0.035
2	$1.83 \cdot 10^{-5}$	$0.151 \cdot 10^{-3}$	0.070
3	$2.02 \cdot 10^{-5}$	$0.152 \cdot 10^{-3}$	0.052
4	$1.57 \cdot 10^{-5}$	$0.102 \cdot 10^{-3}$	0.070
5	$1.45 \cdot 10^{-5}$	$0.125 \cdot 10^{-3}$	0.130
6	$1.18 \cdot 10^{-5}$	$0.090 \cdot 10^{-3}$	0.047

The SAR level and the fields are given in the same units as mentioned above.

As one can see from Table **4** and Fig. **10**, the method based on the measurement of the fields demonstrates a higher efficiency. But acceptable results are obtained only at application of variant 6. Other variants, including placing the third antenna at points E and F (see Fig. **6**), failed to produce satisfactory results.

It should be pointed out that the monitoring signal is a weak signal that serves as that of feedback signal for both the main and auxiliary radiators. Therefore, proper adjustment by both methods requires that no signal from the transmitter impinges on the measuring antenna during the field measurements.

To prevent the field changes under external actions (*e.g.*, an approach of a metallic object), one may use a manual or an automatic circuit. Fig. **11** gives the block diagram of an automatic adjustment circuit, which contains transmitter 1, main radiator A_1 and auxiliary radiator A_2 connected to the transmitter through power divider 2, amplitude controller 3 and phase shifter 4. The amplitude controller and phase shifter provide the initial tuning and the field compensation at a given point. The amplitude controller is usually implemented with a potentiometer, and the phase shifter with a delay line, a low-pass or a high-pass filter. Two circuits consisting of amplitude controllers 13 and 15 and phase shifters 14 and 16 provide two reference signals for the radiators A_2 and A_1, respectively.

Antenna A_3, used for adjustment of the antenna system, receives, in turn, signals from the main and auxiliary radiators and switches between the radiators. The procedure permits to determine the amplitude and phase of the radiated and received signals.

As a result of the external action (*e.g.*, an approach of a metal body) the phases of both radiator signals, received by antenna A_3, change. These phases are in turn compared in phase detector 7 with those of the reference signals. The phase detector produces an error signal proportional to the difference of the said phases. Low-pass filter 8 removes short-term fluctuations of the error signal. The error signal passes through amplifier 9, controls phase shifters 4 and 6 and brings up the optimal phase differences. As it can be seen from Fig. **6**, a feedback circuit is constructed, and it provides a phase self-tuning action similar to that of a phase locked loop (PLL),

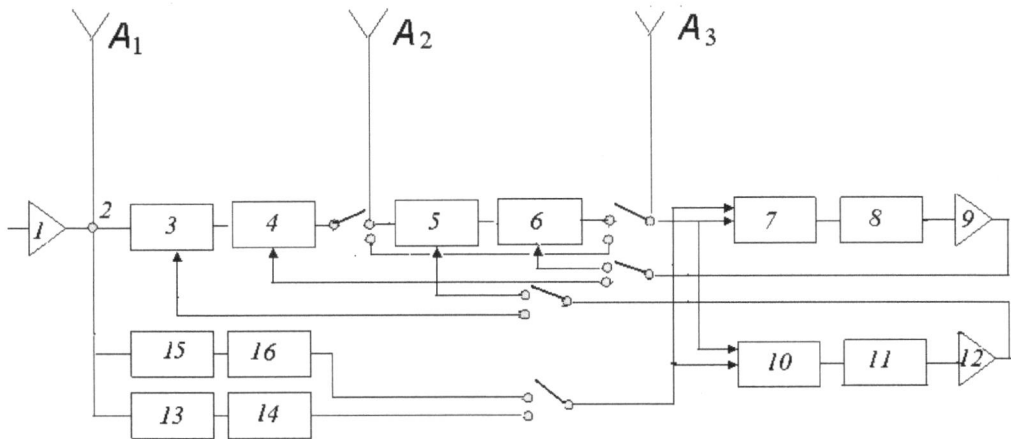

Figure 11: Block diagram of the proposed automatic adjustment circuit based on the constancy of fields.

The second feedback circuit is used for predicting the optimal signal amplitudes. It is similar to an automatic gain control (AGC) circuit. The input signal of radiator A_2 is compared in amplitude by comparator 10 (operational amplifier) with the reference signal. Low-pass filter 11 removes short-term fluctuations of the signal at the comparator output. The error signal passes through the amplifier 12, controls amplitude controller 3 and brings up the amplitudes relationship to the optimal ratio.

As a result two feedback circuits allow optimizing the amplitude relationship and the phases of the emf feeding main radiator A_1 and auxiliary radiator A_2. In contrast to the conventional automatic gain control circuits, the amplitude difference and the phase difference of the two different radiator signals are not zero in this case.

The obtained results show that it is possible to automatically adjust a complicated antenna system under conditions of intensive disturbances. The proposed method based on measurement of the fields demonstrates a higher efficiency. It allows obtaining acceptable practical results under severe disturbance of the antenna system operation even in cases when (as in our example) the zero-field point is not achievable.

7.4. WIDEBAND FIELD COMPENSATION

As said in Section 7.2, cellular communication is not the only area of application where it is necessary to create a region of weak field near the transmitting antenna. The task is vital one, if, for example, a mobile transmitter is located in a vehicle close to users and other passengers. Creating a weak field region in such cases is an efficient technique of protecting against irradiation.

In these cases, requirements to the produced weak field change in part. On the one hand, they often get complicated due to necessity of operation in a broad frequency range. On the other hand, the requirements become less stringent. For example, large dimensions of the transmitter allow increasing the room for mounting auxiliary radiators. Presence of a signal in the direction opposite to user's location sometimes is not required. The requirement for the horizontal pattern to be flat in the direction of user's location is somewhat lifted.

Operation in a broad frequency band gives rise to additional difficulties, since the fields of the main and the auxiliary radiators at the compensation point must have the same magnitude and opposite sign at each frequency. The phase shift can be easily implemented by using a transmitter output stage with balanced output. But, first, these two instances of given radiator type need to have identical characteristics at all frequencies of the operation range. Second, since the main and the auxiliary radiators are located at different distances from the compensation point, and signal velocities in the air and in the feed lines connecting the radiators to the transmitter are different, it is necessary to insert a small phase shifter in one channel. This phase shifter must provide the required delay time throughout the frequency range.

It is not easy to meet both these requirements. To provide the same signal amplitudes, it is necessary to secure identity of the main and the auxiliary radiators or to insert controlled attenuator in one channel. Wideband antennas often have a complicated structure, which uses different conducting and dielectric materials. That is why antenna characteristics are unstable, *i.e.,* different antenna instances have different frequency responses, and so their fields are not equal. It is impossible to construct controllable attenuators in a wide frequency range.

Replacement of the phase shifter with a delay line can provide the required phase, if the delay line is loaded with its wave impedance. Since the radiator impedance is complex and frequency dependent, it is dissimilar to the delay line wave impedance. Placing both radiators at equal distance from the compensation point yields better results.

A simple structure consisting of two radiators placed at the same distance from the compensation point is shown in Fig. **12** (top view). The following notation is used here: 1 is the transmitter top cover, 2 is the user's body, B_1 is the main radiator, B_2 is the auxiliary one, and C is the compensation point. The radiators are two asymmetrical antenna instances of the same type, which have the circular pattern in the horizontal plane and similar amplitude-frequency responses. They are located at the same distance $b = a/\cos\beta$ from point C, where $a=AC$ is the distance from point C to the middle line of the transmitter cover.

Let the auxiliary radiator phase differ from the main radiator phase by $180°$, and the amplitude of the vertically polarized signal $\vec{E}_2$ of the auxiliary radiator be $E_2 = DE_1$, where $\vec{E}_1$ is the amplitude of the main radiator signal, and $D \leq 1$. If $D = 1$, *i.e.,* the amplitudes of the auxiliary signal and the main signal coincide, the sum of the signals in the plane of the structure symmetry is zero, since the signals in the plane are the same in amplitude and opposite in phase. If $D \neq 1$, the summing signal $E = E_1 + E_2 = (1-D)E_1$ is equal to a fixed small share of the main signal and weakly frequency dependent, since the distance from the radiators to the compensation point is the same, and the radiators themselves are similar. No phase shifter is needed in this case.

A disadvantage of the proposed circuit is the absence of the far field signal in the structure symmetry plane (in both directions), when $E_2 = E_1$. But, first, this is not critical in many applications. Second, the angle (gap width), in which radiation signal is nearly zero, is very small. And third, the influence of the ground and neighboring bodies results in a smoothing of the pattern, *i.e.,* both the depth and the width of the gap in the horizontal pattern will be reduced in real conditions.

Estimate the gap width by calculating angle $\Delta\varphi$ between its boundaries, *i.e.,* between the points corresponding to given small value f_1 of the pattern. If $a=25$ cm, and $\beta = 30°$, the signal at angle φ, which consists of two signals with equal amplitude E_1 and opposite phase created by radiators located at distance $d = 2a\tan\beta = 28.9$ cm from each other is equal to $E = 2E_1\sin[(kd/2)\cos\varphi]$, *i.e.,* $E_{max} = 2E_1$. Here, the cylindrical system of coordinates is used.

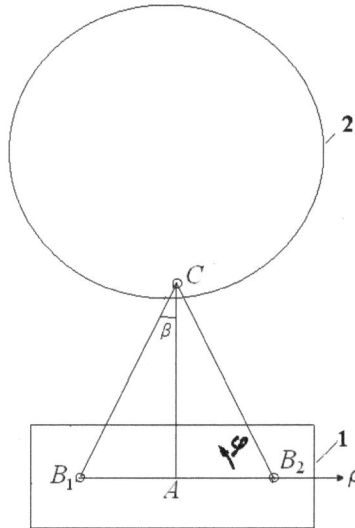

Figure 12: The two-radiator structure.

Since the pattern minimum occurs at $\varphi = \pi/2$, it's more convenient to use coordinate $\varphi_1 = \varphi - \pi/2$, so

$$f_1 = \sin[(kd/2)\sin\varphi_1] \approx kd\varphi_1/2, \tag{7.12}$$

i.e., $\Delta\varphi = 2\varphi_1 = 4f_1/(kd)$. In particular, if $f_1 = 0.1$, the wavelength λ is 30 cm, and $d=28.9$ cm, then $\Delta\varphi = 3.8°$.

Another disadvantage is the fact that, when $D \neq 1$, the irradiation power reduction factor inside the dark spot decreases, as the total field fails to vanish even at the compensation point. But the disadvantage is also inherent in the case of antenna placement at different distance from the compensation point, due to the limited dimensions of the transmitter top cover [47].

Experimental results for the signal magnitude at the compensation point are given in Fig. **13**. Curve 1 corresponds to the signal of the main antenna placed at point A (see Fig. **12**), and curve 2 corresponds to the signal of two antennas placed at points B_1 and B_2. As seen from the results, mounting both radiators at the same distance from the compensation point permits to ensure compensation in a wider frequency band. The auxiliary radiator decreases the signal magnitude at the compensation point by 10-15 dB (3-6 times) in the band from 1.7 to 2.7 GHz.

Using of two auxiliary radiators gives additional advantages. In this case, the amplitude of each auxiliary radiator signal may be smaller than that of the main radiator. One variant of this compensation structure is presented in Fig. **14** (top view) where the same notation is used.

If E_0 is the amplitude of the main radiator signal at the compensation point, the amplitude of each auxiliary radiator signal must be $E_1 = E_2 = E_0/2$. In the cylindrical coordinate system, the origin of which coincides with point A, the ratio of the auxiliary signal to the main signal in direction φ is

$$E_1(\rho)/E_0(\rho) = 0.5\exp\left[jk(a_1 - b_1)\right], \quad E_2(\rho)/E_0(\rho) = 0.5\exp\left[jk(a_1 + b_1)\right], \textbf{(7.13)}$$

where $E_0(\rho)$ is the main radiator field at distance ρ, $E_i(\rho)$ is the field of the auxiliary radiator located at point B_i, and $a_1 \pm b_1$ are the path-length differences of the signals from the main and auxiliary radiators to the observation point. Here, $a_1 = a(1 - \cos\beta)\sin\varphi$, $b_1 = a\sin\beta\cos\varphi$ (these segments are marked in Fig. **14**). The total signal is

$$E(\rho) = E_0(\rho)\left[1 - \cos(kb_1)\exp(jka_1)\right]. \tag{7.14}$$

The ratio of the total and the main signals is equal to

$$\left|E(\rho)\right|/\left|E_0(\rho)\right| = \sqrt{\left[1 - \cos(kb_1)\cos(ka_1)\right]^2 + \cos^2(kb_1)\sin^2(ka_1)}. \tag{7.15}$$

In the antenna symmetry plane the total signal in the far region is

$$E(\rho)/E_0(\rho) = 1 - \exp\left[jka(1 - \cos\beta)\right] = -2j\exp\left[jka(1 - \cos\beta)/2\right]\sin\left[ka(1 - \cos\beta)/2\right],$$

i.e.,

$$\left| E(\rho)/E_0(\rho) \right| = 2\sin\left[ka(1-\cos\beta)/2\right]. \tag{7.16}$$

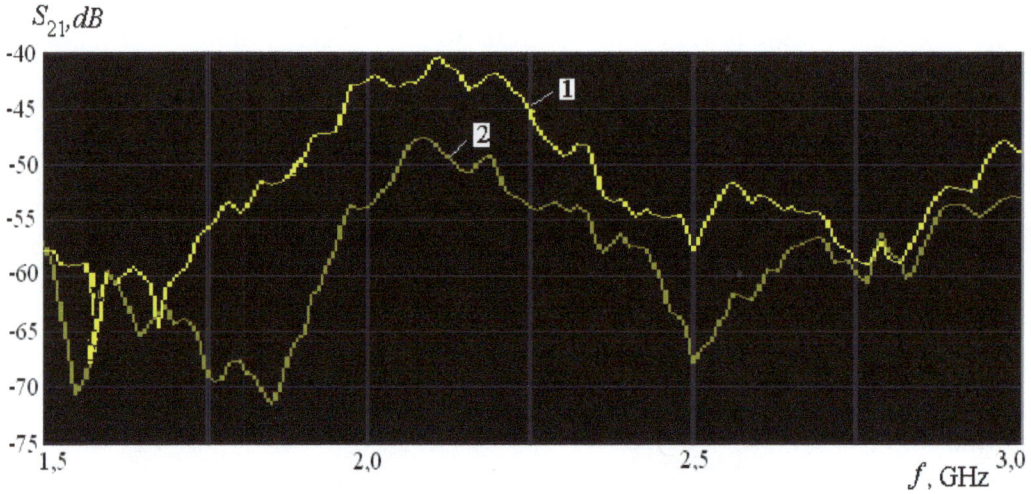

Figure 13: The frequency dependence of the signal at the compensation point, when the main radiator (1) and two radiators (2) are placed at the same distances from the point.

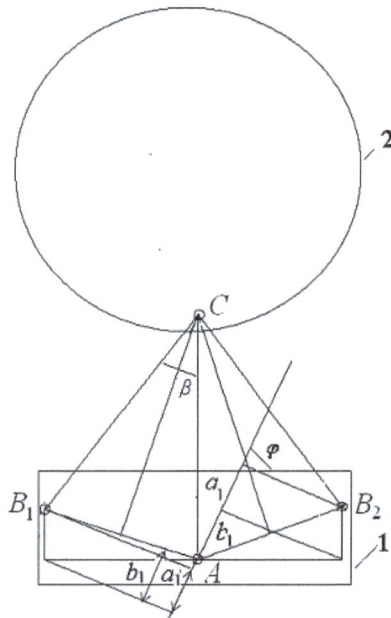

Figure 14: The three-radiator structure.

If, for example, a=25 cm and $\beta = 30°$ or $60°$, we obtain at 1 GHz $|E(\rho)/E_0(\rho)| =$ 0.69 and 1.22, respectively. The total far field in the symmetry plane in both directions is other than zero. It follows from (7.14) that at any φ the total signal in the far region is other than zero, too. This is an inherent advantage of such structure.

The calculated horizontal patterns at frequency 1 GHz of one radiator placed at point A, of two radiators (see Fig. **12**) and of three radiators (see Fig. **14**) located on the cover of infinite dimensions are given in Fig. **15** and marked with symbols 1, 2 and 3, respectively.

The considered compensation structures comprise radiators placed at the same distances from the compensation point. The structures require no use of any phase shifter. The compensation structures, in which flat metal reflectors (mirrors) are used instead of auxiliary radiators, permit to create a weak field in the subspace on the side of the reflectors, where the antenna is located, which is similar to the one generated by three radiators.

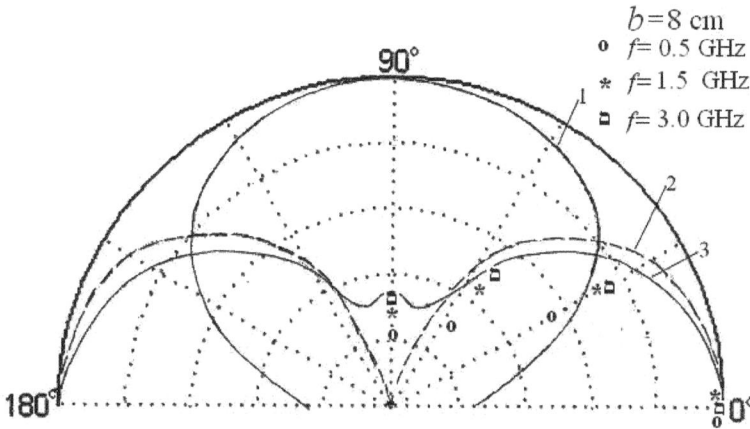

Figure 15: The horizontal patterns of different structures at 1 GHz.

An example of such structure is presented in Fig. **16a** (top view). Here, the following notation is used in addition to that used above: R is the metal reflector, and B is the location of the equivalent radiator. As seen from the Figure, the flat metal plate, in which main radiator A is reflected, is mounted on the transmitter top cover along the structure axis of symmetry (in the vertical plane passing through compensation point C). The reflection (or imaging) of the main radiator signal by the metal plate into the space in front of the reflector is equivalent to the presence of second radiator B located at the same distance behind the plate. This equivalence is clear from Fig. **16b** showing that the path lengths of the reflected signal and the equivalent radiator signal are equal to each other.

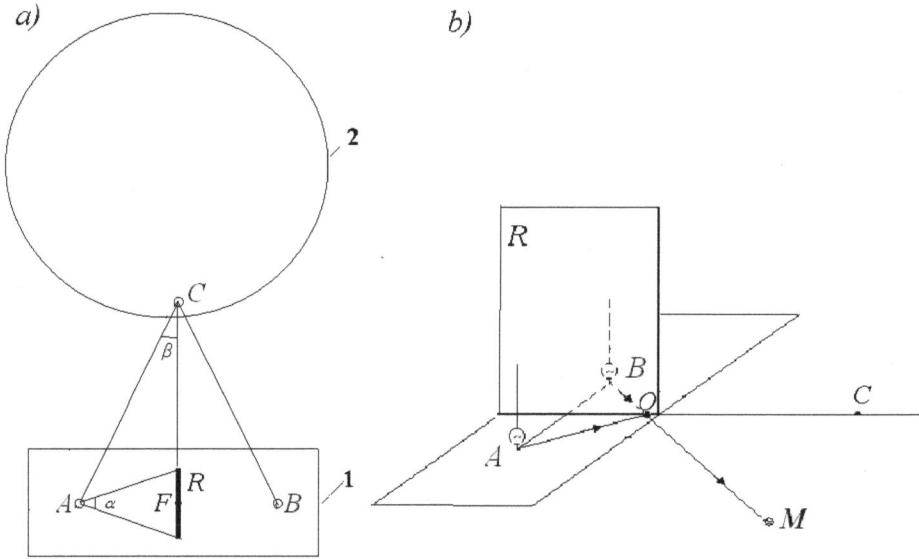

Figure 16: Creating the second signal by a flat mirror (*a*), the *AOM* and *BOM* are equal (*b*).

The reflected signal phase differs from the main radiator phase by $180°$ (see, *e.g.*, [10]), since vector $\vec{E}_1$ of the incident signal is located along the tangent to the reflector surface.

The amplitude of the reflected signal $\vec{E}_2$ is $E_2 = DE_1$, where D depends on magnitude α of the angle, subtended at point A ($D \leq 1$) by the reflector.

If the plate dimensions are infinite (or if they are much larger in all directions than the main radiator dimensions), then $D = 1$, *i.e.,* the amplitudes of the reflected and the main signal coincide, and the total signal in the reflector plane is zero. If $D \neq 1$, total signal $E = E_1 + E_2 = (1-D)E_1$ is a fixed share of the main signal, and the share is independent of frequency, since the distances from both radiators to the compensation point are the same, and the radiators themselves are identical.

On the whole, the performance of the structure is similar to that of the two-radiator structure presented in Fig. **12**. No phase shifter is needed in this structure. An additional and very important advantage of the described structure is that the external actions, *e.g.*, an approach of metal bodies, exert practically no effect on the structure operation, as the main radiator and the reflected signals undergo the same changes upon any external actions. Only the appearance of a metal body between a signal source and the compensation point is an exception and a very rare one at that.

The influence of a mirror (a metal plate) on the antenna input impedance decreases the matching between the antenna and the transmitter. To offset the decreased matching, it is necessary to increase the distance between the antenna and the mirror as well as the mirror width so that the second signal would not decrease. The reflector width has only a weak effect on the antenna input impedance, but essentially increases signal $\vec{E}_2$. Moreover, the mirror influence on the antenna input impedance is constant and can be allowed for a priori. The mirror height must exceed that of the antenna.

Flat reflectors can be manufactured in the form of a light-weight, strong and collapsible construction (if necessary). In such a case, each reflector is implemented as a set of vertical parallel wires, located at a distance of $0.06\,\lambda_{\min}$, where $\lambda_{\min}$ is the minimum wavelength in the frequency range.

Using two reflectors, *i.e.*, creating two equivalent radiators, gives additional advantages. In this case, the amplitude of each reflected signal may be smaller than that of the main radiator signal, and so the requirements to the reflector loosen. A variant of such compensation structure is presented in Fig. **17** (top view), where, as before, B_1 and B_2 denote the equivalent radiators locations. The performance of the structure of Fig. **17** is similar to that of the three-radiator structure presented in Fig. **14**: the distance from all radiators to the compensation point is the same, *i.e.*, field variations due to frequency dependence have no effect on the compensation quality.

As stated above, the amplitude of the reflected signal (vertically polarized) is $E_2 = DE_1$. Here, D depends on the mirror dimensions. If dimensions of the mirror and the radiator are comparable, D is in the first order approximation of the ratio of C_r, the capacitance between the radiator and a rectangular mirror of finite dimensions, to C_∞, the capacitance between the radiator and an infinite mirror

$$D = C_r / C_\infty .$$

In the case of a vertical radiator with a height smaller than that of the mirror, capacitance C_r depends primarily on radiator radius a, mirror width b, and distance h between the radiator and the mirror. In plane-parallel structures, where all wires are infinitely long, the capacitance per unit length between a wire and a metal plate is $C_r = 2\pi\varepsilon/\ln(4h^2/ab)$ - for a plate of width b and $C_\infty = 2\pi\varepsilon/\ln(2h/a)$ - for a plate of infinite width [27]. Here ε is the permittivity. It is easy to see that the necessary value of D can be obtained by selecting the values of b and h (e.g., if $h/b = 1.35$, $h/a = 10$, then $C_r / C_\infty = 0.75$).

Ratio D for plane-parallel structures and for finite structures with heights exceeding the antenna height is approximately the same in magnitude.

An essential drawback of the structure with two reflectors is that the reflectors creating the dark spot in the near region block at the same time prevent the electromagnetic waves in the angular sector, shadowed by them. Here, the signal in the open sector increases approximately twice. This drawback limits the application of such structures.

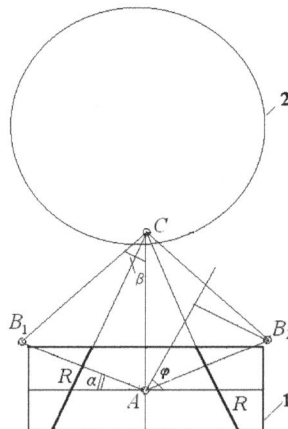

Figure 17: Compensation structure with two reflectors.

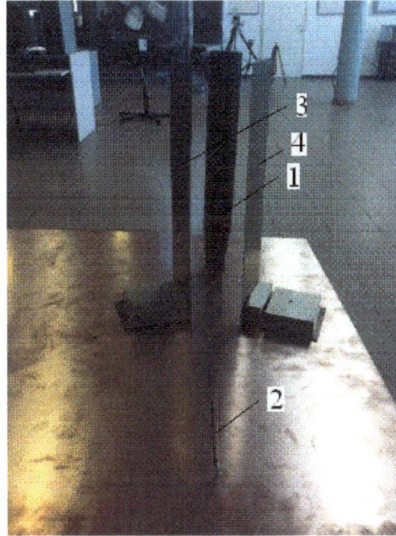

Figure 18: General experimental setup.

The structure with two reflectors was tested experimentally. A general view of the experimental setup is presented in Fig. **18**. The following notation is used in the Figure: 1 is the radiator, 2 is the receiving antenna, placed at the compensation point, and 3 and 4 are flat reflectors, located symmetrically on either sides of the straight line passing through the compensation point and the radiator. The distance between the radiator and the receiving antenna is 25 cm. The reflector is manufactured in the shape of a vertical rectangular plate. Contact with the ground is maintained along the whole width of the reflector, using a horizontal element. In the course of the measurements, each reflector was separately moved along a straight line to the point where the received signal at 2 GHz (close to the band center) decreased by 6 dB. As a result the antenna field at the compensation point in the presence of both reflectors is substantially smaller than the field of the antenna without reflectors. The measurements are performed at two different values of reflectors width *b*, *i.e.,* 8 and 15 cm, and at different angles (10, 15 and 20°) between the axis of the setup symmetry and the lines, along which the reflectors were moved to attain the zero field at the compensation point. In all variants the signal was half as strong as its value when the distance between the compensation point and the near edge of the reflector was 9 cm.

The measurement results are given in Table **5**.

Table 5: Measurement results for the signal magnitudes at the compensation point

f, GHz	Without reflectors	With reflectors of width				Difference			
		8 cm			15 cm	8 cm			15 cm
		$\beta/2 = 10°$	20	15	15	10	20	15	15
0.5	-17	-32.1	-37.4	-33.3	-35.8	15.1	20.4	16.3	18.8
1.0	-19.5	-33.5	-38.3	-35.8	-34.3	14	18.8	16.3	14.8
1.5	-39	-46.4	-38.3	-43.4	-54.9	7.4	-0.7	4.4	15.9
2.0	-28.8	-37.3	-47.2	-40.4	-62.5	8.5	19.4	11.6	33.7
2.5	-27.7	-47.1	-29.4	-42.6	-41.3	19.4	1.7	14.9	13.6
3.0	-36.3	-43.9	-35.1	-47	-42.6	7.6	-1.2	10.7	6.3

The frequency dependence of the signal at the compensation point without reflectors (1) and with reflectors (2) is presented in Fig. **19**, for reflector widths of 8 (*a*) and 15 cm (*b*). As is clear from the table and figures, the reflectors with width 8 cm permit to decrease the signal magnitude without tuning by 10-15 dB for the whole frequency range, and the reflectors with width 15 cm perform even better.

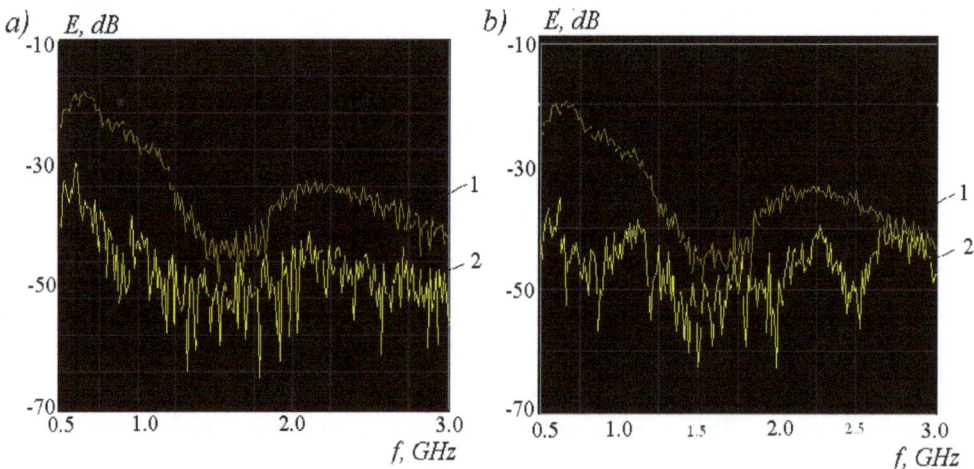

Figure 19: Frequency dependence of signal at the compensation point without reflectors (1) and with reflectors (2) for reflector width 8 cm (*a*) and 15 cm (*b*).

The simulation used the program CST. The structure adjustment (selection of reflector positions, which allows obtaining the minimum signal at point C) was performed at frequency 2 GHz.

In this section, we have analyzed a compensation method for wideband operation. The calculation and experimental results of studying different structures for wideband compensation show the advantages of structures with radiators located at equal distances from the compensation point. Using these structures, one can decrease the irradiation power by a factor of 5 or more in a wide bandwidth. Two auxiliary radiators permit to achieve these results without the deep and wide dips in the horizontal pattern.

CHAPTER 8

Arrays

Abstract: Main circuits of antenna arrays and their characteristics are described. Flat log-periodic structure, reflect arrays and adaptive systems are considered in greater details including the use of algorithms of adaptation and adaptive control of weighting coefficients, which is performed by the adaptive processor.

Keywords: Adaptive array, Adaptive control of weighting coefficients, Adaptive processor, Algorithms of adaptation, Array on the basis of existing set of ship antennas, Broadside radiation, Log-periodic dipole antenna, Flat log-periodic structure close to the self-complementary one, Microstrip antenna, Minimum backward radiation, Multiple-stacked antenna, Pattern, Phase step during reradiating, Principle of the current cut-off, Radiation in the prescribed direction, Reflect array, Relationships of similarity, Theorem of pattern multiplication.

8.1. CHARACTERISTICS OF DIRECTIVITY

As noted in Section 1.1, if sources of electromagnetic field are distributed continuously in some volume V, and a medium surrounding volume V is a homogeneous isotropic dielectric, the solution of equation (1.19) for a harmonic field is given by (1.21). In this expression, R is distance from the integration point to the observation point, which in the far region in the first approximation corresponds to Fig. **1**. If radiators inside the volume have the same direction, then

$$A_p = \frac{A_p(0)}{j_p(0)} \iiint_{(V)} j_p \exp(jkz\cos\theta)dV, \tag{8.1}$$

where $A_p(0)dV$ is the vector-potential of the field, created by an elementary volume with current density $j_p(0)$ located at the coordinate origin. If $E_p(0)dV$ is the field of the volume, then

$$E_p = \frac{E_p(0)}{j_p(0)} \iiint_{(V)} j_p \exp(jkz\cos\theta)dV. \tag{8.2}$$

To exemplify, consider an elementary linear radiator (Hertz dipole) located along the z-axis. It is a filament of length b with current amplitude I, constant along b. In the far region, the filament field is defined by expression (1.7). Substituting it in (8.2), we obtain expression (1.9) for the field of a symmetrical radiator (dipole) located along the z-axis with the center at the coordinate origin (see Fig. **4a**). If, according to (1.8), we consider that the current distribution along the radiator arm is sinusoidal one, we come to expression (1.10) the last factor of which gives the vertical pattern of isolated radiator.

Another example is a linear array situated along the x-axis (Fig. **1a**). In this case, one must replace (8.2) with the sum

$$E = \frac{E_1}{J_1(0)} \sum_{n=1}^{N} J_n(0) \exp\left[jk(n-1)d_1\right], \tag{8.3}$$

where E_1 is the field of the first radiator, $d_1 = d\cos\varphi\sin\theta$ is path difference of beams from the adjacent radiators to the observation point with arbitrary coordinates φ and θ, n is the radiator number, and the array radiators are excited by currents with equal amplitude and linearly growing phase displacement:

$$J_n(0) = J_1(0)\exp\left[-j(n-1)\psi\right]. \tag{8.4}$$

Here, ψ is the phase shift between currents of adjacent radiators. Using the formula for a sum of N terms of geometric progression with ratio $\exp\left[j(kd\cos\varphi\sin\theta - \psi)\right]$, omitting factor $\exp\left[j(N-1)(kd\cos\varphi\sin\theta - \psi)/2\right]$ defining phase characteristic of array, and normalizing the result to 1, we obtain expression for the amplitude characteristic of an array

$$F_N(\theta,\varphi) = \frac{\sin\left[N(kd\cos\varphi\sin\theta - \psi)/2\right]}{N\sin\left[(kd\cos\varphi\sin\theta - \psi)/2\right]}. \tag{8.5}$$

As seen from (8.3), the pattern of a system of identical, equally oriented directional radiators is the product

$$F(\theta,\varphi) = F_1(\theta,\varphi)F_N(\theta,\varphi), \tag{8.6}$$

where $F_1(\theta, \varphi)$ is the pattern of an isolated radiator. Equality (8.6) is referred to as the theorem of pattern multiplication, and $F_N(\theta, \varphi)$ is the array factor.

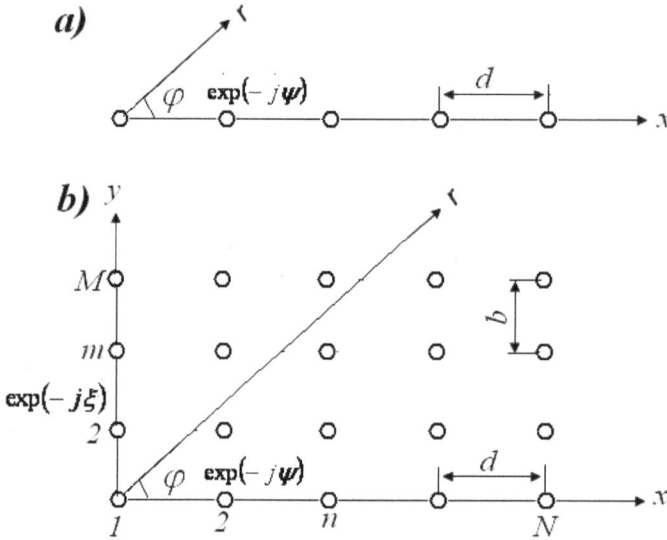

Figure 1: Uniform antenna array: (*a*) – linear, (*b*) – rectangular.

If a system of radiators consists of horizontally-omnidirectional antennas, *e.g.*, of vertical monopoles, its horizontal pattern coincides with the array factor, and the vertical one depends on the corresponding pattern of an isolated radiator. One should emphasize that the array factor also depends on angle θ.

A rectangular array (Fig. **1b**) can be considered as a linear structure consisting of M linear arrays. Therefore, the array factor of the rectangular array is

$$F_{MN} = F_M F_N,\tag{8.7}$$

where F_M is the array factor of a linear array of M radiators situated along the y-axis:

$$F_M(\theta, \varphi) = \frac{\sin\left[M\left(kb\sin\varphi\sin\theta - \xi\right)/2\right]}{M\sin\left[\left(kb\sin\varphi\sin\theta - \xi\right)/2\right]}.\tag{8.8}$$

The electrical characteristics of radiator system at given frequency are conditioned by its phasing mode, *i.e.*, by choice of the phase shift between the currents in radiators. Two phasing modes of antenna arrays are possible: the maximum forward radiation and the minimum backward radiation. In the first mode, the phases of radiator fields are the same at the observation point with a given azimuth; in the second mode, the field is minimum or zero in the direction opposite to the correspondent's direction. The first mode can be set up in any array, while the second one is not feasible always. For example, it is unfeasible in a linear array.

In a rectangular array, the phases of signals coming from the adjacent radiators of one tier to the observation point, located in the same horizontal plane as the array, differ from each other by quantity $\Phi_{12} - \Phi_{11} = kd\cos\varphi - \psi$, and the phase difference from radiators of adjacent tiers is $\Phi_{21} - \Phi_{11} = kb\sin\varphi - \xi$. Here, Φ_{mn} is the phase of signal, coming from the mth radiator of the nth tier. At the observation point with azimuth φ_m, phases of all signals will be the same, if

$$\psi_m = kd\cos\varphi_m, \xi_m = kb\sin\varphi_m = (b\psi_m/d)\tan\varphi_m. \tag{8.9}$$

The conditions are necessary and sufficient for implementation of the first phasing mode.

The second mode can be applied, *e.g.*, in a double-tier array, where the fields of radiators of each tier are added together in phase, and then the fields of different tiers are added together in anti-phase. The condition of no signal in direction φ_0 has the form

$$\psi_0 = kd\cos\varphi_0, \xi_0 = kb\sin\varphi_0 + \pi = \frac{b\psi_0}{d}\tan\varphi_0 + \pi. \tag{8.10}$$

Calculation of the phase shifts in accordance with equalities (8.9) and (8.10) for different angles φ of direction of the maximum radiation (in the first case $\varphi = \varphi_m$, in the second case $\varphi = \varphi_0 + \pi$) clearly demonstrates the difference between the modes (Fig. **2**).

Apart from a rectangular array, the other variant can be implemented on site, *e.g.*, aboard a ship. The variant uses the existing set of ship antennas. As a result, an array of arbitrarily located radiators is formed. Let the current in the base of the

nth radiator be $J_n(0)=\left|J_n\exp(-j\psi_n)\right|$. Fig. **3a** presents phase shift φ_m, ensuring that the field phase of the nth radiator coincides in the far region with the field phase of the radiator located at the coordinate origin. If the condition is met for all radiators, their fields are together in the observation point (in the forward radiation mode).

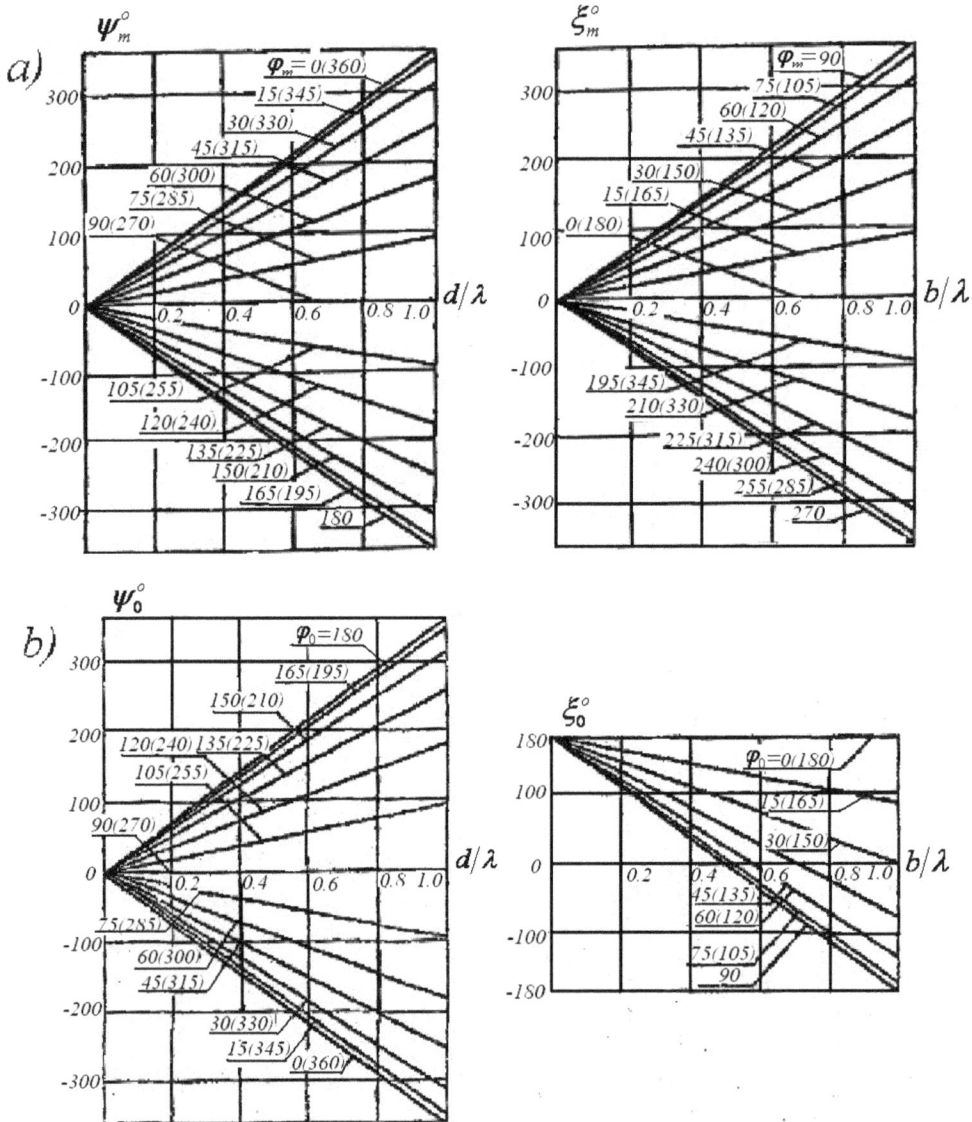

Figure 2: Phase shifts in the modes of maximum forward radiation (*a*) and zero backward radiation (*b*).

The phase of the n*th* radiator field in the far region is

$$\Phi_n = -kr_n - \psi_n,$$

where $r_n = r - x_n \cos\varphi - y_n \sin\varphi$ is the distance from the radiator to the observation point. If $\Phi_n = const(n) = -kr$, then $\psi_n = k(x_n \cos\varphi + y_n \sin\varphi)$. Let $x_n/\lambda = D + d/\lambda$, $y_n/\lambda = B + b/\lambda$, where D and B are integers, d/λ and b/λ are proper fractions. Then

$$\psi_n = D\psi_1 + B\xi_1 + \psi_m + \xi_m. \tag{8.11}$$

Here, $\psi_1 = 2\pi\cos\varphi$ and $\xi_1 = 2\pi\sin\varphi$ are found from Fig. **3b** (depending on φ), ψ_m and ξ_m - from Fig. **2a** or expression (8.9).

The directional performance of radiators is known to be described by the directivity, which is given by the ratio of the maximum of radiation intensity to its average value in the radiation sphere. The parameter represents the number of times, one must increase the power in passing from a directional to omnidirectional (isotropic) antenna under condition of field retention at the receiving point. Since the field strength of a directional antenna is $E = E_m F(\theta,\varphi)$, where E_m is the field in the direction of the maximum radiation, the power of such antenna is an integral of the Poynting vector

$$P_\Sigma = \int\limits_{(s)} \frac{E_m^2 F^2(\theta,\varphi)}{Z_0} dS. \tag{8.12}$$

The integral is taken over the surface of a sphere with great radius r (in the far region), where a surface element is equal to $dS = r^2 \sin\theta d\theta d\varphi$. The radiation power of omnidirectional antenna is

$$P_{\Sigma 1} = \int\limits_{(s)} \frac{E_1^2}{Z_0} dS = \frac{4\pi r^2 E_1^2}{Z_0}, \tag{8.13}$$

where E_1 is the field of omnidirectional antenna, which in accordance to the fields equality condition at the observation point must be equal to $E_1 = E_m(\theta_1,\varphi_1)$. Therefore we obtain for directivity in an arbitrary direction

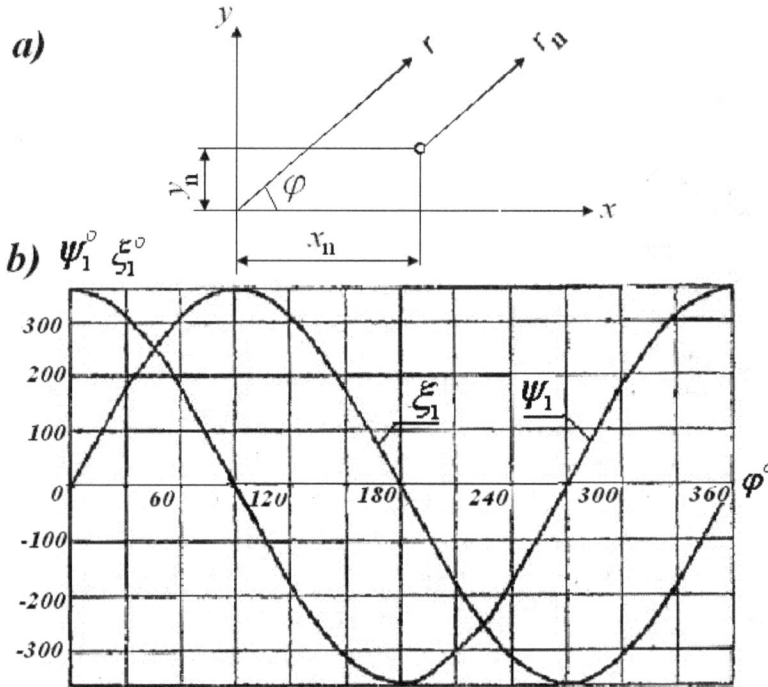

Figure 3: Circuit of the nth radiator placement (a) and dependence of ψ_1 and ξ_1 on φ (b).

$$D_1 = \frac{P_{\Sigma 1}}{P_{\Sigma}} = \frac{4\pi F^2(\theta_1,\varphi_1)}{\int\limits_0^{2\pi} d\varphi \int\limits_0^{\pi} F^2(\theta,\varphi)\sin\theta d\theta}.\tag{8.14}$$

In particular, the directivity in direction of the maximum radiation, when $F(\theta_1,\varphi_1)=1$, is

$$D_m = \frac{4\pi}{\int\limits_0^{2\pi} d\varphi \int\limits_0^{\pi} F^2(\theta,\varphi)\sin\theta d\theta}.\tag{8.15}$$

As one can see from (8.14) and (8.15), quantity D_1 is determined easily, if D_m and $F(\theta_1,\varphi_1)$ are known: $D_1 = D_m F^2(\theta_1,\varphi_1)$.

Figs. **4-6** present the maximum directivity of linear and double-tier arrays consisting of isotropic radiators. Directivity is calculated in accordance with

(8.15) by replacing $F(\theta, \varphi)$ with $F_{MN}(\theta, \varphi)$. The calculated directivity of several arrays at three frequencies at different azimuths of a main lobe is given in Table **1** in the mode of the maximum forward radiation. Processing of calculation results permits to build other characteristics of directivity for the radiator system: the main lobe width, the side lobe level, and the radiation level in the opposite direction. In particular, the half-power width of the main lobe is presented in Fig. 7.

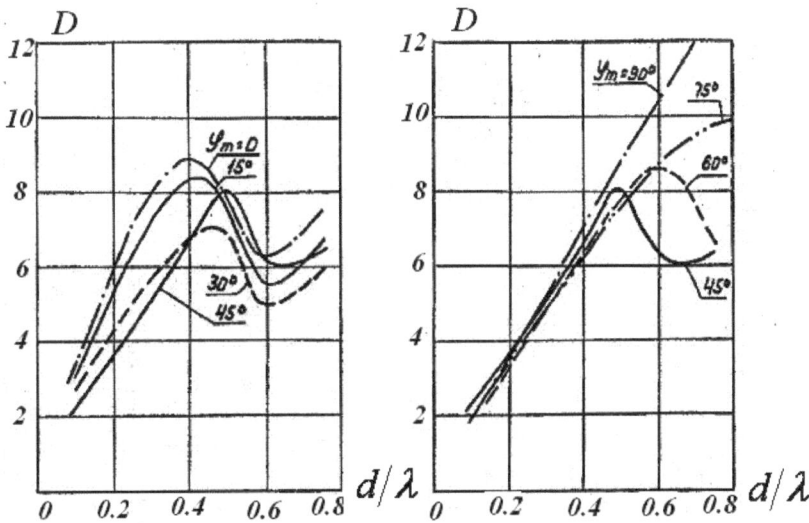

Figure 4: Directivity of a linear array of 8 radiators.

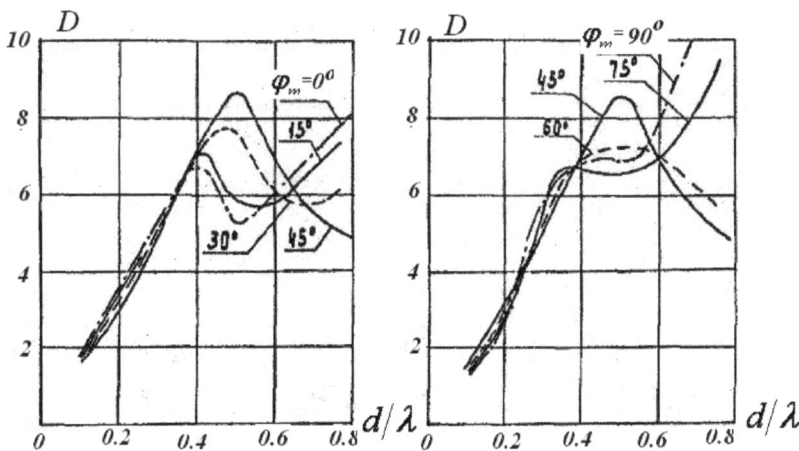

Figure 5: Directivity of a double-tier array of 8 radiators in the mode of the maximum forward radiation.

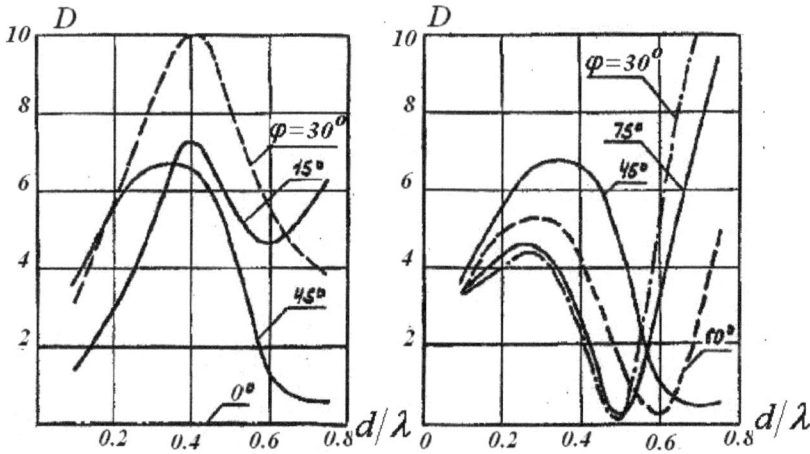

Figure 6: Directivity of a double-tier array of 8 radiators in the mode of the zero backward radiation.

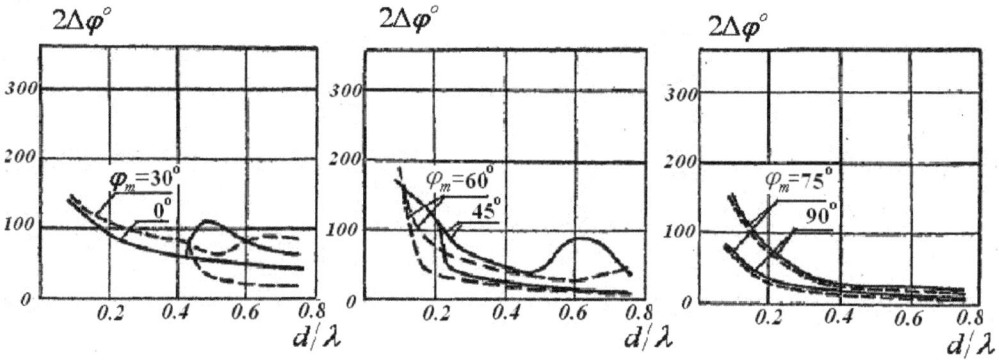

Figure 7: Main lobe width of a linear array of 8 radiators.

If there are several main lobes, two curves are given with the upper curve showing the sum of lobe width.

Table 1: Directivity of antenna arrays in the mode of maximum forward radiation.

Array type	$d/\lambda = 0/2$	0.4	0.6	0.2	0.4	0.6
	in units			in decibels		
Linear, 4 radiators	1.8-3.2	3.1-5.5	2.9-4.9	2.5-5.0	4.9-7.4	4.6-6.9
Linear, 8 radiators	3.3-6.0	6.2-8.9	5.0-10.3	5.2-7.8	7.9-9.5	7.0-10.1
Square	1.8	3.5-4.0	3.5-3.8	2.5	5.4-6.0	5.0-5.8
Double-tier, 8 radiators	2.5-3.5	6.7-7.2	5.7-7.9	4.0-5.4	8.2-8.6	7.-9.0

The analysis of results presented in table and in figures permits to compare the characteristics of different arrays depending on the number of radiators.

8.2. LOG-PERIODIC ARRAYS

A log-periodic dipole antenna (LPDA) provides directional radiation along the longitudinal axis. Also, it retains the pattern shape and has constant input impedance in a wide frequency range, *i.e.,* belongs to the class of frequency-independent antennas, since such antenna implements the principle of electrodynamics similarity. In accordance with the principle, a radiator has the same electrical characteristics at two different frequencies, if its geometric dimensions change, as the frequency changes, in proportion to the wave length. Not only tunable antennas satisfy the principle of electrodynamics similarity, but also those with the shape defined by angular dimensions, *e.g.,* conic radiators of infinite length satisfy the principle. In this case, a change in the scale fails to lead to an antenna change, *i.e.,* the radiator shape and its dimensions in wavelengths are the same at different frequencies.

Relationships of similarity follow from Maxwell equations. For the harmonic field created by an antenna in surrounding space, we can write, accordingly (1.1),

$$curl\vec{H} = (\sigma + j\omega\varepsilon)\vec{E}, \ curl\vec{E} = -j\omega\mu\vec{H} \ . \tag{8.16}$$

Similarly, at another frequency

$$curl\vec{H}' = (\sigma' + j\omega'\varepsilon')\vec{E}', \ curl\vec{E}' = -j\omega'\mu'\vec{H}', \tag{8.17}$$

where $\vec{E}' = k_E\vec{E}, \vec{H}' = k_H\vec{H}, \omega' = k_\omega\omega, \varepsilon' = k_\varepsilon\varepsilon, \mu' = k_\mu\mu, \sigma' = k_\sigma\sigma, l' = k_l l'$,   here, $k_E, k_H, k_\omega, k_\varepsilon, k_\mu, \ k_\sigma, k_l$ are the coefficients, interrelating quantities at different frequencies in the equation, and l is the distance (arbitrary linear coordinate). The said coefficients are called those of modeling, since the use of the principle of similarity allows studying antennas characteristics by means of a model experiment.

Substitute the coefficients into (8.17) and take into account the linearity of quantities and operator $curl\vec{A}$, where $\vec{A}$ is an arbitrary vector. Since the

resulting   equations   are   to   coincide   with   equations   (8.16),   then
$k_H = k_\sigma k_E k_l, k_H = k_\omega k_\varepsilon K_E k_L, k_E = k_\omega K_\mu k_H k_l$.   If $k_\varepsilon = k_\mu = 1$,   then
$k_H / k_E = k_\sigma k_l = k_\omega k_l = 1/(k_\omega k_l)$, i.e.,

$$k_\omega k_l = 1, k_H = k_E, k_\sigma k_l = 1. \tag{8.18}$$

It follows from the expression that the electromagnetic field at different frequencies will be the same, if the electrical dimensions of an antenna (the ratio of linear dimensions to the wavelength) at different frequencies coincide and the material conductivity is in inverse proportion to quantity k_l, i.e., increases with a frequency.

A similar conclusion is valid for model studying antennas. When the geometrical dimensions of a model are smaller by factor N than those of the original, it is necessary to increase the signal frequency and the conductivity of model material N times. Since $k_H = k_E$, i.e., relationship of the currents and voltages remains the same, the resistances of resistors should remains the same, whereas the capacitances of capacitors and the inductances of coils connected in the model should be smaller by factor N. As a rule, the conductivity of model material could not be increased by factor N. So, the active components of an antenna and the model input impedances as well as the characteristics depending on them (*e.g.*, the efficiency, energy factor Q, the gain) differ substantially from each other.

As said earlier, the LPDA belongs to the class of frequency-independent antennas. Of special interest among them are the antennas having the property of the automatic current cut-off. This means that the field at each frequency is radiated by a small segment of the antenna (by the active area), and attenuates quickly outside the boundaries of the area. Here, the coordinates and dimensions of radiated segment are rigidly concerned with the wavelength value. When the frequency changes, the antenna section, radiating the field, (the active area) moves along the antenna. The electrical dimensions of the area, both longitudinal and lateral, remain constant ensuring the invariability of the characteristics.

If the antenna has finite dimensions, its frequency range is finite, but the antenna has, in this finite range, the properties of an infinite antenna. The maximum wavelength is governed by the maximal lateral dimension (by the width) of the antenna, and the minimum one is governed mostly by the accuracy of the structure manufacturing near the excitation point.

LPDA (Fig. **8a**) is a collection of elements (of wires), their dimensions forming a geometric progression with ratio $1/\tau$:

$$R_n/R_{n+1} = l_n/l_{n+1} = \tau. \tag{8.19}$$

Accordingly, the antenna electrical characteristics are repeated at frequencies forming the geometric progression with the same ratio. It means that the directional characteristics and input impedance of the antenna prove to be periodic functions of logarithm of frequency f, *i.e.,* when the electrical characteristics are plotted as a function of $\ln f$, their values repeat with period equal to $\ln \tau$. Hence the antenna name.

An indispensable condition of a weak frequency dependence of antenna characteristics is their varying little within one period. To meet the condition, the frequency period itself must be small. But this is insufficient.

The LPDA shown in Fig. **8a** consists of two structures situated in one plane. Each structure is shaped as a straight wire, with linear conductors attached at right angles to it alternately from the left and from the right. Their lengths increase with the growing distance from the excitation point, following the law of geometric progression. Such antenna is a simplified and modified variant of a flat log-periodic structure shown in Fig. **8b**, which is the self-complementary structure, *i.e.,* consists of metal plates and slots coinciding with each other in shape and dimensions. The input impedance of a flat infinite self-complementary structure is purely active, independent of the frequency, and equal to 60π (see Chapter 4). Construction of a log-periodic antenna in the form of a self-complementary or similar structure ensures a little variation of electrical characteristics of the antenna within one frequency period.

Each of two structures, forming a LPDA (see Fig. **8a**), differs from the structure, forming an arm of a flat log-periodic antenna (see Fig. **8b**), with replacements of the metal sector 1' with the longitudinal wire 1, of metal strip 2' situated along the arc of a circumference with lateral wire 2, tangent to the arc, and of the slot strip 3' with the interval between the lateral wires. Such construction is essentially simpler to implement and, at the same time, close to the original one in the electrical properties.

Rotation of one of antenna metal structures (of one arm) about the *y*-axis (see Fig. **8b**) through angle π and placement of both structures in one plane allow providing unidirectional radiation. It is convenient to interpret the unidirectional log-periodic antenna shown in Fig. **8a** as a linear array of symmetrical radiators that have monotonically changing length and are excited by a two-wire feeder. A generator is connected in the feeder from the shorter radiators.

A convenient implementation of an antenna design, which is requires no special balun, is shown in Fig. **9**. The cable is laid inside one of tubes forming a two-wire feeder (distribution line).The cable sheath forms an integral unit united with the tube, and its inner conductor is connected to the second tube at the antenna vertex.

Consider the active area of LPDA with the view of explaining the principle of its operation. The area consists of dipoles with the arm length close to $\lambda/4$. In their input impedance, an active component is predominant, and the reactive component is small. In actual practice, the number of dipoles forming the active area can be as high as five. To simplify assume only three dipoles, with the arm length of central one beyng $\lambda/4$. The upper arms of dipoles are seen from Fig. **8a**

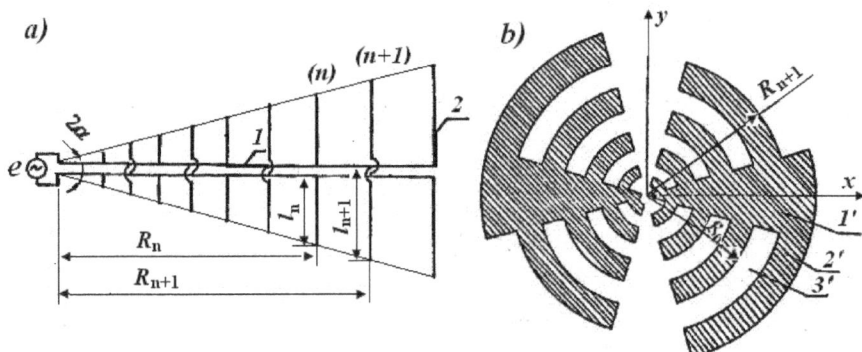

Figure 8: Log-periodic dipole antenna (*a*) and flat structure (*b*). 1 and 1' are distribution lines, 2 is a dipole, 2' is a metal strip, 3' is a slot

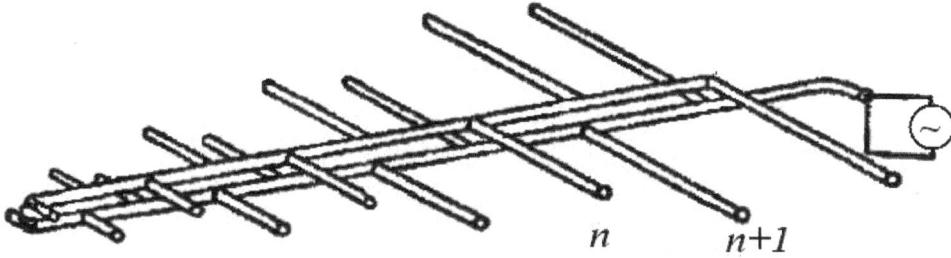

Figure 9: Construction of LPDA.

to be connected alternately to one or another conductor of the distribution line, which is equivalent to a crossing of conductors on the line segments between the dipoles. With allowance for the crossing the current in the larger dipole leads in phase the current in the resonance radiator, and the current in the shorter dipole lags behind the current in the resonance radiator, *i.e.,* larger dipole acts as a reflector and the shorter one as a director. Therefore, the fields of individual radiators are added in the direction toward the feed point (to the side of shorter dipoles) and cancel each other in the opposite direction.

The waves in the distribution line reflected from the dipoles of the active area cancel each other to a considerable extent, since the reactive components of the input impedances of shorter and large dipoles are opposite in sign. This explains a high matching level of the active area with the distribution line. In addition the electrical length of the line from the feed point to the active area remains unchanged with the changing frequency. Therefore, the active area impedance referred to the antenna input is the same at different frequencies as well.

The dipoles located outside the active area are excited weakly due to the great reactive impedance. The short dipoles at the structure beginning practically fail to radiate, since the fields established by them add almost in anti-phase because of line wires crossing and the proximity of dipoles to each other (as compared with the wavelength). As a result, the *EM* wave along this segment of distribution line has almost no attenuation, *i.e.,* the distribution of currents and voltages at the line segment between the feed point and the active area is close to that of the traveling wave mode. The short dipoles act as capacitances shunting the distribution line and thereby decreasing slightly its wave impedance. The long dipoles situated behind the active area radiate weakly as well, since, first, their input impedance is great and, second, the power of the *EM* wave at that line segment drops substantially as a result of attenuation in the active area.

The method of LPDA calculation [49] is based on antenna presentation in the form of a parallel connection of two multiport networks (Fig. **10**), one of which describes the dipole system and is defined by matrix $[Z_A]$ of mutual impedances, and the other describes the distribution line with matrix $[Y_l]$ of admittances. For each n*th* dipole of the structure with current J_{An} and input impedance Z_{An}, the following equation is valid

$$J_{An}Z_{An} = (J - J_{An})/Y_{\ln}.$$

(8.20)

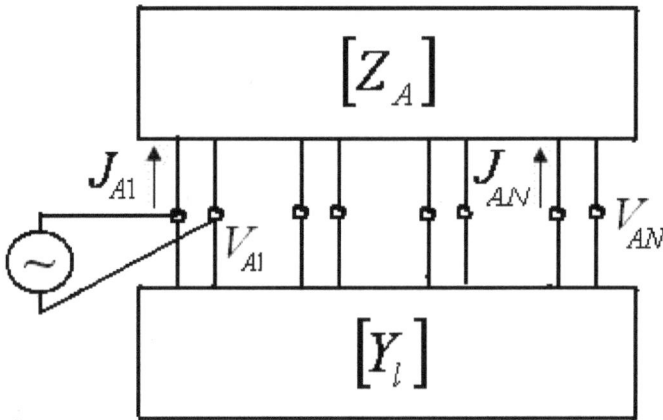

Figure 10: Equivalent circuit of LPDA.

Here, $Y_{\ln}$ is the admittance of the distribution line segment, to which the dipole is attached, and J is the impressed current excited at given junction point (or, rather, a line connecting this segment with a dipole). From here,

$$J_{An}(Z_{An}Y_{\ln} + 1) = J.$$

Accordingly, a matrix equation with respect to column-vector $[J_A]$ of dipole input current is written in the form

$$([E] + [Y_l][Z_A])[J_A] = [J],$$

(8.21)

where $[E]$ is the identity matrix, $[J]$ is the column-vector of currents exciting the lines connecting multiport networks with each other. Since the only exciting source is the generator of current

J_0 connected in the first connecting line, then $[J] = \begin{vmatrix} J_0 \\ 0 \\ \dots \\ 0 \end{vmatrix}$. Solving equation (8.21),

we find the column-vector $[J_A]$, and then matrix $[V_A] = [Z_A][J_A]$ of the voltages at the dipoles inputs. The first element of the matrix is the voltage at the input of the shortest dipole, which is equal to the input impedance of antenna, if the exciting current J_0 is equal to 1.

In [49], the elements of the matrix $[Z_A]$ are calculated, in fact, by the induced emf method. Later on, to obtain more exact results, the matrix elements were calculated by means of the integral equation solution with the help of the Moment Method. The difference between the approximate and the exact method is particularly noticeable, if the LPDA consists of thin radiators or has a wide angle at the antenna vertex. The energy in this antenna propagates along the distribution line beyond the boundaries of the active area and excites the long dipoles.

Log-periodic antennas have rather large overall dimensions. To lessen their lateral dimensions, it is expedient to shorten the longest dipoles, using loads of different kinds or concentrated impedances. Attempts to decrease longitudinal dimensions of an antenna failed, since violation of geometric progression relationships and increase of the number of dipoles cause, as a rule, sharp deterioration of the electrical characteristics and produce insignificant decrease of overall dimensions.

8.3. REFLECT ARRAYS

Lately, reflect arrays became widespread in the capacity of a flat equivalent of a parabolic reflector. The calculation method of the array is based on the reciprocity theorem described in Section 3.5. For example, an array of microstrip radiators is such an array.

The simplest microstrip antenna is a rectangular metal plate of length L and width b situated on a dielectric substrate above a metal plane (Fig. **11a**). Length L of the plate is about $\lambda_1/2$, where λ_1 is the wavelength in the substrate material. A simplified model of a microstrip antenna is a planar dipole with a sinusoidal current distribution (Fig. **11b**) and dimensions coinciding with those of the microstrip antenna. The propagation constant of the current is close to $k_1 = k\sqrt{\varepsilon_r}$, the propagation constant of a wave traveling in the substrate material (ε_r is the relative permittivity of the substrate).

The equivalent circuit of the antenna operating in the receive mode is shown in Fig. **12a**. Here Z_A is the antenna impedance in the transmit mode. In this mode, the current distribution along a planar radiator is similar to that along an impedance electric dipole excited at its center. In the first approximation, the input impedance of the radiator when $L/2 < 0.3\lambda_1$ is

$$Z_A = R_\Sigma - jW_A \cot(k_1 L/2), \tag{8.22}$$

where R_Σ is the radiation resistance, W_A is the wave impedance of the planar dipole, the value of which is equal to the doubled wave impedance of a strip line. The wave impedance of a strip line is $W = 120\left[\pi t/\left(b\sqrt{\varepsilon_r}\right)\right]$, where t is the thickness of the substrate [50]. If $L/2 < 0.3\lambda_1$, the radiation resistance can be calculated by formula $R_\Sigma = 20k^2 l_e^2$, where $l_e = (2/k_1)\tan(k_1 L/4)$ is the effective length of the antenna. The values of the input impedance, corresponding to $L/2 > 0.3\lambda_1$, can be calculated with the use of the Moment Method.

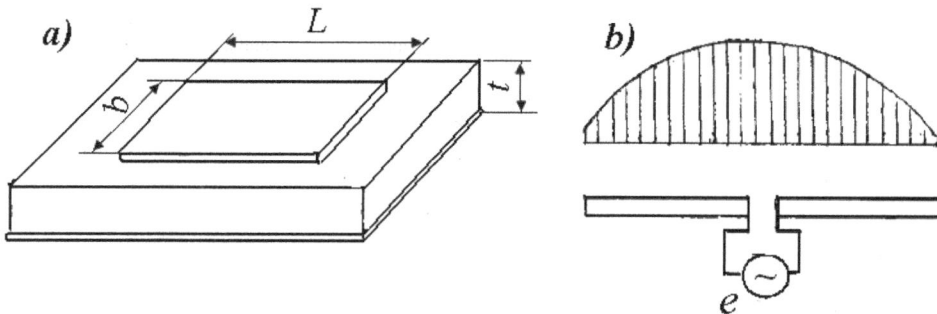

Figure 11: General view of a rectangular microstrip antenna (*a*) and it's excitation at the center of the metal plate (*b*).

The amplitude and phase of the current generated in a reradiator depend on the load impedance. For the circuit shown in Fig. **12b**,

$$I_A = e/Z_A = e/|Z_A| \exp\{j \cot^{-1}[(W_A/R_\Sigma)\cot(kL/2)]\}. \tag{8.23}$$

Quantity $\cot^{-1}[(W_A/R_\Sigma)\cot(kL/2)]$ is the phase increment of the current running in the antenna relative to the phase of the incident field. In accordance with the reciprocity theorem, the increment is equal to the phase increment of the reflected field relative to the phase of the current running in the antenna. Hence, the phase step during reradiation is

$$\varphi_1 = 2\tan^{-1}[(W_A/R_\Sigma)\cot(kL/2)]. \tag{8.24}$$

The value of the step is zero for a tuned antenna, negative for an elongated antenna, and positive for a shortened antenna. Increase of dipole radius a lowers its wave impedance W_A and decreases the phase step (Fig. **13**).

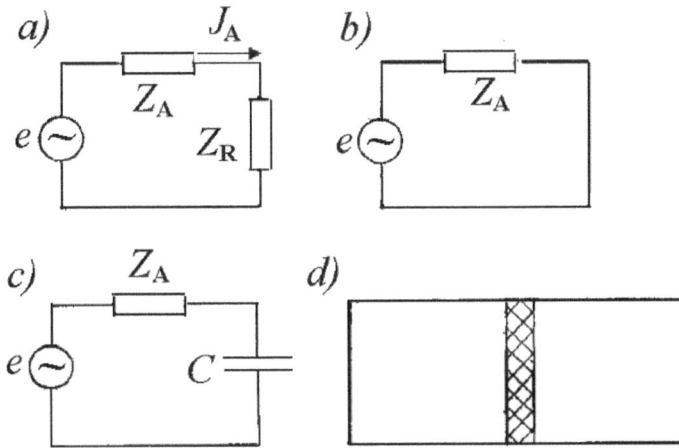

Figure 12: Equivalent circuits of a receive antenna for load Z_R (*a*), zero load (*b*), and a capacitive load formed by the slot filled with a dielectric (*c, d*).

The approximate method for calculating the phase step of the field reradiated by an element of a reflection array is based on the reciprocity theorem and the theory of dipoles. The method is simple and efficient. After only slight sophistication, this method can be used to analyze loaded antennas. For the circuit shown in Fig. **12c**,

$$I_A = \frac{e}{Z_A - (j/\omega C)} = e \Big/ \sqrt{|Z_A| + (1/\omega C)^2} \, \exp\left\{ j \tan^{-1}\left[\frac{1}{R_\Sigma}\left(\frac{1}{\omega C} + W_A \cot\frac{kL}{2} \right) \right] \right\},$$

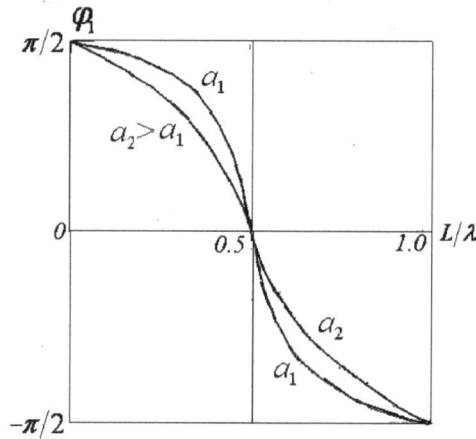

Figure 13: Field phase step during reradiating as a function of the antenna length.

i.e.,, the phase step during reradiating is

$$\varphi_2 = 2\tan^{-1}\left[\frac{1}{R_\Sigma}\left(\frac{1}{\omega C} + W_A \cot\frac{kL}{2} \right) \right].$$ (8.25)

Here ω is the circular frequency of the signal.

Apply the above results to calculation of the phase step of a microstrip antenna during the reradiating. To sum the signals from secondary radiators in the direction perpendicular to the array plane, the phases of the signals should be equal.

Assume the primary feed coordinates to be x_0, y_0 and z_0 (see Fig. **13a**). Then, the phase step in the ith reradiator, needed to compensate the phase difference in the direction of the x-axis, must be equal to $\xi_i = k\left[\sqrt{x_0^2 + (y_i - y_0)^2 + (z_i - z_0)^2} - x_0 \right]$. The choice of geometric dimensions of the ith reradiator allows it to receive phase step $\varphi_{1i} = \xi_i$.

Fig. **14** plots the field phase step φ_1 created at 60 GHz by a micro strip antenna situated on a substrate with a thickness of 0.254 mm and a relative permittivity of

2.22 as a function of the antenna length. The results were obtained by the proposed technique. Curves *1* and *2*, respectively, correspond to antenna widths of 0.3 and 2.3 mm.

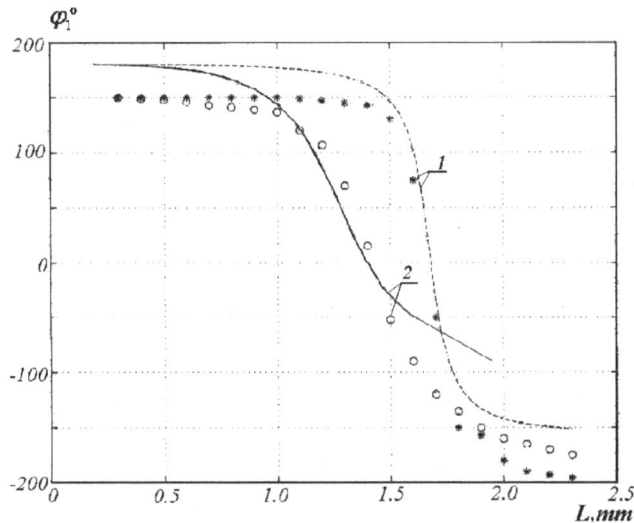

Figure 14: Phase steps φ_1 for micro strip antennas with lengths of 0.3 (curve *1*) and 2.3 mm (curve*2*). Closed and open circles show the results obtained in [51].

The rigorous method for calculating the step value was described in [51]. It relies on the analysis of an infinite periodic array of identical elements illuminated by a plane wave, *i.e.,*, on solution of the analysis problem in the spectral domain and on the Floquet's theorem. The open and closed circles in Fig. **14** correspond to the results presented in [51]. As seen from the Figure, the correspondence is satisfactory.

Along with simple microstrip antennas, multilayer (multiple-stack) microstrip antennas can be used in reflection arrays. If the field phase step arising during reradiating in a single-layer antenna is less than 360° (the phase increment arising with a sufficiently thick substrate, which ensures smooth phase variation during plate length changing and a wider frequency band, does not exceed 300°), the maximum phase step, attainable, *e.g.*, in a two-stack antenna, is 540°.

The design of a two-stack micro strip antenna is shown in Fig. **15**. In this antenna, two rectangular metal plates (of lengths L_1 and L_2 and widths b_1 and b_2, respectively) are separated from each other and from the metal plane by a dielectric substrate. The upper plate is smaller than the lower one. Plate length L_2 is larger than $\lambda_1/2$, where λ_1 is the wavelength in the substrate material.

Characteristics of a multiple-stack antenna can be calculated by the method described above. The phase step of the field generated at 12.5 GHz by a two-stack microstrip antenna containing substrates with a thickness of 3 mm and a relative permittivity of 1.03 is shown in Fig. **16** as a function of antenna length L_2. The antenna is a square. The calculating curves were obtained with the use of the proposed technique. The open and closed circles in the figure indicate the results of [52].

In order to set the maximum of the radiation pattern of an antenna array in the prescribed direction, the phases of the radiator fields must be equal in this direction. Hence, if this angle varies in plane xOz, the phases must vary linearly as functions of coordinate z_i (see Fig. **13b**): $\psi_i = k(z_1 - z_i)\cos\theta$. To simplify the control procedure, the field phases should be controlled by an electric signal.

Among the circuits of microstrip antennas, circuits with loads are most suitable for continuous control of the phase of the reradiated field. In order to connect the load impedance in series into the circuit of a receiving antenna, a slot can be cut in the central part of the planar radiator across its long sides (Fig. **12d**). In this case, a slot filled with a dielectric forms the simplest load: a capacitor with capacitance C (see Fig. **12c**) and reactance $1/j\omega C$. The capacitor capacitance can be controlled readily *via* variation of the permittivity of a special material placed between the capacitor plates during application of voltage between the plates.

Consider the circuit presented in Fig. **12c**. The phase step during reradiation is determined by (8.25). The tangent of angle φ_2 depends on two terms. The first term, $\alpha = 1/(R_\Sigma \omega c)$ is related to the presence of a capacitor at the radiator center while the second term, $\beta = (W_A/R_\Sigma)\cot(k_1 L/2)$ is related to the deviation of the radiator length from the resonant value. The second term can be used to compensate of the phase difference caused by the differences in distances r_i between primary feed 1 and reradiators 2 (see Fig. **13a**), while the first term can be used for turning the radiation pattern. The total phase step in the ith reradiator should be $\varphi_{2i} = \xi_i + \psi_i$. Here, if reactances $1/(j\omega C_i)$ of capacitors C_i, corresponding to $\varepsilon_r = 1$ vary linearly with coordinate z_i, then, applying equal voltages to the capacitors filled with the same dielectric, we find that the reactances of these capacitors retain a linear dependence on this coordinate. This statement is equally valid for the first term in the expression for the above tangent.

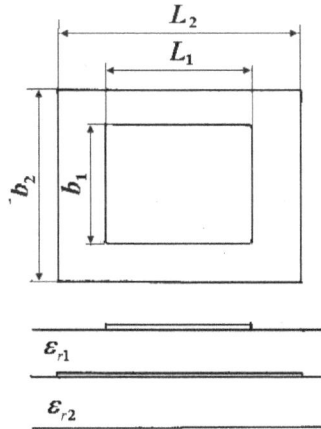

Figure 15: Two-stack microstrip antenna.

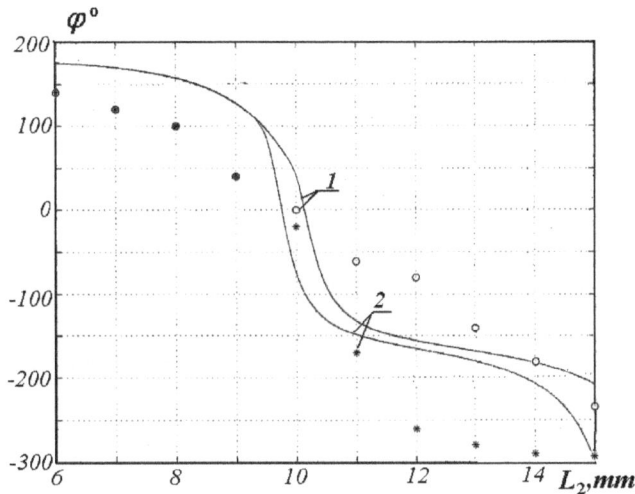

Figure 16: Phase steps φ for a two-stack micro strip antennas with L_1/L_2 =0.6 (*1*) and 0.8 (*2*). The open and closed circles present the results obtained in [52].

Note, however, that the second term takes different values for different reradiators because it depends on distance r_i. Therefore, phase φ_2 will vary nonlinearly after application of the same voltage to all capacitors. Note that even if the second term is absent, *i.e.,* $\tan\psi_i = 1/(R_\Sigma\omega C_i)$ and the angular turning the radiation pattern requires a linear dependence of phase ψ_i on coordinate z_i, the value $\tan\psi_i$ and, accordingly, capacitance C_i do not possess this property. Therefore, in the general case, the angular displacement of the radiation pattern requires application of an individual voltage to each capacitor.

The case when angle θ of the maximum radiation of the antenna array is close to $\pi/2$, *i.e.,* $\alpha\langle\langle\beta$, requires a special analysis. In this case, by expanding the function $\varphi_2 = \tan^{-1}(\alpha + \beta)$ into the Taylor series, we obtain

$$\varphi_2(\alpha + \beta) = \tan^{-1}\beta + \alpha/(1 + \beta^2).$$

Here, the term $\dfrac{\alpha}{1+\beta^2} = \left[R_\Sigma\omega C\left(1 + \dfrac{1}{R_\Sigma^2}W_A^2\cot^2\dfrac{k_1 L}{2}\right)\right]^{-1}$ is proportional to

reactance $1/(j\omega C)$ of capacitor C. Hence, phases of the fields generated by secondary radiators will vary linearly with the coordinate upon application of equal voltages to the capacitors filled with the same dielectric.

If the direction of the maximum radiation differs substantially from the perpendicular to the array plane, different voltages can be applied to several groups of antennas in order to bring the law of variation of phase φ_2 along the antenna closer to a linear function. Note that the number of these voltages can be substantially less than the total number of radiators.

During calculation of the capacitance formed by a slot cut in the plate of a microstrip antenna, it should be taken into consideration that this capacitance consists of two terms: capacitance C_1 between thin planar plates and capacitance C_2 of the planar capacitor formed between the plate edge surfaces.

8.4. ADAPTIVE ARRAY

During receiving a radio signal, possible filtration of interference, *i.e.,* separation of useful signal and simultaneous suppression of interfering signals, is of important significance. The spatial filtration is one of most efficient methods of fighting interference. To apply it, one must know the direction of the useful signal arrival so that the main lobe of the reception pattern could be oriented on the source of the signal, and nulls of the pattern – on sources of interferences and disturbances.

The array of receive antennas, forming a unified system with receiving equipment can act as an efficient spatial filter. Adjustment of such system is performed with the help of special weighting devices (attenuators), forming the required pattern, and adaptive controlling circuit (feedback loop), which uses an iterative procedure to automatically choose optimal parameters of the system and then automatically adapts to changing conditions. For this reason, the described antenna system is called an adaptive one.

An advantage of adaptive processing is the fact that the suppression of interference, as a rule, involves no decrease of the useful signal. The automatic control of parameters is of special importance where constant factors, degrading the antenna performance, act often in concert with variable factors that exist, for example, aboard ship: running rigging, motion of various steel ropes under the action of wind or pitching and rolling, rotation or tuning of nearby antennas, the weather effects, *etc.*

An adaptive antenna system operates in the situation, when the spectrum of the useful signal and the direction of its arrival are known, whereas the field structure of the source incorporating noise and interference and the direction towards the source are not. The system uses an artificially introduced reference signal that is produced in the receiver and has the spectral characteristics and azimuth coinciding with those of the useful signal, known approximately.

The principle of beam forming in the adaptive antenna system with the help of weighting devices is clear from Fig. **17**. When multiplying output signals of array elements by weighting coefficients, the latter can be selected to secure that the main lobe undergoes almost no change (*i.e.,* that the magnitude of the received useful signal remains the same), and the direction of zero reception coincides with that towards the interference source. A possible variant to implement required weighting coefficient W is using a circuit two parallel channels at each element output with system adjustment in amplitude and phase delay by $\pi/2$ in one channel only. Such element is named as a circuit with quadrature channels. Introduction of the phase delay equal to $\pi/2$ is unnecessary, yet useful, since it allows obtaining a close magnitude of the weighting coefficient in the adjacent channel.

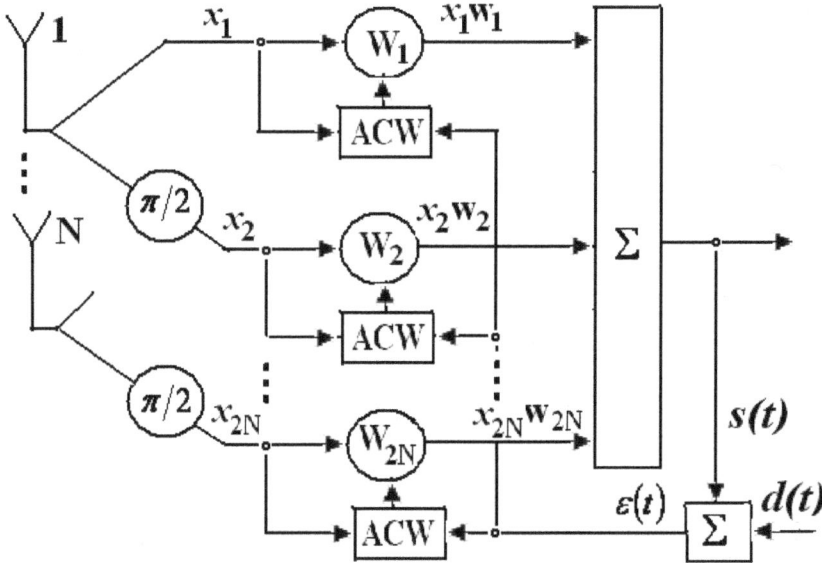

Figure 17: Antennas array with adaptive control of weighting coefficients.

The adaptive control of weighting coefficients (ACW) is performed with the help of controlling circuit – adaptive processor (AP). It automatically adjusts the weights by the iterative procedure in accordance with the chosen algorithm. Error signal $\varepsilon(t)$ is used as a controlling one in adjustment circuits for weighting coefficients. The error signal is equal to the difference between reference signal $d(t)$ (close to required output signal) and actual signal $s(t)$ at the adder output:

$$\varepsilon(t) = d(t) - s(t), \tag{8.26}$$

where output signal $s(t)$ is the sum of signals $x_n(t)$ with weighting coefficients W_n:

$$s(t) = \sum_{n=1}^{2N} x_n(t) W_n . \tag{8.27}$$

Here, N is the number of antennas, $2N$ is the number of weighting coefficients. Quantity $x_n(t)$ is to take into account the phase delay, equal to $\pi/2$ for even values of n.

Three adaptation algorithms are applied: 1) differential or greatest steepness, 2) least mean squared error, 3) random search [53]. The first two are based on the steepest descent method.

The adaptation process starts with a set of several arbitrary coefficients. Then, if the steepest descent method is used, the gradient of error function is measured, and the weighting coefficients are set such that the error function will change in the direction opposite to the gradient. The procedure is repeated to ensure that the error decreases and the weighting coefficients approach the optimal values. If the differential method is used, the gradient is evaluated directly according to the error function derivatives. In the least mean squared error method, the error value is squared, and the derivatives of the square are used to calculate the gradient. The random search method includes the measurement of the mean squared error before and after an arbitrary change of the weighting coefficients and the comparison of the results to decide whether to accept the change, if the error has decreased, or to discard it otherwise.

The least mean squared error algorithm ensures either the most rapid convergence to the same value of the error in all cases or the least error in the same operation time. Implemented in practice easier than other one, the algorithm bases on using of feedback and adjusting each weighting coefficient following the law

$$dW_n/dt = \mu\overline{\partial\varepsilon^2(t)/\partial W_n}, \tag{8.28}$$

where μ is a negative constant governing the convergence rate and the system stability, and the overline denotes the mathematical expectation. In this case, the number of arithmetical operations is linear in the number of weighting coefficients, that is, far less than in the direct calculation of the coefficients with the help of covariance matrix, where the number of operations is proportional to the third power of the number of weighting coefficients.

Rather than calculate the gradient of the mean squared error, which requires a great number of statistical samples, it is expedient to use the gradient of a single sample of the squared error (the gradient estimate), *i.e.,* to replace derivative $\partial \varepsilon^2(t)/\partial W_n$ with derivative $\partial \varepsilon^2(t)/\partial W_n$. Then the law of feedback takes the shape

$$dW_n/dt = -2\mu x_n(t)\varepsilon(t).\tag{8.29}$$

One can show expected value of the gradient estimate to be the gradient, *i.e.,* the gradient estimation is unbiased.

If the number of iterations increases without limit, the mathematical expectations of the weighting coefficients converge to the Wiener solution, for which gradient

$$\overline{\nabla \varepsilon^2(t)} = \sum_{n=1}^{2n} \overline{\partial \varepsilon^2/\partial W_n}$$ vanishes. But the convergence is secured only in the case,

when constant μ lies within certain limits. A practically convenient restriction (although stricter than necessary) is inequality

$$-1/P < \mu < 0,\tag{8.30}$$

where P is total power of input signals.

In accordance with (8.29), input signal $x_n(t)$ and error signal $\varepsilon(t)$, *i.e.,* the difference between reference signal $d(t)$ and actual signal $s(t)$ at the adder output, are fed into the processor. The error signal would have performed best, if the error signal had rather included the required output signal instead the reference one. But the latter is unavailable in the receive antenna, and we have to resort to the reference signal, close to the required one. For this reason, the main lobe of array in the process of adaptation orients in the direction specified by the reference signal. The amplitude response of the antenna system in the frequency band of the reference signal becomes uniform, and the phase response becomes linear.

Lest the reference signal distort the useful signal, two manners of adaptation, single-mode and dual-mode, are developed and used. In the dual-mode adaptation (Fig. **18**) only one processor is used, *i.e.,* it is more economical. As seen from the

Figure, the reference generator signals (RG) outputs two signals. One signal goes as the reference signal $d(t)$ to circuit for the processing. The second (control) signal imitates the useful signal arrival from the given direction. It goes through the circuits of delay δ_n to inputs of array channels. Delays δ_n are chosen so that the received input signals are identical to the signals coming from the given direction.

In the first mode, with the switch set to position I, the control signals are fed to the inputs of adaptive processor channels, and the processor adjusts the weighting coefficients so that the output signal does not differ from the reference signal, *i.e.,* turns the pattern main lobe in the given direction. In the second mode, with the switch set to position II, the signals from array elements (*i.e.,* from the surrounding space) are fed to the inputs of adaptive processor channels, and the reference and control signals are removed, lest they distort the external signal. Since there is no reference signal, *i.e.,* $d(t) = 0$, all received signals are suppressed.

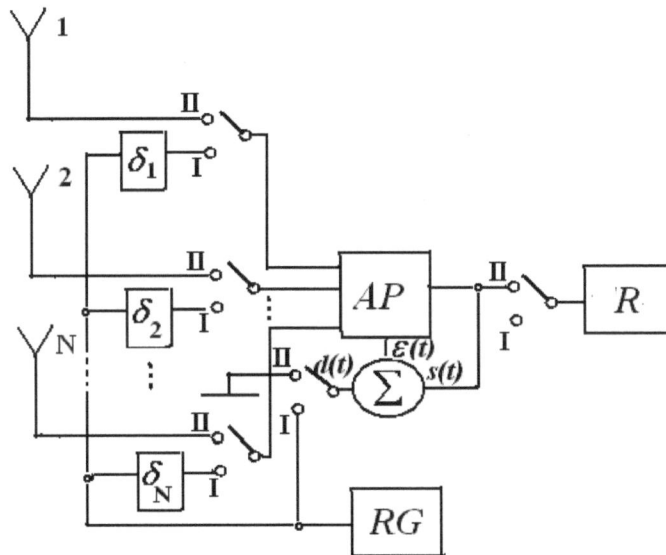

Figure 18: The structure circuit of dual-mode adaptation.

Sustained operation in the second mode leads to the self-clinching of the system, when all weighting coefficients tend to zero. But, if the modes rapidly alternate and the weights vary little during operation in each mode, the required direction of the main lobe is retained (at the operation in the first regime), and the power of interference is reduced to minimum (mostly, in the second mode). The useful signal in the second mode (switch is set to position II) arrives at the receiver input (R).

Digital simulation adaptive processing of signals confirms the procedure convergence and shows that this is an efficient method of the spatial filtration of interference with the useful signal retained. The experimental testing of adaptive system confirming its efficiency has simultaneously revealed the danger of suppressing the system operation by interference with a frequency close to that of the useful signal as well as necessity of protection, for example, by means of modulating the useful and reference signals with pseudo noise code.

Send Orders for Reprints to reprints@benthamscience.net

CHAPTER 9

Particular Antennas and Problems

Abstract: The chapter treats particular antenna problems, namely, transparent antennas, ship antennas, rectangular loop, and ground resistance. A rigorous method to calculate a transparent antenna based on solving the integral equation for the current permits to develop the antenna design. An analysis of electrical characteristics of a rectangular loop shows how its pattern differs from that of a circular loop. An application of the theorem on the oscillating power enables to simplify and define more precisely the magnitudes of the monopole ground resistance. Section dedicated to ship antennas deals with specific difficulties in their designing and service, the influence of superstructures and other metal bodies on their electrical characteristics, and the creation of the main ship antenna with inductance-capacitance load.

Keywords: Antenna located flush with metal construction, Azimuth dependence on the near field, Electrical characteristics, Energy conservation law, Flat metal triangle, Ground resistance, Inductance-capacitance load, Integral equation for the current in a flat transparent antenna, Invisible antenna, Leontovich impedance boundary condition, Linear self-capacitance of a film, Low efficiency, Main ship antenna, Oscillating power, Propagation constant, Rectangular loop, Sheet resistivity, Ship antennas, Stray capacitance between the antenna and the mast, Thick superstructures, Transparent film, Upper and lower feeding, Weak dependence of input impedance on radiator length.

9.1. TRANSPARENT ANTENNAS

The present chapter is dedicated to some specific issues of antenna engineering. In particular, it concerns the progress in creation of transparent antennas. Manufacture of transparent conducting films became the base of such progress.

Special films fabricated as *ITO* (Indium-Tin-Oxygen) coating on high-quality glass substrates are electrically conductive and optically transparent and have high sheet resistance homogeneity that permits one to use them as flat antennas for mobile communication and other applications. As an example, Fig. **1**, quoted from [54], presents the dependence of the *ITO* and *FTO* (Fluorine-Tin-Oxygen) film transmittance on its sheet resistivity R_{sq}, specified in Ohms per square section, for light (visible) wavelength 550 nm. As seen from the Figure, the transmittance increases with the film resistivity growth, and, in the case of *ITO*, sufficiently high transmittance (near 95%) is ensured, provided film resistivity R_{sq} is greater than 5 Ohm/□.

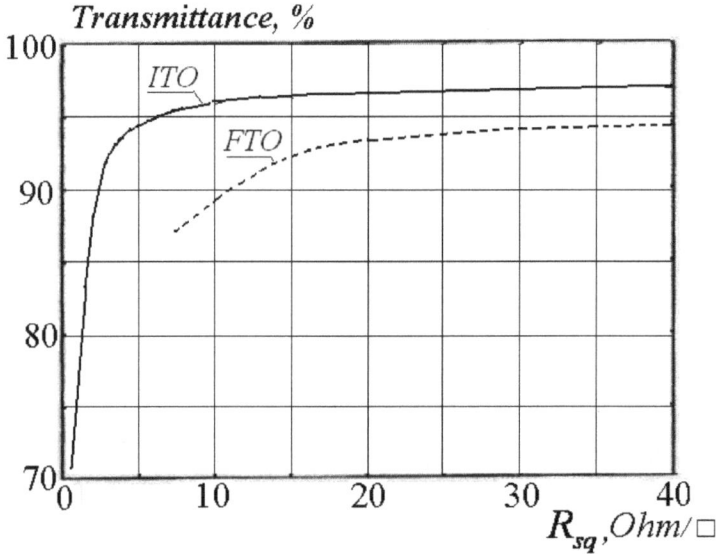

Figure 1: Film transmittance as a function of its sheet resistivity R_{sq}.

According to Leontovich impedance boundary condition, if the thickness of a metal film is greater than the penetration depth, the film loss resistance per square section is

$$R_{sq} = 34.41\sqrt{\mu_r/(\sigma\lambda)} \text{ Ohm/□,} \tag{9.1}$$

where μ_r is the relative permeability of the metal, σ is its specific conductivity for constant current (in units S/m), and λ is the wavelength (m). The specific conductivities of copper and aluminum, often used in antennas, are $5.8 \cdot 10^7$ and $3.5 \cdot 10^7$ S/m, respectively, and so their loss sheet resistances are $4.52 \cdot 10^{-3}/\sqrt{\lambda}$ and $5.92 \cdot 10^{-3}/\sqrt{\lambda}$, respectively. The conductivity of the *ITO* films used in transparent antennas is substantially lower than that of aluminum and copper. For example, sheet resistivity R_{sq} of the *CEC005P* film is about 4.5 Ohm to square. In this case, the sheet resistivity is the resistance of a film square section with thickness equal to penetration depth δ given by

$$\delta = 1/\sqrt{\pi f \mu \sigma} \,. \tag{9.2}$$

Since typically $\mu_r = 1$, absolute permeability is $\mu = \mu_0 = 4\pi \cdot 10^{-7}$ F/m. If R_{sq} is known, one can calculate specific conductivity σ from (9.1):

$$\sigma = 34.41^2 / \left(\lambda R_{sq}^2\right). \tag{9.3}$$

If thickness d of a metal film is less than penetration depth δ, the film sheet resistance is

$$R_0 = R_{sq}\delta / d \,. \tag{9.4}$$

Using expressions (9.2)-(9.4), we find: $\sigma = 195f$, $\delta = 1.14 \cdot 10^{-3}/f$, $\delta/d = 3677/f$ (for $d = 0.31 \cdot 10^{-6}$ m), $R_0 = 16549f$, where frequency f is given in GHz. At 1 GHz (λ =0.3 m), $R_0 = 16549$ Ohm. Obviously, resistance R_0 of a transparent film, *e.g.*, *ITO*, is much higher (by several orders of magnitude) than that of copper, namely, $R_{0C} = R_{sqC}\delta / d = 8.25 \cdot 10^{-3}\sqrt{f}$.

To verify the conformity of the actual sheet resistivity of the transparent films to the calculated results presented above, measurements of the sheet resistivity were performed. The measurements were carried out, as shown schematically in Fig. **2a**, on a mock-up consisting of a transparent film of thickness $0.62 \cdot 10^{-6}$ m and two flat metal cones connected to either side of the film. Length L and width b of the transparent film are 2 and 7.5 cm, respectively.

The measurements included two stages. The first stage (calibration) was accomplished on the mock-up without the transparent film (Fig. **2b**), and the second stage (measurement) dealt with the total model (Fig. **2c**). Its results are given in Table **1**. Here, Z_C and Z_f are the measured values of the impedance in the calibration and measurement modes, respectively. Difference $Z_f - Z_C$ is the transparent film impedance, given by a parallel connection of resistance R and reactance X:

$$Z_f - Z_C = \frac{RjX}{R + jX} = \frac{RjX(R - jX)}{R^2 + X^2} = \frac{RX^2}{R^2 + X^2} + j\frac{R^2 X}{R^2 + X^2},$$

and X/R is equal to the ratio of resistive and reactive components of the film impedance: $Re(Z_f - Z_C)/Jm(Z_f - Z_C) = X/R$.

Since $Re(Z_f - Z_C) = \dfrac{RX^2}{R^2 + X^2}$, then $R = Re(Z_f - Z_C)\dfrac{R^2 + X^2}{X^2}$, so, using ratio X/R, we find

$$R = Re(Z_f - Z_C)\frac{1 + [Re(Z_f - Z_c)/Jm(Z_f - Z_C)]^2}{[Re(Z_f - Z_c)/Jm(Z_f - Z_C)]^2}.$$

As noted above, the measurements were performed on a film $0.62 \cdot 10^{-6}\, m$ thick, *i.e.*, twice that used in (9.4). Since the currents run on both the top and bottom sides of the film, the value of R must be halved one more time. It was shown earlier that the sheet resistance of the film at 1 GHz is equal to 16549 Ohm. Thus, with due account of the model dimensions, *i.e.*, a double-thick film, its width, and the two film surfaces, the measured resistance should be equal to $R = R_0 L/(4b) \approx 1103$ Ohm. That is close to the measured values presented in the last row of Table **1**.

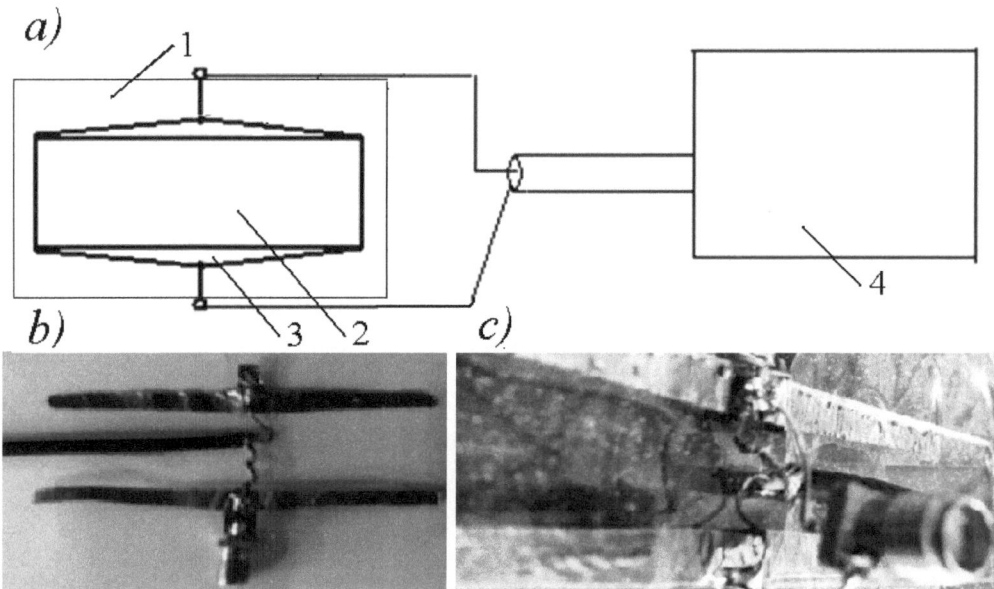

Figure 2: Sheet resistivity measurements: (*a*) –measurement circuit, (*b*) – mock-up without film, (*c*) – total mock-up. (1) – glass substrate, (2) –microstrip coating, (3) –transparent film

Knowing the current structure of antenna, in particular, knowing the current distribution law along the antenna axis, is of great importance for understanding of different physical processes in the antenna. Write down the equation for the current in a flat antenna in accordance with the integral equation for the current in a thin cylindrical antenna with nonzero (impedance) boundary conditions at its surface.

The mentioned equation (2.36) for current $J(z)$ of the symmetrical radiator in the first order approximation with respect to χ has the form

$$\frac{d^2 J(z)}{dz^2} + \left(k^2 - U\right)J(z) = -4\pi j\,\omega\varepsilon K(z)$$

(9.5)

Here, z is the coordinate along the antenna axis, $k = 2\pi/\lambda$ is the wave propagation constant along a perfectly conducting metal surface, $U = j2\omega\varepsilon\chi Z/a$, $K(z) = e\delta(z)$ is the impressed electromotive force, χ is the small parameter of thin antennas theory ($\chi \approx 1/\Omega$, where Ω is the parameter used by Hallen), ω is the circular frequency, and Z is the surface impedance of the antenna, which is in this case equal to $R_0 f$. Here it is taken into account that the symmetrical radiator consists of two wires. It is easy to see that the equation is identical to the first equation of (2.37).

The long line, equivalent to an impedance dipole, is shown in Fig. **4** of Chapter 2.

Make a transition from the cylindrical antenna to a flat one. For that purpose, we shall replace the linear capacitance (per unit length) of a cylindrical monopole in the expression for the small parameter χ with the linear capacitance of a flat plate. As shown in Section 2.5, the linear capacitance of a cylindrical monopole is the self-capacitance of a long wire of radius a. It is equal to $C_1 = \pi\varepsilon/\ln(2L/a) = 4\pi\varepsilon\chi$. Therefore, it is related to small parameter χ by $\varepsilon\chi = C_1/4\pi$.

The linear capacitance of a flat monopole is the linear self-capacitance of the plate with length $2L$ and width b, namely,

$$C_{r1} = 8\varepsilon/\left[sh^{-1}(2L/b) + (2L/b)sh^{-1}(b/\langle 2L\rangle)\right]$$

(9.6)

(see, for example, [27]).

It follows from (9.5) that the propagation constant of the wave along the antenna is a complex quantity:

$$\gamma^2 = k^2 - U = k^2 - j\omega C_{r1} R_0 f / (2b).\qquad(9.7)$$

If $k^2 << |U|$, then, as $\omega = 2\pi f \cdot 10^9$, we obtain

$$\gamma \approx \gamma_0 (1 - j),\qquad(9.8)$$

where $\gamma_0 = 10^4 f\sqrt{5\pi C_{r1} R_0 / b}$. In accordance with (9.8), the current outgoing from the feed point decays exponentially.

Table 1: The measurements and calculations results

f	1.0	1.001
Z_C	-261.11-j536.51	-511.20-j998.91
Z_f	256.61-j108.89	200.40-j116.73
$Z_f - Z_C$	517.72+j427.62	711.60+j882.17
X/R	1.21	0.807
R	871	1805

To calculate the wave impedance of the equivalent line, one should replace inductive impedance $j\omega\Lambda$ per unit length with loss resistance $R_0 f$ per unit length,

$$W_1 = \sqrt{R_0 f / (j\omega C_{r1})} = W_{10}(1 - j),\qquad(9.9)$$

where $W_{10} = \sqrt{R_0 \cdot 10^{-9} / (4\pi C_{r1})}$. The linear capacitance and resistance are constant along the equivalent length, *i.e.,* the line is uniform. Since the current quickly decays along the line, the line is an infinite long line from the viewpoint of the input impedance, and its input impedance is equal to the wave impedance:

$$Z_1 = W_1.\qquad(9.10)$$

In a particular case, when $L=2$, $b=7.5$ cm, $C_{r1} = 0.57 \cdot 10^{-10}$ F/m, then at frequency 1 GHz, $\gamma_0 = 140.5$, *i.e.,* the current decreases by a factor of e at distance 0.7 cm from the feed point. The distance varies with increasing frequency in inverse proportion to it. It is clear that the coordinate of the point where the current decreases by a factor of e can be used as the upper limit of integration and also as the monopole length, *i.e.,* the monopole should be shorter than 1 cm.

Squared propagation constant γ^2 is equal to the sum of two terms: k^2 and $U = -k_1^2$. If $k^2 \langle\langle k_1^2$, then $\gamma^2 \approx -k_1^2$. Consider different variants of ratios k^2 and k_1^2. If k_1^2 is substantially greater than k^2, then, neglecting k^2, we can write the following equation for the current in the first order approximation with respect to χ as:

$$\frac{d^2 J(z)}{dz^2} + k_1^2 J(z) = -4\pi j \omega \varepsilon K(z).$$ (9.11)

The equation solution shows that the current distribution along the antenna has the form of an exponential curve that decays from the generator towards the antenna free end. The greater the value of γ_0, the sharper the decay curve is, and the smaller the necessary length and radiator efficiency are. As shown earlier, γ_0 depends on the shape and the width of the plate. But, if the film conductivity is small, the antenna radiation resistance and efficiency are small regardless of the plate shape and width, *i.e.,* the level of its matching with the signal source is low.

If the surface resistivity is small so that k_1^2 is substantially smaller than k^2, the equation for the current in the first order approximation with respect to χ will have the form:

$$\frac{d^2 J(z)}{dz^2} + k^2 J(z) = -4\pi j \omega \varepsilon K(z).$$ (9.12)

In this case, the current distribution along the radiator has, in the first order approximation, the sinusoidal character: $J(z) = J(0)\sin k(L-|z|)/\sin kL$.

It is reasonable to describe the implications of the surface resistivity being great or small. In the first case, k_1^2 is substantially greater than k^2, while in the second case the opposite relationship holds true. Parameters k_1^2 and k^2 depend on frequency to the same extent. For k to exceed k_1 by a factor of 4-5, it is necessary, as it follows from (9.7), to decrease R_0 300-400 times at the expense of increasing the material conductivity and the transparent film thickness. At present time, the requirements are unlikely to be met.

Fig. **3** presents variants of asymmetrical flat radiators, with the ground having the shape of an infinite metal sheet: *a*) a narrow rectangular plate, *b*) a wide rectangular plate, *c*) a self- complementary antenna. The radiators were compared with each other and also with a thin cylindrical monopole of circular cross section of radius *a* (variant *d*).

It was shown above that the linear capacitance is in the case of a rectangular plate of width 7.5 cm equal to $C_1 = 0.57 \cdot 10^{-10}$ F/m. In the case of a self-complementary antenna, angle φ at the vertex of the triangle adjacent to the generator is equal to 90°, and $C_1 = 0.37 \cdot 10^{-10}$ F/m. Dividing the radiator into short segments, solving an individual equation for each segment, and tailoring the segments currents at the boundaries, one can show that the self-complementary antenna is for high sheet resistivity films (when $k^2 \langle\langle k_1^2$) undoubtedly worse than rectangular-plate one.

Compare different antennas variants at $k^2 \rangle\rangle k_1^2$. In this case, the losses power in a rectangular flat monopole of width *b* is equal to

$$P = J^2(0)R_0 / \left(b \sin^2 kL \right) \int_0^L \sin^2 k(L-z)dz ,$$

i.e., the loss resistance is

$$R_{loss} = \frac{P}{J^2(0)} = \frac{R_0 L}{2b \sin^2 kL} \left[1 - \frac{\sin 2kL}{2kL} \right].$$

If $L = \lambda/4$, $L = b$, then $R_{loss} = 0.5R_0$.

In the case of the self-complementary dipole, its width depends on coordinate *z*, namely,

$b = 2z$. Accordingly, $R_{loss} = \dfrac{R_0}{2\sin^2 kL}\displaystyle\int_{\Delta}^{L}\sin^2 k(L-z)\dfrac{dz}{z}$, where Δ is a small value tending to zero, *i.e.*,

$$R_{loss} = \frac{R_0}{4\sin^2 kL}\cdot\left[\ln(L/\Delta) - \cos 2kL(Ci2kL - Ci2k\Delta) + \sin 2kL(Si2kL - Si2k\Delta)\ \right]. \qquad (9.13)$$

Functions *Cix* and *Six* are the integral cosine and sine functions. Since Δ is small, then $Ci2k\Delta \approx C - \ln 2k\Delta$ ($C=0.5772\ldots$is the Euler's constant), and $Si2k\Delta \approx \Delta$. For $kL = \pi/2$ we obtain $R_{loss} = R_0(\ln\pi - C + Ci\pi)/4$, *i.e.*, $R_{loss} = 0.164R_0$. Comparison between the obtained results shows that the self-complementary antenna is for small sheet-resistivity films better than the rectangular one in terms of loss. Even if $b=L$ (b is to the mean width of the self-complementary antenna), the loss in the antenna of constant width is thrice higher

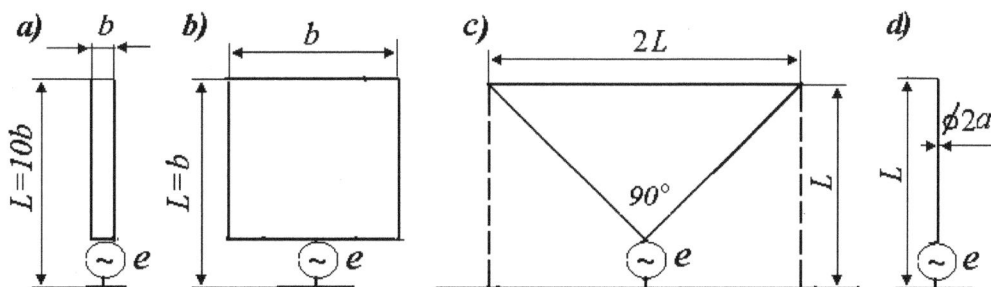

Figure 3: Variants of asymmetrical antennas: *a*) narrow rectangular plate, *b*) wide rectangular plate, *c*) self-complementary antenna, *d*) cylindrical monopole.

The detailed analysis shows that flat antennas with existing transparent conductive coatings have low efficiency. The digital simulation with the program CST confirms the qualitative coincidence of the calculations and simulation results. In particular, it confirmed that the rectangular-plates antenna is for high sheet resistivity films better than the self-complementary one.

Calculations of transparent antennas permit one to make the following conclusions:

1) Flat antennas with existing transparent conductive coatings have low efficiency.

2) Dependence of their electrical characteristics on dimensions and frequency has no marked resonant behavior. Owing to the fast current decay along the radiator, the characteristics weakly depend on the radiator length.

3) The efficiency of a rectangular antenna with the use of existing films is substantially higher than that of the self-complementary one.

4) The measurements of the film impedance comply with the tabulated values and confirm the calculated results of the sheet resistivity.

5) To increase the antenna efficiency, it is expedient to somewhat increase its width and to establish a uniform current distribution along a radiator cross section, one should use a flat metal cone (triangle), with vertex and base are connected to the feeding cable and to the radiator base, respectively. The length of the triangle base is equal to the width of the radiator formed by the film. The triangle can be realized as a printed circuit of high conductivity. The triangle width must not exceed $\lambda_{min}/4$ lest the matching the antenna elements be disturbed.

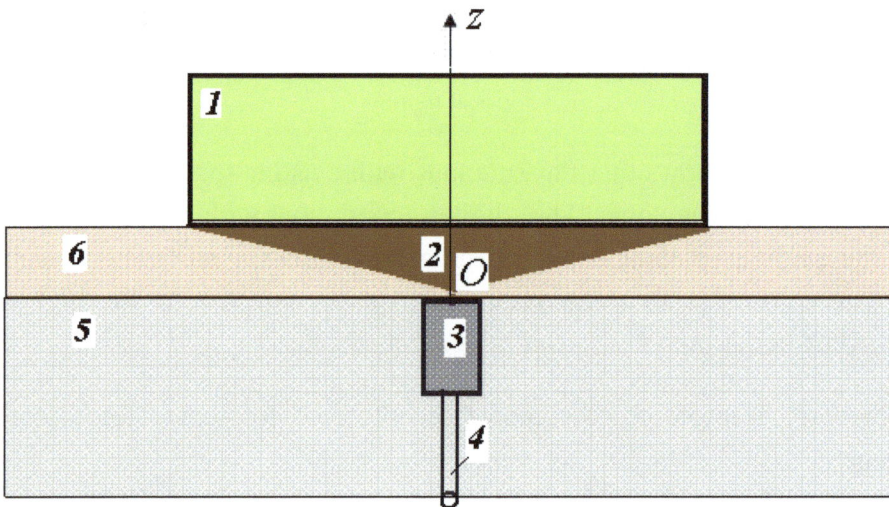

Figure 4: Transparent antenna with a flat metal cone. (1) – transparent radiator, (2) – metal triangle, (3) – balun, (4) – cable, (5) – ground, (6) - opaque material

The triangle can be placed behind an opaque material, *e.g.,* beyond the lower edge of a car window pane. In general, it is better to implement an asymmetrical radiator (monopole), as it is easier to shelter the feeding triangle beyond the window pane edge (in contrast to dipole). The general view of such antenna, based on these conclusions, is presented in Fig. **4**.

The shape of the triangle determines its wave impedance. If the wave impedances of the triangle and the film are close in magnitude, the signal reflection from the boundary of two sections is minimum, the current distribution along the structure is smooth, and the whole structure "triangle-film" is close to uniform one.

The wave impedance of two flat triangles with vertex angle 2α situated at an angle $180°$ to each other is given in [38]. In an asymmetrical variant, it is half as much and equal to

$$W_2 = 60\pi\, K(n)/K\left(\sqrt{1-n^2}\right), \tag{9.14}$$

where $K(n)$ is the total elliptic integral of the first kind of argument n where $n = (1-\sin\alpha)/2$.

Table **2** gives the calculation results of ratio $K(n)/K\left(\sqrt{1-n^2}\right)$ and quantity W_2 for the structure of two flat triangles as a function of angle α. They show W_2 to increase as α decreases. Here, quantity W_2 tends to quantity W_1. The input impedance of the total antenna is

$$Z_2 = W_2\, \frac{Z_1/W_2 + j\tan kl}{1 + jZ_1\tan kl/W_2},$$

where l is the triangle height. The reflectivity is

$$\rho = \sqrt{\left[(R_2 - W_c)^2 + X_2^2\right]/\left[(R_2 + W_c)^2 + X_2^2\right]},$$

where W_c is the cable wave impedance. The standing wave ratio (SWR) is equal to $SWR = (1+\rho)/(1-\rho)$.

If the vertical z-axis coincides with the axis of the structure symmetry, and its origin with the feed point (see Fig. **4**), we write currents $J_1(z)$ and $J_2(z)$ along the upper (thin film) and lower (metal triangle) sections of the antenna in the form

$$J_1(z) = J_1(l)\exp(-\gamma z), \quad J_2(z) = J_2(0)\sin k(l + l_e - z)/\sin k(l + l_e), \qquad (9.15)$$

where $l_e = (1/k)\tan^{-1}(W_2/Z_1)$, $J_1(l)$ and $J_2(0)$ are the currents in the base of the upper and lower sections, respectively, with $J_1(l) = J_2(l) = J_2(0)\sin kl_e/\sin k(l + l_e)$. The antenna effective height is found from the expression

$$h_e = h_{e1} + h_{e2}, \qquad (9.16)$$

where

$$h_{e1} = \frac{1}{J_2(0)}\int_l^{l+L} J_1(l)\exp(-\gamma z)dz = \frac{\sin kl_e}{\gamma \sin k(l + l_e)}\left[\exp(-\gamma l) - \exp(-\gamma\langle l + L\rangle)\right] \text{ and}$$

$$h_{e2} = \int_0^l \frac{\sin k(l + l_e - z)}{\sin k(l + l_e)}dz = \frac{\cos kl_e - \cos k(l + l_e)}{k \sin k(l + l_e)}$$

are the effective heights of the upper and lower sections of the antenna, respectively. The radiation resistance and efficiency are $R_\Sigma = 40(kh_e)^2$ and $\eta_\Sigma = R_\Sigma/(R_\Sigma + R_2)$.

Further on, the characteristics of two antenna variants of width 7.5 and 2 cm are considered. In both variants, the rectangle and the triangle are equal to 2 and 0.5 cm high, respectively. Antennas were excited by the generators with the output resistance 75 and 50 Ohm.

Table 2: Wave impedance of two flat triangles

α°	40	30	20	12	10	8	5	1
$K(n)/K(\sqrt{1-n^2})$	0.51	0.57	0.64	0.69	0.71	0.72	0.74	0.77
W_2		96	107	121	130	134	136	139	145

The electrical characteristics of antennas calculated with the Program CST were compared with measurement results of the full-scale models. Calculated values of input impedance Z_A with allowance for short insert at point 3 (see Fig. **4**), the total efficiency η (with allowance for matching) and gain G of both antenna models are given in Table **3**, at frequencies from 2 to 8 GHz. Calculated curves for reflectivity ρ, the standing-wave ratio and the total efficiency are presented at Figs. **5** to **7**. The antenna patterns in the horizontal and vertical planes are given in Figs. **8** and **9**. Experiment (points) showed on the whole, compliance with these results.

Table 3: Electrical characteristics of antennas models

f, GHz	Width 7.5 cm			Width 2 cm		
	η	G	Z_A, Ohm	η	G	Z_A,Ohm
2	0.06	3.1	1.5-j5.5	0.008	3.0	3.1-j97
3	0.23	3.3	5.6+j46.5	0.05	3.0	3.0-j54
4	0.09	3.7	19.5-j123.5	0.21	3.0	4.1-j28.5
5	0.10	5.0	4.4-j13.3	0.57	3.0	6.8-j9.2
6	0.52	4.4	8.5+j19.6	0.79	3.1	11.4+j8.1
7	0.47	3.8	42.4+j83.7	0,79	3.2	18.3+j23.0
8				0.65	3.5	30.5+j41.3

It is seen from the tables and figures that the curves for the variant of antenna 7.5 cm wide have discontinuous form, and this effect is an essential disadvantage when the antenna is to operate in a wide frequency range. The causes of the effect are the large dimensions of the metal triangle on order of the wave lengths and reflection of the current wave from the interface of the transparent film and the metal strip. The absence of the disadvantage is an important asset of the narrower model.

The transparent metal radiator has been analyzed basing on the integral equation for the current to reveal that the use of existing films with high resistivity causes the radiator current to decay rapidly from the feed point following the exponential law. For this reason, the radiator has low efficiency, and its characteristics weakly depend on the radiator length. Creating a uniform distribution of currents along the radiator cross section by means of a metal triangle permits to obtain a sufficiently efficient antenna, the undoubted advantage of which is its wide frequency band, making it suitable for UWB communications. It is expedient to use operation frequencies starting with 3-4 GHz. The poor performance of existing transparent films prohibits obtaining high levels of matching and efficiency at low frequencies, while the increasing the antenna dimensions fails to lead to any positive practical effect. But one can create at the said frequencies an invisible antenna with high electrical characteristics on the base of transparent films.

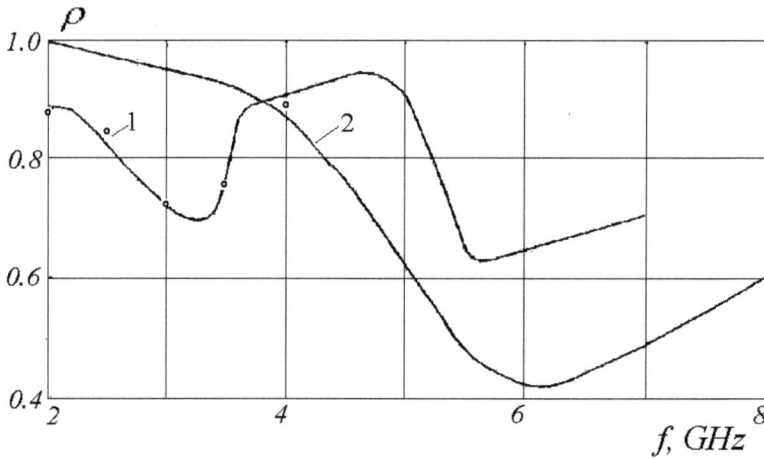

Figure 5: Reflectivity of the antennas with widths 7.5 (1) and 2 (2) cm.

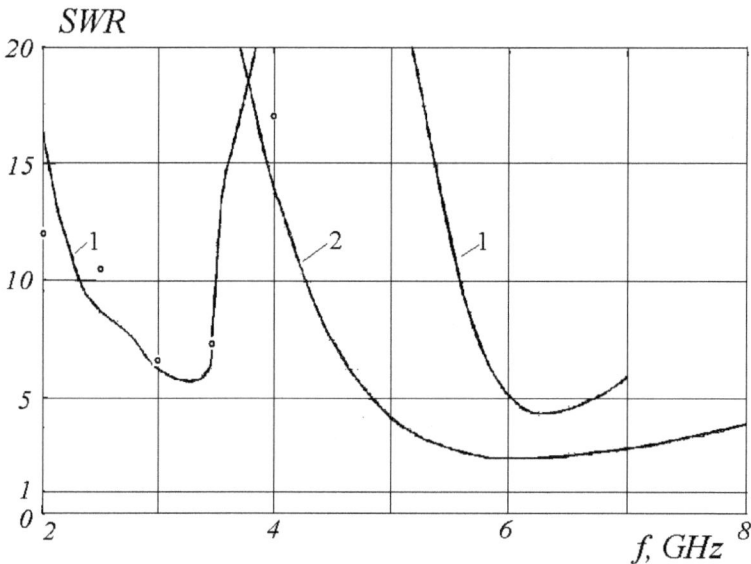

Figure 6: Standing-wave ratio of antennas with width 7.5 (1) and 2 (2) cm.

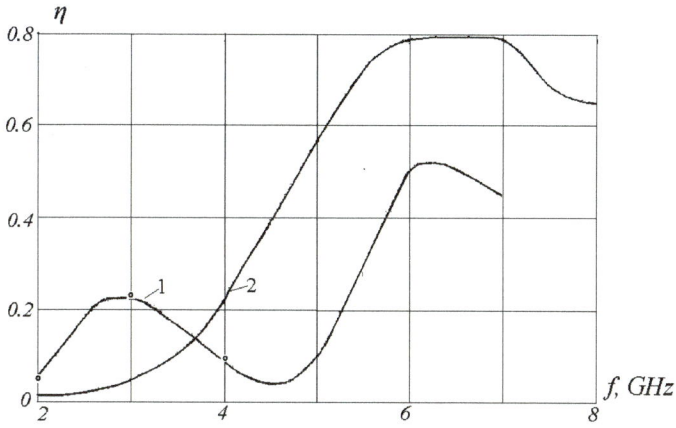

Figure 7: Efficiency of antennas with widths 7.5 (1) and 2 (2) cm.

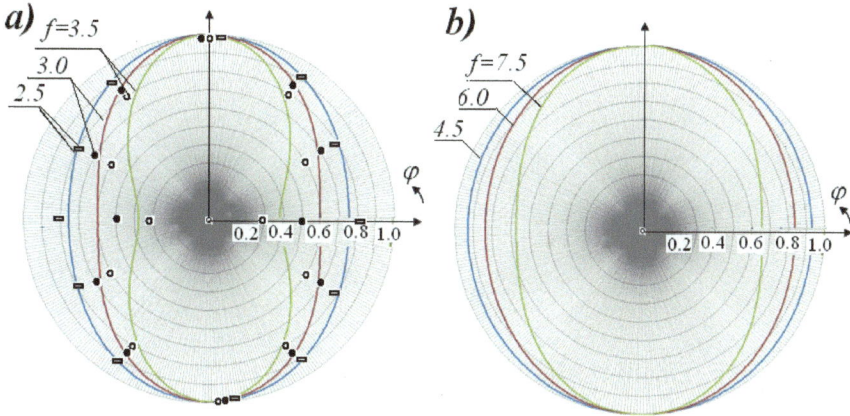

Figure 8: Patterns in the horizontal plane of antennas with widths 7.5 (*a*) and 2 (*b*) cm

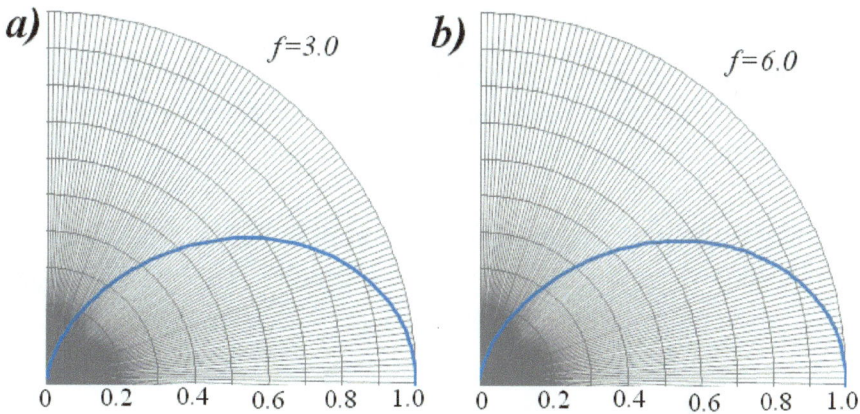

Figure 9: Patterns in the vertical plane of antennas with widths 7.5 (*a*) and 2 (*b*) cm.

9.2. GROUND LOSS RESISTANCE

Calculation of loss in ground was performed for a long time in accordance with a procedure based on assumption that the ground conductivity is high. For this reason, magnetic field H_φ on the ground surface (Fig. **10a**) is practically no different from magnetic field $H_{\varphi 0}$ of an antenna located on perfectly conducting ground, which is equal to the surface current density in the ground

$$j_\rho(\rho) = H_{\varphi 0}(\rho), \tag{9.17}$$

where the surface current is of the radial nature.

If R_0 is the unit-area resistance of the ground surface, the resistance of a circle with radius ρ and a width $d\rho$ is

$$dR_g = R_0\, d\rho/(2\pi\rho).$$

The loss power in the circle is

$$dP_g = \left(2\pi\rho\left|H_{\varphi 0}\right|\right)^2 dR_g,$$

so, the loss resistance related to the antenna base is

$$R_g = \frac{1}{\left|J(0)\right|^2}\int_a^b dP_g = \frac{2\pi R_0}{\left|J(0)\right|^2}\int_a^b \left|H_{\varphi 0}\right|^2 \rho\, d\rho. \tag{9.18}$$

Here $J(0)$ is the current in the antenna base, and $R_0 = 1/(\delta\sigma) = 11\pi/\sqrt{\sigma\lambda}$, where length δ is the depth of the current penetration into the ground, and σ is the ground conductivity.

Quantity a, which is the lower limit of integration in (9.18), is equal to an antenna radius or a radius of ground with infinite conductance. The natural choice of a upper limit is the infinity. But then the integral becomes divergent. Actually, using the monopole model in the form of the straight filament, we obtain for the antenna without load from (1.32)

$$H_{\varphi 0} = j\frac{J(0)}{2\pi\rho\sin kL}\left[\exp\left(-jk\sqrt{\rho^2 + L^2}\right) - \cos kL\exp(-jk\rho)\right]. \tag{9.19}$$

Accordingly, the integrand is

$$\left|H_{\varphi 0}\right|^2\rho = \frac{J^2(0)}{4\pi^2\rho\sin^2 kL}\left[1 + \cos^2 kL - 2\cos kL\cos k\left(\sqrt{\rho^2 + L^2} - \rho\right)\right].$$

If $b_2 \gg b_1 \gg L$, integral $\displaystyle\int_{b_1}^{b_2}\left|H_{\varphi 0}\right|^2\rho d\rho = \frac{J^2(0)(1 - \cos kL)^2}{4\pi^2\sin^2 kL}\ln\frac{b_2}{b_1}$ increases

without limit with the increasing upper limit, *i.e.,* quantity R_g increases without limit too.

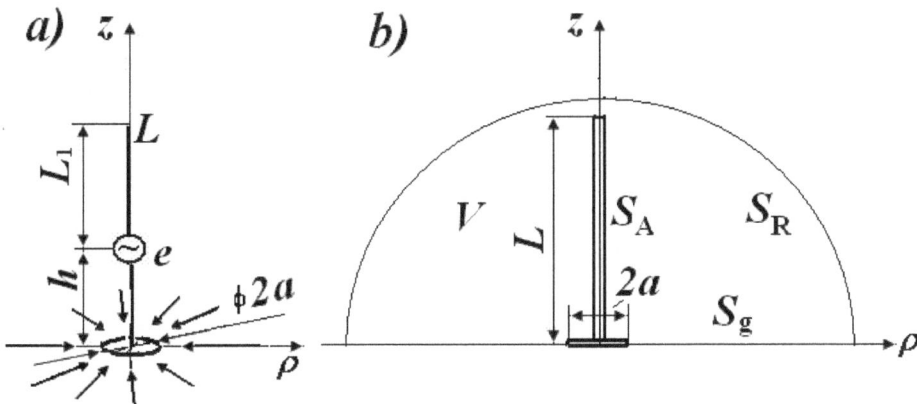

Figure 10: Surface currents in the ground (*a*) and integration area (*b*).

In connection with this, it was proposed to adopt $\lambda/2$ as the upper limit of the integral. Firstly, the inductive current component, which decays following the law $1/\rho^2$ and causes the ground loss, predominates inside a zone with radius $\lambda/2$. Secondly, the current component, which decays following the law $1/\rho$ and is taken into account at calculation of the radiation resistance, predominates outside of the zone. Arbitrariness of such choice of the upper limit is, broadly speaking, obvious.

In [56], the following expression for antenna additional impedance caused by the finite conductance of the ground was derived

$$Z_g = -\frac{2\pi}{J^2(0)}\int_a^\infty E_\rho H_{\varphi 0}d\rho,$$ (9.20)

where

$$H_{\varphi 0} = -\frac{1}{2\pi}\frac{\partial}{\partial\rho}\int_0^L J(z)\frac{\exp\left(-jk\sqrt{\rho^2+z^2}\right)}{\sqrt{\rho^2+z^2}}dz,$$ (9.21)

E_ρ is the radial component of the electrical field on the ground surface, and $J(z)$ is the current along the antenna.

The expression (9.21) is obtained in [56] in an intricate manner, using direct and inverse Fourier-Bessel transforms. A similar result can be obtained with the help of the theorem on the oscillating power (see Section 1.3). Apply the said theorem to volume V bounded by semi-sphere S_R of great radius R, ground surface S_g, and antenna surface S_A (Fig. **10b**). Let R tend to infinity. Since the energy inside the volume is constant in steady-state regime,

$$\int_{(V)} \vec{E}\vec{j}dV = \int_{(S=S_R+S_A+S_g)} \left[\vec{E},\vec{H}\right]_n dS + j\omega\int_{(V)}\left(\mu\vec{H}^2 + \varepsilon\vec{E}^2\right)dV,$$ (9.22)

where $\vec{j}$ is the impressed current density, n is outward normal to surface S.

The left-hand side in (9.22) is the oscillating power related to the energy of external sources, *i.e.,* to the antenna:

$$P_I = \int_0^L E_z J(z)dz = J^2(0)\left(Z_{A0} + Z_g\right).$$ (9.23)

Here, Z_{A0} is the antenna input impedance in the case of a perfectly conducting ground. Its active component at lossless in the antenna wires and external bodies is equal to the radiation resistance. It is assumed that the antenna radiation resistance is independent of the ground conductance, since the procedure presupposes that the ground conductance is high, and the electromagnetic field coincides with the field of antenna situated on the ground with infinite conductance.

The right-hand side of (9.22) is

$$P_{II} = P_1 + P_2 + P_3 + 2j\omega W , \qquad (9.24)$$

where $\quad P_1 = \int_{S_R} \left[\vec{E}, \vec{H}\right]_n dS_R , \quad P_2 = \int_{S_A} \left[\vec{E}, \vec{H}\right]_n dS_A , \quad P_3 = \int_{S_g} \left[\vec{E}, \vec{H}\right]_n dS_g \quad$ and

$$W = \frac{1}{2} \int_V \left(\mu \vec{H}^2 + \varepsilon \vec{E}^2\right) dV$$

are the oscillating parts of the radiation power, of the loss power in the antenna wires, part of the loss power in the grounds and of the electromagnetic field energy in the volume V, respectively.

Since electromagnetic fields at high and infinite ground conductance are actually the same, assume that

$$J^2(0)Z_{A0} = P_1 + P_2 + 2j\omega W , P_3 = J^2(0)Z_g ,$$

where from

$$Z_g = \frac{1}{J^2(0)} \int_{S_g} \left[\vec{E}, \vec{H}\right]_n dS_g = -\frac{2\pi}{J^2(0)} \int_0^\infty E_\rho H_\varphi d\rho . \qquad (9.25)$$

Here, E_ρ and H_φ are the field components on the surface of a perfectly conducting ground.

Expression (9.25) differs from (9.20) only by replacement of $H_{\varphi 0}$ with H_φ. As is seen from (9.25), the statement on insertion of the magnetic field on the surface of perfectly conducting ground into the integrand is erroneous. The error is caused by an incorrect interpretation of expression (9.21). Since $J(z)$ is the current along the actual antenna, H_φ is the magnetic field on the surface of the ground having the finite conductance.

The impedance boundary conditions on the ground surface are

$$Z_0 = -E_\rho / H_\varphi, \tag{9.26}$$

where Z_0 is the surface impedance. The sign minus in this expression is caused by the direction of current density $\vec{j} = [\vec{n}, \vec{H}]$ along the radius to the coordinate origin. For the ground with great conductance

$$Z_0 = R_0(1 + j), \tag{9.27}$$

i.e.,

$$Z_g = \frac{2\pi R_0(1 + j)}{J^2(0)} \int_a^\infty H_\varphi^2 \rho d\rho. \tag{9.28}$$

In particular,

$$R_g = \operatorname{Re} Z_g = \frac{2\pi R_0}{J^2(0)} \int_a^b \left(H_{\varphi 1}^2 - H_{\varphi 2}^2 - 2H_{\varphi 1} H_{\varphi 2} \right) \rho d\rho, \tag{9.29}$$

where $H_{\varphi 1} = \operatorname{Re} H_\varphi$, $H_{\varphi 2} = \operatorname{Im} H_\varphi$. The current is in accordance with (1.8) considered as a real quantity. The difference between the old and new procedures is clear from (9.18) and (9.29). If $H_\varphi = H_{\varphi 0}$, then at $b_2 \rangle\rangle b_1 \rangle\rangle L$

$$\int_{b_1}^{b_2} \left(H_{\varphi 10}^2 - H_{\varphi 20}^2 - 2H_{\varphi 10} H_{\varphi 20} \right) \rho d\rho = J^2(0)(1 - \cos kL)^2 F / (2\pi \sin kL)^2,$$

where $F = -\int_{b_1}^{b_2} \left(\frac{\cos 2k\rho}{\rho} + \frac{\sin 2k\rho}{\rho} \right) d\rho = -\sqrt{2} \int_{b_1}^{b_2} \frac{\cos 2k\rho}{\rho} d\rho \le \frac{1}{kb_1\sqrt{2}}$, *i.e.,* the integral in (9.29) converges.

It is useful to emphasize that expression (9.29) follows from the theorem on the oscillating power. One can also obtain expression (9.18) with the upper limit of the integral equal to ∞ with help the help of the theorem on the complex power. As noted before, the reactive power has no physical sense. The error proved to be a decisive one here.

9.3. FIELD OF A RECTANGULAR LOOP

A widely used type of antennas is the loop, which can be used, *e.g.,* as the main or additional antenna of a cellular phone. The loop antenna theory usually considers the field in the far region, which is created by a circular loop. However, the field of a square or a rectangular loop antenna is rather different from that of circular loop one. It is known only that if the square loop dimensions are small against the wavelength, the far field of such loop in the plane of the loop in the direction perpendicular to side of the square coincides with the field of a circular loop of the same area. Advantages of square and rectangular loop antennas make their use attractive, especially in small-sized devices, and the comprehensive analysis of their performance in the near field becomes more important. For example, it is obvious that the loop shape influences the field in the near field, if loop sizes are small, but finite in comparison to the wavelength and distance to the observation point. This section continues the examination of parameters of a rectangular loop antenna in the near field and compares them to those of a circular loop.

The diagram of a rectangular loop is presented in Fig. **11a**. The loop of size $a \times b$ lies in the plane xOy, and its center coincides with the origin. It is assumed, that the loop sizes are small ($a, b \langle\langle \lambda$), and the amplitude and phase of current I along a loop wire do not vary. The azimuth of observation point P is φ_1, azimuths of loop tops are $\varphi_{01} = \tan^{-1}(b/a)$, $\varphi_{02} = \pi - \varphi_{01}$, $\varphi_{03} = \pi + \varphi_{01}$, $\varphi_{04} = -\varphi_{01}$, respectively. Radius-vector r of the radiation point situated on a loop wire depends on angle φ and is

$$r = \begin{cases} a/(2\cos\varphi), & \varphi_{04} \le \varphi \le \varphi_{01} \\ b/(2\sin\varphi), & \varphi_{01} \le \varphi \le \varphi_{02} \\ -a/(2\cos\varphi), & \varphi_{02} \le \varphi \le \varphi_{03} \\ -b/(2\sin\varphi), & \varphi_{03} \le \varphi \le \varphi_{04} \end{cases}. \tag{9.30}$$

Azimuth component E_φ of the loop electrical field is

$$E_\varphi = -j\omega A_\varphi, \tag{9.31}$$

where $A_\varphi = \dfrac{\mu I}{4\pi} \int\limits_{(l)} [\exp(-jkR)/R] \sin \psi \, dl$ is the corresponding component of the

vector potential, μ is permeability, R is the distance from observation point (R_0, θ, φ_1) to integration point $(r, \pi/2, \varphi)$, ψ is an angle of the current vector with the direction from the coordinate origin to the observation point (straight-line segment R_0), and $k = 2\pi/\lambda$. Expression (9.31) takes into account that the loop current is closed, *i.e.*, $div \vec{A} = 0$.

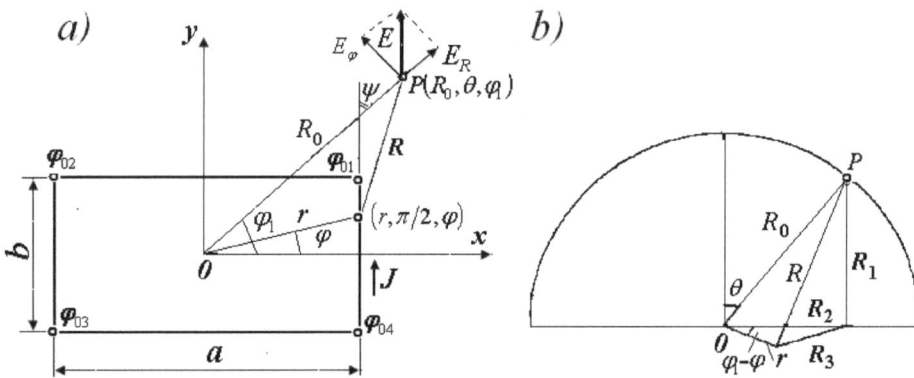

Figure 11: Rectangular loop geometry: a – in the loop plane, b –axonometric view.

As seen from Fig. **11a**, current I at integration point $(r, \pi/2, \varphi)$ creates at observation point P both radial and azimuth components of the electrical field. Thus, unlike the circular loop, there is no other point (symmetrical with respect to the direction from the coordinate origin to the observation point) on the rectangular loop wire, where the current creates the radial component with the same magnitude, but opposite sign at the observation point. It means that, in addition to component E_φ, the electrical field has also component E_ρ. Nevertheless, as well as in the case of a circular loop, field components E_φ and H_θ predominate in an antenna far region and

$$H_\theta = \frac{1}{\mu} curl_\theta \vec{A} = \frac{1}{j\omega\mu R_0} \frac{\partial}{\partial R_0}\left(E_\varphi R_0\right)$$ (9.32)

From Fig. **11b**,

$$R^2 = R_1^2 + R_3^2,$$

where $R_1 = R_0 \cos\theta$, $R_3^2 = R_2^2 + r^2 - 2rR_2 \cos(\varphi - \varphi_1)$, $R_2 = R_0 \sin\theta$,

i.e.,

$$R^2 = R_0^2 + r^2 - 2rR_0 \sin\theta \cos(\varphi - \varphi_1). \tag{9.33}$$

Rewrite the expression (9.33) as

$$R = R_0 \sqrt{1 + x}, \tag{9.34}$$

where $x = -(2r/R_0)\sin\theta \cos(\varphi - \varphi_1) + r^2/R_0^2 \langle\langle 1.$

Representing function $\sqrt{1 + x}$ as a sum of an ascending power series and confining ourselves to the first four terms, the get $\sqrt{1 + x} = 1 + x/2 - x^2/8 + x^3/16 - \dots$ Accordingly,

$$R = R_0 - r\sin\theta \cos(\varphi - \varphi_1) + r^2/(2R_0)\left[1 - \sin^2\theta \cos^2(\varphi - \varphi_1)\right] +$$
$$+ r^3/(2R_0^2)\sin\theta \cos(\varphi - \varphi_1)\left[1 - \sin^2\theta \cos^2(\varphi - \varphi_1)\right] \tag{9.35}$$

Using the smallness of $R - R_0$, expand fraction $\exp(-jkR)/R$ into Taylor series in the vicinity of point $R = R_0$:

$$\frac{\exp(-jkR)}{R} = \frac{\exp(-jkR_0)}{R_0} + (R - R_0)\frac{\partial}{\partial R_0}\left[\frac{\exp(-jkR_0)}{R_0}\right] +$$
$$+ \frac{1}{2}(R - R_0)^2 \frac{\partial^2}{\partial R_0^2}\left[\frac{\exp(-jkR_0)}{R_0}\right] + \frac{1}{6}(R - R_0)^3 \frac{\partial^3}{\partial R_0^3}\left[\frac{\exp(-jkR_0)}{R_0}\right] + \dots \tag{9.36}$$

Substituting the derivatives into (9.36) and reducing similar terms

$$f(\varphi) = \frac{\exp[-jk(R - R_0)]}{R/R_0} = f_1(\varphi) + jk(1 + \alpha)\sin\theta[f_2(\varphi)\cos\varphi_1 + f_3(\varphi)\sin\varphi_1] +$$
$$+ (k^2/2)\left[\alpha(1 + \alpha) - \frac{1}{2}(1 + 3\alpha + 3\alpha^2)\sin^2\theta\right]f_4(\varphi) - (k^2/2)(1 + 3\alpha + 3\alpha^2)\sin^2\theta.$$

$$\cdot \left[f_5(\varphi) \cos 2\varphi_1 + f_6(\varphi) \sin 2\varphi_1 \right] + j \left(k^3 \alpha / 2 \right)\!\left(1 + 3\alpha + 3\alpha^2 \right) \sin \theta \left[f_7(\varphi) \cos \varphi_1 + f_8(\varphi) \sin \varphi_1 \right] -$$

$$- j \left(k^3 / 8 \right)\!\left(\frac{1}{3} + 2\alpha + 5\alpha^2 + 5\alpha^3 \right) \sin^3 \theta \left[f_9(\varphi) \cos 3\varphi_1 + f_{10}(\varphi) \sin 3\varphi_1 + 3 f_7(\varphi) \cos \varphi_1 + 3 f_8(\varphi) \sin \varphi_1 \right] \quad \textbf{(9.37)}$$

where

$$f_1(\varphi) = 1,\ f_2(\varphi) = r \cos\varphi,\ f_3(\varphi) = r \sin\varphi,\ f_4(\varphi) = r^2,\ f_5(\varphi) = r^2 \cos 2\varphi,$$
$$f_6(\varphi) = r^2 \sin 2\varphi,\ f_7(\varphi) = r^3 \cos\varphi,\ f_8(\varphi) = r^3 \sin\varphi,\ f_9(\varphi) = r^3\!\left(4\cos^3\varphi - 3\cos\varphi \right)$$
$$,\ f_{10}(\varphi) = r^3\!\left(3\sin\varphi - 4\sin^3\varphi \right),$$

and $\alpha = 1/(jkR_0)$. It is easy also to show, that on the loop sides

$$\sin\psi = \begin{cases} \cos\varphi_1 \\ \sin\varphi_1 \\ -\cos\varphi_1 \\ -\sin\varphi_1 \end{cases} \quad dl = \begin{cases} rd\varphi / \cos\varphi = ad\varphi/\left(2\cos^2\varphi\right) & \varphi_{04} \le \varphi \le \varphi_{01} \\ bd\varphi/\left(2\sin^2\varphi\right) & \varphi_{01} \le \varphi \le \varphi_{02} \\ ad\varphi/\left(2\cos^2\varphi\right) & \varphi_{02} \le \varphi \le \varphi_{03} \\ bd\varphi/\left(2\sin^2\varphi\right) & \varphi_{03} \le \varphi \le \varphi_{04} \end{cases} \quad \textbf{(9.38)}$$

Substituting (9.30), (9.37) and (9.38) into expression for A_φ, we get

$$A_\varphi = \frac{\mu I \exp(-jkR_0)}{4\pi R_0} \int\limits_{(\varphi)} f(\varphi) \sin\psi\, dl = \frac{\mu I \exp(-jkR_0)}{4\pi R_0}\, jkSF(f) \sin\theta, \quad \textbf{(9.39)}$$

where $\quad jkSF(f)\sin\theta = \dfrac{a}{2}\cos\varphi_1\!\left(\int\limits_{\varphi_{04}}^{\varphi_{01}} - \int\limits_{\varphi_{02}}^{\varphi_{03}} \right)\dfrac{f(\varphi)d\varphi}{\cos^2\varphi} + \dfrac{b}{2}\sin\varphi_1\!\left(\int\limits_{\varphi_{01}}^{\varphi_{02}} - \int\limits_{\varphi_{03}}^{\varphi_{04}} \right)\dfrac{f(\varphi)d\varphi}{\sin^2\varphi},$

and $S = ab$ is

the loop area. Denote

$$F_i = jkSF(f_i)\sin\theta. \quad \textbf{(9.40)}$$

Then from (9. 37)

$$jkSF(f)\sin\theta = F_1 + jk(1+\alpha)\sin\theta(F_2\cos\varphi_1 + F_3\sin\varphi_1) + (k^2/2)[\alpha(1+\alpha)-(1/2)(1+3\alpha+3\alpha^2)\sin^2\theta]F_4 -$$
$$-(k^2/2)(1+3\alpha+3\alpha^2)\sin^2\theta(F_5\cos2\varphi_1 + F_6\sin2\varphi_1) + j(k^3\alpha/2)(1+3\alpha+3\alpha^2)\sin\theta(F_7\cos\varphi_1 + F_8\sin\varphi_1)-$$
$$- j(k^3/8)(1/3+2\alpha+5\alpha^2+5\alpha^3)\sin^3\theta(F_9\cos3\varphi_1 + F_{10}\sin3\varphi_1 + 3F_7\cos\varphi_1 + 3F_8\sin\varphi_1).$$

The integration of expressions (9.40) gives

$$F_1 = F_4 = F_5 = F_6 = 0, F_2 = ab\cos\varphi_1, F_3 = ab\sin\varphi_1, F_7 = (a^3b/4)[1+b^2/(3a^2)]\cos\varphi_1,$$
$$F_8 = (ab^3/4)[1+a^2/(3b^2)]\sin\varphi_1, F_9 = (a^3b/4)(1-b^2/a^2)\cos\varphi_1, F_{10} = -(a^3b/4)(1-a^2/b^2)\sin\varphi_1,$$

and, accordingly,

$$F(f) = 1 + \alpha + k^2/8[(a^2+b^2/3)\cos^2\varphi_1 + (b^2+a^2/3)\sin^2\varphi_1]\alpha(1+3\alpha+3\alpha^2)-$$
$$- (k^2/8)(a^2\cos^2\varphi_1 + b^2\sin^2\varphi_1)(1/3+2\alpha+5\alpha^2+5\alpha^3)\sin^2\theta. \tag{9.41}$$

Quantity A_φ is found from (9.39); quantity E_φ is

$$E_\varphi = 30k^2 IS \frac{\exp(-jkR_0)}{R_0} F(f)\sin\theta. \tag{9.42}$$

If the loop sizes are infinitesimally small in comparison with the wavelength, then

$$E_\varphi = 30k^2 IS(1+\alpha)\frac{\exp(-jkR_0)}{R_0}\sin\theta. \tag{9.43}$$

The expression coincides with the known one for the azimuth field component of the elementary circular loop with radius is small against the wavelength. The circular loop area in it is replaced with the rectangular loop area.

Comparing the formulas (9.42) and (9.43), it is easy to see, that in (9.43) $F(f)=1+\alpha$, i.e., only the first two terms are retained from expression (9.41). The remaining terms are proportional to the squared ratios of loop linear size to the wavelength and higher powers of α (to inverse powers of the distance to the observation point). The field magnitude in a near field is seen from (9.41) to depend on the observation point azimuth.

The rectangular (noncircular) loop shape gives rise to radial field component

$$E_R = -j\omega A_R, \tag{9.44}$$

where $A_R = \dfrac{\mu I}{4\pi}\displaystyle\int_{(l)} \dfrac{\exp(-jkR)}{R}\cos\psi\,dl$ is the radial component of the vector potential, and

$$\cos\psi = \begin{cases} \sin\varphi_1 & \varphi_{04} \le \varphi \le \varphi_{01} \\ -\cos\varphi_1 & \varphi_{01} \le \varphi \le \varphi_{02} \\ -\sin\varphi_1 & \varphi_{02} \le \varphi \le \varphi_{03} \\ \cos\varphi_1 & \varphi_{03} \le \varphi \le \varphi_{04}. \end{cases} \tag{9.45}$$

The expression for A_R subject to (9.45) assumes the form

$$A_R = \frac{\mu I \exp(-jkR_0)}{4\pi R_0} jkSG(f)\sin\theta, \tag{9.46}$$

where

$$jkSG(f)\sin\theta = \frac{a}{2}\sin\varphi_1\left(\int_{\varphi_{04}}^{\varphi_{01}} - \int_{\varphi_{02}}^{\varphi_{03}}\right)\frac{f(\varphi)d\varphi}{\cos^2\varphi} - \frac{b}{2}\sin\varphi_1\left(\int_{\varphi_{01}}^{\varphi_{02}} - \int_{\varphi_{03}}^{\varphi_{04}}\right)\frac{f(\varphi)d\varphi}{\sin^2\varphi}, \tag{9.47}$$

and function $f(\varphi)$ is consistent with the formula (9.37). Denote

$$G_i = jkSG(f_i)\sin\theta, \tag{9.48}$$

then the expression for $jkSG(f)\sin\theta$ is similar to that for $jkSF(f)\sin\theta$ with quantity F_i replaced with G_i. The integration of equation (9.48) gives

$$G_1 = G_4 = G_5 = G_6 = 0, G_2 = ab\sin\varphi_1, G_3 = -ab\cos\varphi_1, G_7 = (a^3b/4)[1 + b^2/(3a^2)]\sin\varphi_1,$$
$$G_8 = (ab^3/4)[1 + a^2/(3b^2)]\cos\varphi_1, G_9 = a^3b/4(1 - b^2/a^2)\sin\varphi_1, G_{10} = -(a^3b/4)(1 - b^2/a^2)\cos\varphi_1,$$

and, accordingly,

$$G(f) = \left(k^2/4\right)\sin 2\varphi_1 \left[\alpha\left(1/3 + \alpha + \alpha^2\right)\left(a^2 + b^2\right) - \left(1/8\right)\left(1/3 + 2\alpha + 5\alpha^2 + 5\alpha^3\right)\left(a^2 + 3b^2\right)\sin^2 \theta\right] \quad \textbf{(9.49)}$$

Quantity A_R is found from (9/46); whereas

$$E_R = 30k^2 IS \frac{\exp(-kR_0)}{R_0} G(f)\sin\theta. \tag{9.50}$$

As seen from (9.50), the radial field component contains only terms that are proportional to the squared ratios of loop linear size to the wavelength, *i.e.,* it is small in comparison to the azimuth component. In addition, the radial component substantially depends on the azimuth. The component vanishes in the directions that are perpendicular to the rectangle sides. Indeed, in this case, each loop conductor point, creating field dE_R at the observation point, opposes a point symmetric with respect to that perpendicular, the current at the point creating field $-dE_R$ at the observation point. Peaks of E_R occur at angle $\pi/4$ with respect to the said directions.

Fig. **12** shows the calculated results for field E_φ. They are depending on a distance R_0/λ between the rectangular loop center and the observation point (in wavelengths). The loop dimensions are assumed equal to $a = 2b = 1/2k$, the observation point azimuth is $\varphi_1 = \pi/2$, and angle θ varies from $\pi/6$ to $\pi/2$. The field magnitude is reduced to its value for $\theta = \pi/2$, $R_0/\lambda = 0.1$.

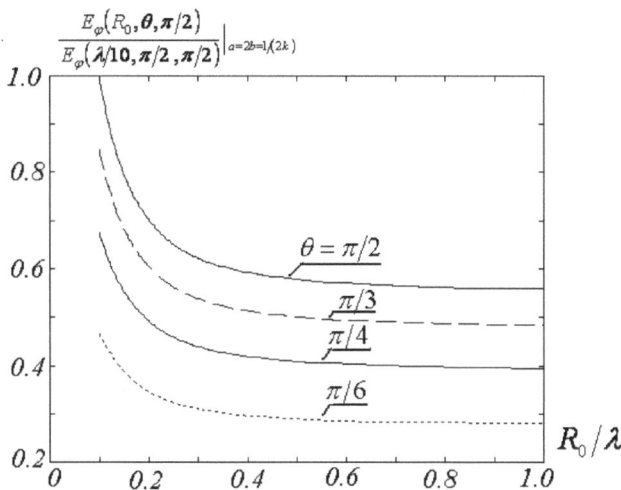

Figure 12: Relative azimuth components as a function of distance for various values of angle θ.

Fig. **13** presents curves for E_φ of the similarly shaped loops ($a = 2b$), but with different areas (in wavelengths), and $\varphi_1 = \pi/2, \theta = \pi/2$. The fields of different area loops are the same in the far region ($IS = const$), and they are reduced to the field of a loop with $a = 1/2k$ and $R_0/\lambda = 0.1$. Curve 1 was calculated by the approximate formula (9.44), that is, for a loop, with dimensions are infinitesimally small in comparison with the wavelength. The loop field in the near region calculated by the exact formula can be seen to differ substantially from that by the approximate one.

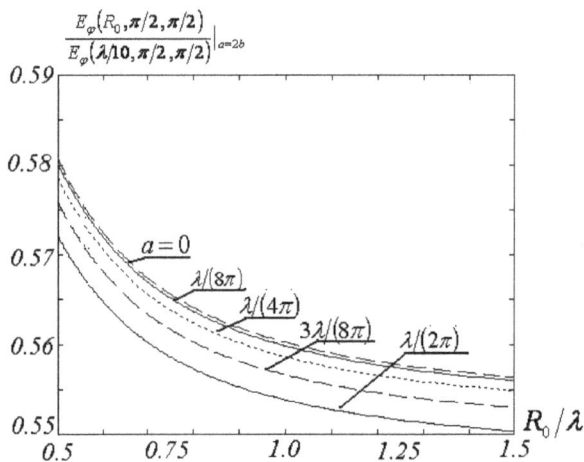

Figure 13: Relative azimuth components for the loops of different shapes and dimensions.

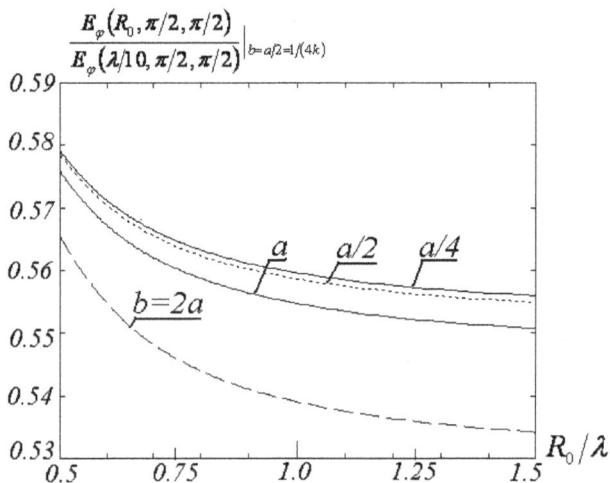

Figure 14: Relative azimuth components for the same a and the same fields in the far region.

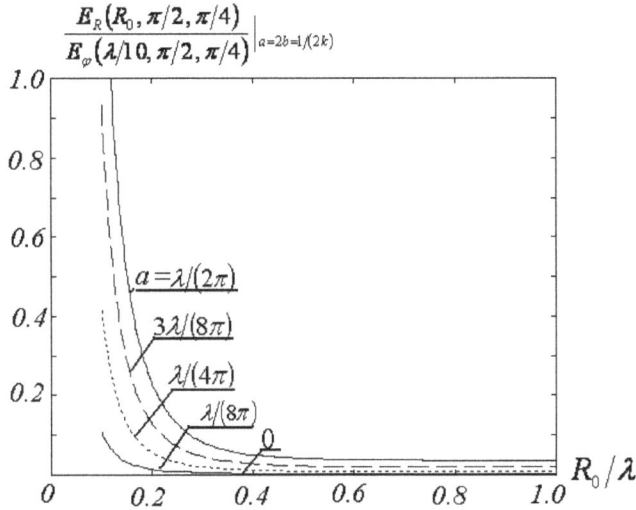

Figure 15: Relative radial components for loops with different sizes and the same fields in the far region.

Fig. **14** gives similar curves for various aspect ratios. The fields of different loops are the same in the far region ($IS = const$), and they are reduced to the field of a loop with $b = a/2 = 1/(4k)$ and $R_0/\lambda = 0.1$.

Fig. **15** presents the results for radial field component E_R. Loop dimensions are $a = 2b$, observation point azimuth is $\varphi_1 = \pi/4$, and the angle θ is $\pi/2$. The magnitudes of E_R for loops with the same fields E_φ in the far zone ($IS = const$) are reduced to that of E_φ for a loop with $a = 1/2k$ and $R_0/\lambda = 0.1$. As can be seen, E_R is small in comparison with E_φ, and vanishes in case of an elementary loop with dimensions infinitesimally small in comparison with the wavelength.

The results demonstrate that dimensions and shape of a loop have a substantially effect on its field magnitude, especially in the near field.

9.4. SHIP ANTENNAS

The section is dedicated to separate issues of designing ship antennas. They are representatives of numerous groups of antennas deployed on mobile objects and as such have specific features of their own being placed under constrained conditions in close proximity to metal bodies of different shapes and dimensions.

An antenna of medium-frequency (hectometer) waves is known as the main ship antenna. It must ensure the tuning of the main and emergency transmitters, and its efficiency in operation with the main transmitter must be sufficient in order to establish the electric field with strength $50 \, \mu V/m$ at distance 150 miles. With allowance for details of medium-frequency waves propagation, the antenna must create vertically polarized electromagnetic waves with the maximum radiation direction along the ground surface. The pattern in the horizontal plane must be close to the circular one.

For this reason, the hectometer antenna is an asymmetrical vertical radiator. As a rule, its height is small in comparison with the wavelength, *i.e.,* the radiation resistance is low, which leads to low efficiency. Accordingly, the underlying problem in the development of new antennas is the increase of its effective length. Therefore, wide use is made of antennas with capacitance loads at the upper end permitting to improve (to make more uniform) a current distribution along the antenna, to increase their effective length and radiation resistance (Fig. **16**). The load is performed as a system of horizontal wires, which are stretched between masts either in one plane, or along generatrices of a circular cylinder.

Such an antenna is mostly base-fed. Its circuit corresponds to Fig. **16a**, and the current distribution is given in Fig. **16b**. But one can, in principle use top-fed upper antennas (Fig. **17**), when the transmitter connects to a wire located inside the mast, with the upper end of the wire integrated with the horizontal load (see Fig. **17a**). The circuit is equivalent to connection of the exciting emf at the top (vertex) of the antenna – between the vertical radiator (the outer mast surface) and the load (a horizontal sheet or circular cylinder) – see Fig. **17b**. If the wire cannot be laid inside the mast, the variant with a shielded wire is feasible (see Fig. **17c**)

The current distribution along a top-fed antenna is given in Fig. **17d**. In this case, the current antinode is in the radiator base. The current varies weakly near the antinode, and the effective length is in the first approximation equal to the geometric length of antenna, *i.e.,* it is longer than in the case of base-fed antenna. But one has towards the antenna base (to the point of the exciting emf connection). To weaken the effect, the ratio of the mast and the internal wire diameters needs to be increased.

In a base-fed antenna, the mast acts as a support and creates stray capacitance between the antenna and the ground, which causes that carries decrease of the radiation resistance. In the general case for analysis of the mast effect can rely on the CST program. In the particular case, when the antenna represents a vertical wire without load, which is located in parallel to the mast and has the same length (Fig. **18**), one can use an explicit technique based on the folded radiator theory. As seen from the Figure, the radiating wire and the mast form short-ended at the ground and open-ended above the folded radiator with different wire diameters. By analogy with Section 3.4, one can divide the radiator into two auxiliary circuits: a linear radiator of height L with an equivalent radius a_e and open-ended at the end long line with wave impedance W_l. The input admittance of the antenna near the mast

$$Y_A = 1/Z_l + p^2/Z_m(a_e). \tag{9.51}$$

Here, $Z_l = -jW_l \cot kL$ is the input impedance of the long line, Z_m is input impedance of the linear radiator, $p = C_{11}/(C_{11} + C_{22})$ is the in-phase current share in the excited wire, C_{11} and C_{22} are the self-capacitances of excited and passive wires.

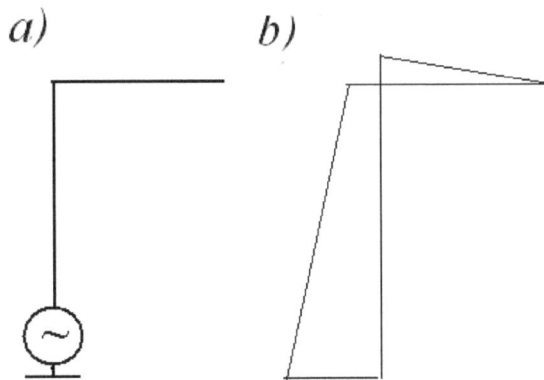

Figure 16: Inverted L antenna (a) and the current distribution along it (b).

Expression (9.51) is valid, if distance b from the antenna to the mast is small against the antenna length and the wavelength. If these values are commensurable, it is expedient to use the induced emf method, in accordance with which the input impedance of the excited radiator near the passive one is

$$Z_A = Z_{11} - Z_{12}^2/Z_{22} , \qquad (9.52)$$

where Z_{11} and Z_{22} are the self-impedances of both radiators, and Z_{12} is their mutual impedance.

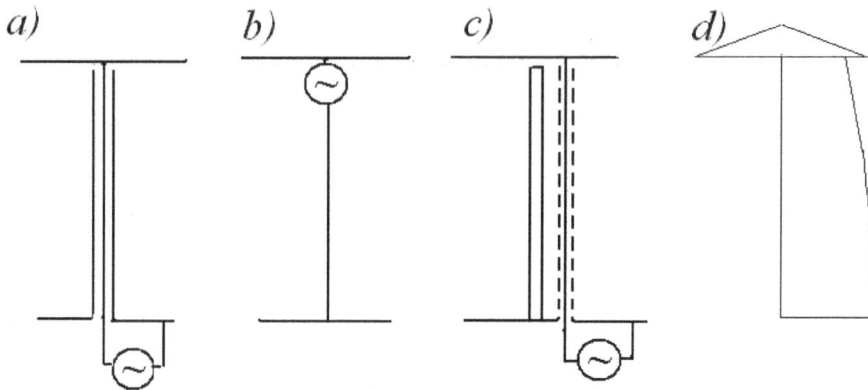

a)   b)   c)   d)

Figure 17: Top-fed antenna with the upper feeding: (*a*) – a wire inside the mast, (*b*) - equivalent radiator, (*c*) – a shielded wire, (*d*) –the current distribution.

Fig. **19** gives an example of radiation resistances R_Σ of wire antennas 6.52 and 13 m long (the wire radius is $a_1 = 3.7 \cdot 10^{-3}$ m) at frequency 460 kHz plotted against distance b between the antenna and the mast for different radii a_2 of the mast. To compare, quantity R_{11} of the single antenna radiation resistance is plotted. Experimental values for the mast of radius $a_2 = 0.4$ m are shown with dots. The good coincidence of experimental results with calculation confirms the validity of obtained results.

It is seen from the figure that active component R_A of the antenna input impedance drops sharply as the distance between the antenna and the mast decreases. Horizontal loads of the wires weaken the influence of the mast. But in this case as well, it is necessary to move the antenna away from the mast as far as possible (from 4 to 8 m, depending from the mast height).

Low radiation resistance and lower efficiency are not only drawbacks of wires antennas. To them one must add a wide variation range of the input impedance, which makes it difficult to specify the antennas type and complicates the onboard equipment. Besides, an antenna curtain can break as a result of a storm or ice formation, the antenna can hinder cargo handling, it requires mounting a second mast unnecessary onboard contemporary ships. For this reason antenna-masts have found their use as the main ships antennas. At first, three variants of such antennas appeared: with guy ropes and guy wires, self-supporting and mast-mounted. The first variant was dropped soon after because of a great area occupied by the antenna. The second variant was at first too expensive. Often, it is difficult to erect a cumbersome self-supporting antenna-mast. So, it is expedient to use an antenna with inductance-capacitance load, installable on the existent mast (Fig. **20**).

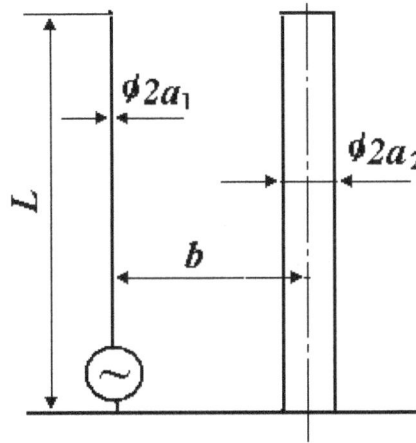

Figure 18: The relative disposition of the radiator and the mast.

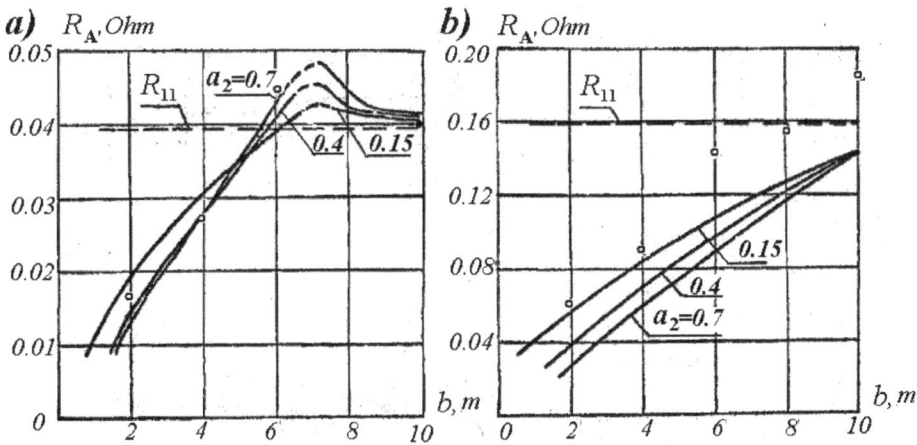

Figure 19: Radiation resistance of wire antenna with height 6.52 m (*a*) and 13 m (*b*) against the distance to the mast.

The circuit presented in Fig. **20a** corresponds to the variant with open vertical wire and can be used onboard operational ship. The antenna is base-fed and differs from inverted *L* antenna only by the type of load. Top-fed variants are possible also. The load is produced in the shape of a vertical structure, which is the mast extension. The mast supports the load, so for a given mast height, the geometric height of antenna increases, *i.e.,* the effective height of antenna increases and other characteristics are improved, too.

The antenna load (see Fig. **20b**) consists of four whip antennas connected in the base by a conducting ring and double-coiled volume spiral inserted in series with it. The system of four whip antennas is equivalent to a thick metal radiator with low wave impedance and high capacitance. The spiral increases the antenna electrical length. Both elements decrease the reactive component of the input impedance and increase the effective height. Use of tumbling whip antennas permits to decrease, if necessary, the total height of the structure. The man access to the load elements at the time of ships lying in a port and the calm weather is foreseen. Lightning spark gap is installed at the point of antenna wire lead-in to the radio room.

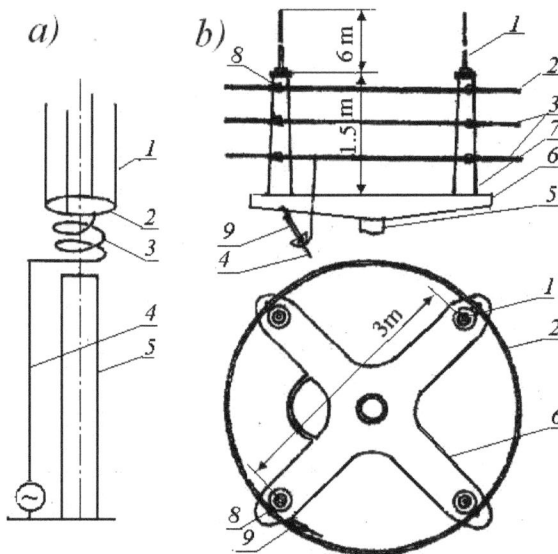

Figure 20: Antenna with inductance-capacitance load: (*a*) – circuit, (*b*) – load structure, (1) - tumbling whip antenna, (2) - conducting ring, (3) - double-coiled volume spiral, (4) - open vertical wire, (5) - mast, (6) – ground, (7) - column, (8) – base insulator, (9) – rod insulator.

Figure 21: Antenna-mast with inductance-capacitance load on the upper bridge of cargo ship *Konstantin Shestakov.*

The antenna-mast with inductance-capacitance load was proposed in 1966 and improved in 1970 [57]. The antenna prototype was mounted onboard cargo ship *Konstantin Shestakov* of displacement 3500 ton (Fig. **21**). Antenna is placed on the ship's upper bridge, on the mast of height 9.5 m and of diameter 0.3 m. The static antenna capacitance is 442 pF, the natural wave length is 240 m, and the active component of the input impedance at frequency 400 kHz is 4.3 Ohm.

In later years, the antennas in accordance with a similar circuit were constructed in other countries (Fig. **22**).The capacitance load in them is made as a systems of whip radiators installed at the antenna top, and the inductance load is made as a coil or spiral connected in series with the radiating wire. They include Norwegian antenna AS9 (in tumbling version AS9ST) and antenna 938G-1 by *Collins*, USA. They are built as self-supporting structures of fiberglass and have the total height about 15.3 m. No provision exists for their placement on a common ship mast.

The development of described antennas resolved the problem of ship radio communication in the range of medium-frequency waves. Meanwhile, solution of the problem in the short-wave range met with little success. Of most interest were the multi-radiator antennas and antennas with capacitance loads.

The structure of the first kind is described in Section 1.5 (Fig. **7b**), and the procedure of calculation is given in Section 5.4. The antenna consists of the central radiator and the side radiators, mounted around the central one along the circular cylinder generatrices and connected with it in the base. The number and length of the side radiators should be selected so that antenna could operate in the entire frequency range.

A multi-wire antenna consisting of radiators without concentrated loads has high electric characteristics in the vicinity of series resonances of isolated radiators. But it ensures the required TWR level and the necessary pattern shape over the entire frequency range, only if the antenna diameter is increased to 1 m and greater. For this reason, it is expedient to use the central radiator with the concentrated complex impedance consisting of parallel-connected resistor and inductance coil.

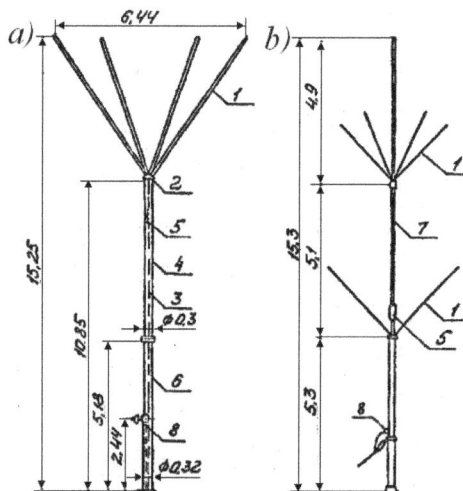

Figure 22: Self-supporting antenna-masts with inductance-capacitance load: (*a*) – 938G-1, (*b*) – AS9. (1) – whip radiator of fiberglass with copper sennit, (2) – bronze cap, (3) – central copper conductor, (4) – fiberglass mast, (5) – inductance coil, (6) – aluminum tube, (7) – fiberglass tube, (8) – antenna lead-in.

A mode close to traveling wave one is established on the lower segment of the central radiator and raises the minimum matching level in the gaps between the normal zones. The combination of the multi-wire structure and the complex load permits to obtain acceptable performance in the range from 6 to 25 MHz with the help of antenna of height 12 m and diameter 0.3 m. The antenna uses absorber BAA-3-2 as a resistor. Unfortunately, the antenna efficiency at certain frequencies drops below 50%.

The antenna with capacitance loads has the minimum loss. But, the capacitance loads cannot provide for a tunable linear antenna, which ensures PF not less than 0.5 and TWR not less than 0.3 over entire short-wave range. To adjust the antenna, it is more than enough to change the loads magnitudes, for example, by means of switches. Unfortunately, the work of creating such antenna has not been carried through.

An important feature of any mobile object, in particular, a ship, is confined area for antenna placement, in view of that antennas are installed near differently-shaped metal structures, such as masts, superstructures, pipes, *etc.* Analysis of their influence on the antenna performance is complicated even in the simplest cases. The superstructure can be considered as an additional passive radiator. Solving the set of Kirchhoff equations for the totality of radiators, we can find the current in each one. The difficulty lies in large lateral dimensions of the superstructure, *i.e.,* the calculation of its self-impedance and mutual impedances between it and other radiators, which is based on the thin antenna theory, yields too rough approximation. The analysis method based on replacing the metal body with a system of thin wires [23, 41] is more efficient.

Begin with a single superstructure of a regular shape. Fig. **23** gives the circuit of the wire structure, equivalent to a thick superstructure shaped as a circular cylinder of finite length, next to which a whip antenna is placed. It is assumed that the ground surface is perfectly conducting, and the structure is symmetrical with respect to the surface. The circular cylinder is replaced with the wire structure of eight equidistant wires located along the cylinder generatrices and the radii of their ends. Since the antenna excites mainly the longitudinal current component in the superstructure, then, to simplicity the calculations the lateral circle wires are disregarded. The wire and the whip antenna diameters are assumed equal. The coordinate origin coincides with the superstructure center. The dimensions at the figure are given in meters.

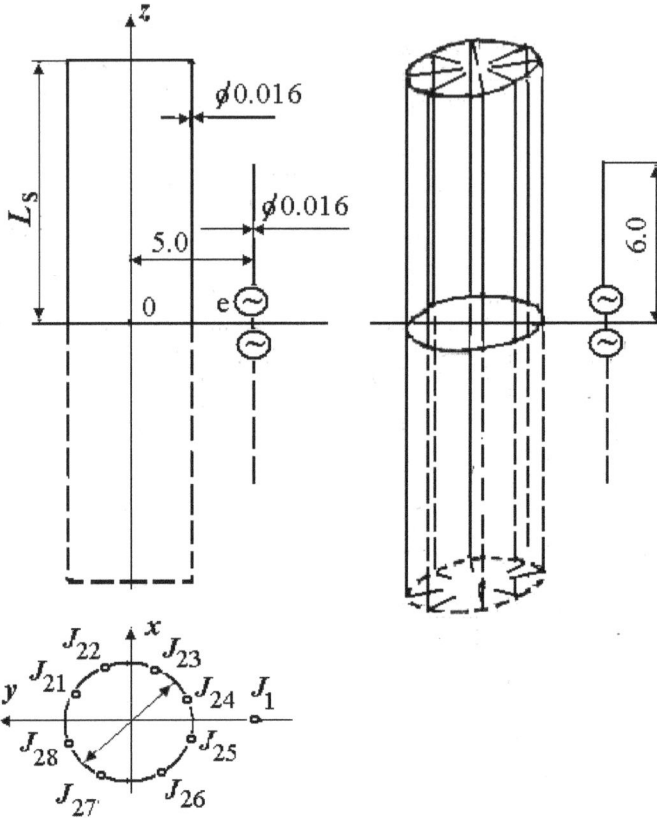

Figure 23: Whip antenna near a cylindrical superstructure.

Table **4** presents input impedance Z_A and directivity D_m (reduced to a half-wave dipole) of the 6-m whip antenna near a superstructure of diameter 5 m and of heights 6 and 20 m at three high frequencies. To compare the parameters are given for the case without the superstructure ($L_S = 0$). The calculations show that the antenna characteristics to a high degree depend on the superstructure presence and on its height. The radiation resistance decreases (excepting the region of parallel resonance). The radiation in the direction to the superstructure decreases sharply (except for lower frequencies and the structures, shorter than the whip antenna). The currents in the wires of the superstructure equivalent on either side of the whip antenna differ substantially both in amplitude and in phase. The calculated pattern of the antenna located near superstructures of different height both in the horizontal plane and in the vertical plane are in fairly good agreement with the experiment.

The said agreement indirectly confirms obliquely the validity of the wire structure choice. As an additional test the characteristics of the whip antenna situated near a superstructure with than in vertical ones. For example, in the superstructure of the height 6 m at frequency 6 MHz they differ by a factor of order 10^5.

Table 4: Characteristics of a whip antenna near a superstructure

f, MHz	Z_A, Ohm			D_m		
	$L_S = 0$	6	20	0	6	20
6.0	6.0-j353	3.7-j351	1.8-j350	1.86	2.46	4.32
12.5	39.7+j20.9	24.7+j38	17.0+j28.2	2.00	5.39	5.32
19.0	289+j424	356+j505	185+i448	2.35	4.76	6.80

The calculation permits to find the minimal superstructure height, at which the antenna characteristics coincide with those of an antenna located near an infinitely high superstructure. This height exceeds the antenna height approximately by a quarter of the wavelength.

In actual practice, the superstructure shape is other than cylindrical. Comparison of influence of superstructures differently shaped, equally distant from the antenna axis, of close dimensions, *e.g.,* with circular or polygonal cross sections, shows that the influence on the antennas behavior remains basically the same.

Abundance of variants of antenna placement onboard ships calls for its type-design practice in order to predict their performance with allowance for environment, if possible. Namely, it is necessary to include the nearby antennas of the great dimensions, *e.g.,* antennas-masts and radiolocation antennas, which distort the antennas characteristics. As a rule, the problem reduces to determining the performance of antennas placed near two superstructures. The additional superstructure causes additional decrease of the radiation resistance and increase of directivity. The above procedure was used also for calculating characteristics of a linear array consisting of whip antennas situated near a cylindrical superstructure, and a convergent double-tier array near two superstructures.

It is worth emphasizing that the wire structure used in the calculation as the equivalent of a metal structure must agreement with the physical object of the problem. For example, it is expedient to replace structure wires along the expected current lines. That permits to decrease substantially the number of wires and, accordingly, the amount of calculation at the same calculation accuracy (or, *vice versa*, to increase the accuracy of results at the same amount of calculation).

Different authors quote different values of the minimum number of wire structure conductors to secure the equivalence of influencing metal body and its model. This value determines the number of segments, into which the wire system is divided. The variance is caused by solving different problems and using different basis functions. As shown here, when calculating electrical characteristics of antennas located near ship's superstructures, the distance between wires should not exceed $0.08\,\lambda$. Similar results were obtained in calculations of characteristics of whip antennas situated on the edge of ship deck or on the yard and of symmetrical radiator placed along the axis of trough with finite length. The latter problem arises, if the antenna is located near a metal structure and is built flush with it, *i.e.,* does not rise over it. Another variant of the same problem occurs, if the radiator is placed in a dielectric capsule floating along the water surface.

REFERENCES

[1]   R. W. P. King, "The linear antenna – eighty years of progress," *Proc. IEEE*, vol. 55, No. 1, 1967, pp. 2-16.

[2]   J. C. Maxwell, "A dynamical theory of the electromagnetic field," *Proc. Royal Soc. (London)*, vol. 13, 1864, p. 531.

[3]   H. Hertz, "Uber sehr schnelle electrische Schwingungen," *Wiedemanns Ann. Phys.,* vol. 31, 1887, p. 421.

[4]   H. Hertz, "Die krafte electrischer Schwingungen behandelt nach der Maxwellshen theorie," *Wiedemanns Ann. Phys., v*ol. 36, 1888, p. 1.

[5]   J. A. Stratton, *Electromagnetic Theory.* New York: McGraw-Hill, 1941.

[6]   J. H. Poynting, "On the transfer of energy in the electromagnetic field," *Phil. Trans. Roy. Soc. (London)*, vol. 2, 1884, p. 343.

[7]   G. A. Lavrov, *Mutual effect of linear radiators*, Sviyaz, Moscow, 1975 (*in Russian*).

[8]   M. L. Levin, "About one new method of finding the thin antenna characteristic reactance," *Izvestiya AN SSSR*, ser. phys., vol. 11, No. 2, 1947, pp. 117-133 (*in Russian*)..

[9]   G. Z. Aizenberg, *Short-Wave Antennas*, Sviyazizdat, Moscow, 1962 (*in Russian*).

[10]   C. A. Balanis, *Antenna Theory: Analysis and Design*, Wiley & Sons, New York, 2005.

[11]   J. Aharoni, *Antennae*, Oxford, 1946.

[12]   M. A. Leontovich, and M. L. Levin, "On the theory of oscillations excitation in the linear radiators," *Journal of Technical Physics*, vol. 14, No. 9, 1944, pp. 481-506 (*in Russian*).

[13]   M. I. Kontorovich, "Some remarks in connection with the induced emf method," *Radiotechnika*, 1951, No, 2, pp. 3-9 (*in Russian*).

[14]   J. D. Kraus, *Antennas,* McGraw-Hill, Boston, 1988.

[15]   L. A. Vainshtein, *Electromagnetic Waves*, Sovetskoye Radio, Moscow, 1988 (*in Russian*).

[16]   E. Hallen, "Theoretical investigations into the transmitting and receiving qualities of antennae," *Nova Acta Regiae Soc. Sci. Upsaliensis*, ser. IV, vol. 11, No. 4, 1938, pp. 1-44.

[17]   H. C. Pocklington, "Electrical oscillations in wire," *Cambridge Philosophical Society Proc.,* London, vol. 9, 1897, pp. 324-332.

[18]   R.W.P. King, *Theory of Linear Antennas*, Harvard University Press, Cambridge, Mass., 1956.

[19]   B. M. Levin, *Monopole and Dipole Antennas for Ship Radio Communication*, Abris, St.-Petersburg, 1998 (*in Russian*).

[20]   B. D. Popovic, "Theory of cylindrical antennas with lumped impedance loadings," *The Radio and Electronic Engineer*, vol. 43, No. 3, 1973, pp. 243-248.

[21]   R. Mittra (ed.), *Computer Techniques for Electromagnetics,* Pergamon Press, New York, 1973.

[22]   A. R. Djordjevic, B. D. Popovic, and M. B. Dragovic, "A method for rapid analysis of wire antenna structures," *Archiv fur Electrotechnic (W. Berlin)*, vol. 61, No. 1, 1979, pp. 17-23.

[23]   J. H. Richmond, "A wire-grid model for scattering by conducting bodies," *IEEE Trans. Antennas Propagat.*, vol. AP-14, No.6, 1966, pp. 782-786.

[24]   N. N. Mirolubov, M. V. Kostenko, M. L. Levinstein, and N. N. Tichodeev, *Methods of Electrostatic Field Calculation,* Visshaya Shkola, Leningrad, 1963 (*in Russian*).

[25]   R. L. Carrel, "The characteristic impedance of two infinite cones of arbitrary cross section", *IEEE Trans. Antennas Propagat.*, vol. AP-6, No. 2, 1958, pp. 197-201.

[26] H. Buchholz, *Elektrische und Magnetische Potentialfelder,* Berlin, 1957 (*in German*).

[27] Yu. Ya. Iossel, E. S. Kochanov, and M. G. Strunsky, *Calculation of Electrical Capacitance*, Leningrad, Energoisdat, 1981 (*in Russian*).

[28] A. A. Pistolkors, *Antennas*, Sviyazizdat, 1947 (*in Russian*).

[29] B. L. J. Rao, J. E. Ferris, and W. E. Zimmerman, "Broadband characteristics of cylindrical antennas with exponentially tapered capacitive loading," *IEEE Trans. Antennas Propagat.*, vol. AP-17, No. 2, 1969, pp. 145-151.

[30] D. Himmelblau, *Applied Nonlinear Programming*, McGraw-Hill, 1972.

[31] ANSI C95.1-1990, *Final Draft: American National Standard Safety Levels with Respect to Human Exposure to Radio Frequency Electromagnetic Fields, 300 kHz to 100 GHz,* New York, IEEE Press, 1990.

[32] M. Bank, and B. Levin, "The development of the cellular phone antenna with a small radiation of human organism tissues," *IEEE Antennas Propagat. Magazine*, vol. 49, No. 4, 2007, pp. 65-73.

[33] A. A. Pistolkors, "General theory of diffraction antennas," *Journal of Technical Physics*, vol. 14, No. 12, 1944, pp. 693-701 (*in Russian*)

[34] A. A. Pistolkors, "Theory of the Circular Diffraction Antenna," *Proc. IRE*, vol. 36, No. 1, 1948, pp. 56-60.

[35] Y. Mushiake, "The input impedance of a slit antenna," in *Joint Convention Record of Tohoqu Sections of IEE and IECE in Japan*, 1948, pp. 25-26.

[36] Y. Mushiake, "Self-Complementary Antennas," *IEEE Antennas and Propagation Magazine,* vol. 34, 1992, No. 6, 1992, pp. 23-29.

[37] H. G. Booker, "Slot aerials and their relation to complementary wire aerials (Babinet's principle)," *The Journal of the Institution of Electrical Engineers*, Part IIIA, 1946, No. 4, pp. 620-626.

[38] V. H. Rumsey, *Frequency Independent Antennas,* Academic Press, New York, 1966.

[39] G. Korn, and T. Korn, *Mathematical Handbook for Scientists and Engineers*, McGraw-Hill, New York, Toronto, London, 1961.

[40] Ya.N. Feld, *Foundations of Slot Antennas Theory*, Moscow, Sovetskoye Radio, 1948 (*in Russian*).

[41] J. Perini, and D. J. Buchanan,."Assessment of MOM techniques for shipboard applications," *IEEE Transactions on Electromagnetic Compatibility*, vol. EMC-24, No.1, 1982, pp.32-39.

[42] IEEE. Recommended Practice for Determining the Peak Spatial-Average Specific Absorption Rate (SAR) in the Human Head from Wireless Communications Devices: Measurement Techniques. IEEE Standards Coordinating Committee 34 (1528TM). 2003.

[43] C. Valenti, "NEXT and FEXT models for twisted-pair North American loop plant," *IEEE Journal Select. Areas Commun.,* vol. 20, No. 5, 2002, pp. 893-900.

[44] D. Cochrane, "Passive cancellation of common-mode electromagnetic interference in switching power converters," M.S. thesis, Virginia Polytechnic Inst. State Univ., Blacksburg, 2001.

[45] B. D. Popovic, and J. V. Surutka, "Cylindrical cage antenna," *Nachrichtentechn. Fachber,* No. 45, 1972, p. 51.

[46] T. T. Wu, and R. W. P. King, "The cylindrical antenna with non-reflecting resistive loading," *IEEE Trans. Antennas Propagat.,* vol. AP-13, No. 3, 1965, pp. 369-373.

[47] B. Levin, M. Bank, M. Haridim, and V. Tsingauz, "Dimensions of a dark spot produced by the compensation method," in *Proc. 9th Intern. Symp. EMC*, 2010, Wroclaw, pp. 317-322.

[48]   B. M. Levin, and V. G. Markov, *Method of Complex Potential and Antennas*, Ship Electrical Engineering and Communication, St.-Petersburg, 1997 *(in Russian)*.

[49]   R. L. Carrel, "The design of log-periodic dipole antennas," *IRE Intern. Convention Record,* vol. 9, part 1, 1961, pp. 61-75.

[50]   R. C. Johnson, *Antenna Engineering Handbook,* McGraw-Hill, New York, 1993.

[51]   W. Menzel, D. Pilz, and M. Al-Tikriti, "Millimeter-wave folded reflector antennas with high gain, low loss and low profile," *IEEE Antennas Propagat. Magazine*, vol. 44, No. 3, 2002, pp. 24-29.

[52]   F.-C. E. Tsai, and M. E. Bialkowski, "Designing a 161-element Ku-band micro strip reflect array of variable size patches using am equivalent unit cell waveguide approach," *IEEE Trans. Antennas Propagat.*, vol. AP-51, No. 10, 2003, pp. 2953-2962.

[53]   B. Widrow, and J. M. McCool, "A comparison of adaptive algorithms based on the methods of steepest descent and random search," *IEEE Trans. Antennas Propagat.*, vol. AP-24, No. 5, 1976, pp. 615-637.

[54]   N. Guan *et al.*, "Antennas of Transp. cond. films," *PIERS Online*, No. 1, 2008, p. 116.

[55]   Janke, Emde, and Losch, *Tafeln Hoherer Funktionen,* Verlagsgezellschaft, Stuttgart, 1960 *(in German)*.

[56]   J. R. Wait, and W. A. Pope, "The characteristics of a vertical antennas with a radial conductor ground system," *Applied Scientific Research*, vol. 4, No. 1, 1954, pp. 177-195.

[57]   M. V. Vershkov, B. M. Levin *et al.*, "Antenna- mast," Patent of SSSR 328824, March 31, 1970, Bulletin of Inventions, 1973, No. 45.

INDEX

www.ingramcontent.com/pod-product-compliance
Lightning Source LLC
Chambersburg PA
CBHW050809220326
41598CB00006B/164